Springer Series in Statistics

Advisors:
P. Bickel, P. Diggle, S. Fienberg, K. Krickeberg,
I. Olkin, N. Wermuth, S. Zeger

Springer
New York
Berlin
Heidelberg
Hong Kong
London
Milan
Paris
Tokyo

Springer Series in Statistics

Andersen/Borgan/Gill/Keiding: Statistical Models Based on Counting Processes.
Atkinson/Riani: Robust Diagnotstic Regression Analysis.
Berger: Statistical Decision Theory and Bayesian Analysis, 2nd edition.
Borg/Groenen: Modern Multidimensional Scaling: Theory and Applications
Brockwell/Davis: Time Series: Theory and Methods, 2nd edition.
Chan/Tong: Chaos: A Statistical Perspective.
Chen/Shao/Ibrahim: Monte Carlo Methods in Bayesian Computation.
David/Edwards: Annotated Readings in the History of Statistics.
Devroye/Lugosi: Combinatorial Methods in Density Estimation.
Efromovich: Nonparametric Curve Estimation: Methods, Theory, and Applications.
Eggermont/LaRiccia: Maximum Penalized Likelihood Estimation, Volume I: Density Estimation.
Fahrmeir/Tutz: Multivariate Statistical Modelling Based on Generalized Linear Models, 2nd edition.
Farebrother: Fitting Linear Relationships: A History of the Calculus of Observations 1750-1900.
Federer: Statistical Design and Analysis for Intercropping Experiments, Volume I: Two Crops.
Federer: Statistical Design and Analysis for Intercropping Experiments, Volume II: Three or More Crops.
Ghosh/Ramamoorthi: Bayesian Nonparametrics.
Glaz/Naus/Wallenstein: Scan Statistics.
Good: Permutation Tests: A Practical Guide to Resampling Methods for Testing Hypotheses, 2nd edition.
Gouriéroux: ARCH Models and Financial Applications.
Gu: Smoothing Spline ANOVA Models.
Györfi/Kohler/Krzyżak/ Walk: A Distribution-Free Theory of Nonparametric Regression.
Haberman: Advanced Statistics, Volume I: Description of Populations.
Hall: The Bootstrap and Edgeworth Expansion.
Härdle: Smoothing Techniques: With Implementation in S.
Harrell: Regression Modeling Strategies: With Applications to Linear Models, Logistic Regression, and Survival Analysis
Hart: Nonparametric Smoothing and Lack-of-Fit Tests.
Hastie/Tibshirani/Friedman: The Elements of Statistical Learning: Data Mining, Inference, and Prediction
Hedayat/Sloane/Stufken: Orthogonal Arrays: Theory and Applications.
Heyde: Quasi-Likelihood and its Application: A General Approach to Optimal Parameter Estimation.
Huet/Bouvier/Gruet/Jolivet: Statistical Tools for Nonlinear Regression: A Practical Guide with S-PLUS Examples.
Ibrahim/Chen/Sinha: Bayesian Survival Analysis.
Jolliffe: Principal Component Analysis.
Kolen/Brennan: Test Equating: Methods and Practices.

(continued after index)

J.K. Ghosh R.V. Ramamoorthi

Bayesian Nonparametrics

With 49 Illustrations

J.K. Ghosh
Statistics-Mathematics Division
Indian Statistical Institute
203 Barrackpore Trunk Road
Kolkata 70035
India

R.V. Ramamoorthi
Statistics and Probability
Michigan State University
A431 Wells Hall
East Lansing, MI 48824
USA

Library of Congress Cataloging-in-Publication Data
Ghosh, J.K.
Bayesian nonparametrics / J.K. Ghosh, R.V. Ramamoorthi.
p. cm. — (Springer series in statistics)
Includes bibliographical references and index.
ISBN 0-387-95537-2 (alk. paper)
1. Bayesian statistical decision theory. 2. Nonparametric statistics. I. Ramamoorthi, R.V. II. Title. III. Series.
QA279.5 .G48 2002
519.5′42—dc21 2002026665

ISBN 0-387-95537-2 Printed on acid-free paper.

Printed in the United States of America.

9 8 7 6 5 4 3 2 1 SPIN 10884896

Typesetting: Pages created by the authors using a Springer TEX macro package.

www.springer-ny.com

Springer-Verlag New York Berlin Heidelberg
A member of BertelsmannSpringer Science+Business Media GmbH

to our wives

Ira and Deepa

Preface

This book has grown out of several courses that we have given over the years at Purdue University, Michigan State University and the Indian Statistical Institute on Bayesian nonparametrics and Bayesian asymptotics. These topics seemed sufficiently rich and useful that a book length treatment seemed desirable.

Through the writing of this book we have received support from many people and we would like to gratefully acknowledge these. Our early interest in the topic came from discussions with Jim Berger, Persi Diaconis and Larry Wasserman. We have received encouragement in our effort from Mike Lavine, Steve McEachern, Susie Bayarri, Mary Ellen Bock, J. Sethuraman and Shanti Gupta, who alas is no longer with us.

We have enjoyed many years of collaboration with Subashis Goshal and much of our joint work finds a place in this book. Besides, he looked over an earlier version of the manuscript and gave very useful comments. The book also includes joint work with Jyotirmoy Dey, Roy Erickson, Liliana Dragichi, Charles Messan, Tapas Samanta and K.R.Srikanth. They have helped us with the proof, as have others. In particular, Tapas Samanta played an invaluable role in helping us communicate electronically and Charles Messan with computations.

Brendan Murphy, then a graduate student at Yale, gave us very useful feed back on an earlier version of Chapter 1. We also benefited from many suggestions and criticisms from Jim Hannan on the same chapter. We like to thank Nils Hjort both for his interest in the book and comments.

Dipak Dey made Sethuraman's unpublished notes available to us and these notes helped us considerably with Chapter 3.

When we first thought of writing a book, it seemed that we would be able to cover most, if not all, of what was known in Bayesian nonparametrics. However the last few years have seen an explosion of new work and our goals have turned more modest. We view this book as an introduction to the theoretical aspects of the topic at the graduate level. There is no coverage of the important aspect of computations but given the interest in this area we expect that a book on computations will emerge before long.

Our appreciation to Vince Melfi for his advice in matters related to Latex. Despite it, our limitations with Latex and typing skills would be apparent and we seek the readers' indulgence.

Contents

Introduction: Why Bayesian Nonparametrics—An Overview and Summary

Bayesians believe that all inference and more is Bayesian territory. So it is natural that a Bayesian should explore nonparametrics and other infinite-dimensional problems. However, putting a prior, which is always a delicate and difficult exercise in Bayesian analysis, poses special conceptual, mathematical, and practical difficulties in infinite-dimensional problems. Can one really have a subjective prior based on knowledge and belief, in an infinite-dimensional space? Even if one settles for a largely non-subjective prior, it is mathematically difficult to construct prior distributions on such sets as the space of all distribution functions or the space of all probability density functions and ensure that they have large support, which is a minimum requirement because a largely nonsubjective prior should not put too much mass on a small set. Finally, there are formidable practical difficulties in the calculation of the posterior, which is the single most important object in the output of any Bayesian analysis.

Nonetheless, a major breakthrough came with Ferguson's [61] paper on Dirichlet process priors. The hyperparameters $\alpha(\mathbb{R})$ and $\alpha(\cdot)$ of these priors are easy to elicit, it is easy to ensure a large support, and the posterior is analytically tractable. More flexibility was added by forming mixtures of Dirichlet processes, introduced by Antoniak [4].

Mixtures of Dirichlet have been very popular in Bayesian nonparametrics, especially in analyzing right censored survival data. In these problems one can combine analytical work with Markov Chain Monte Carlo (MCMC) to calculate and display various posterior quantities in real time. By choosing $\alpha(\cdot)$ equal to the exponential distribution and by tuning the parameter $\alpha(\mathbb{R})$, one can make the analysis close to classical analysis based on a parametric exponential or close to classical nonparametrics. However, the whole range of $\alpha(\mathbb{R})$ offers a whole continuum of options that are not available in classical statistics, where typically one either does a model based parametric analysis or use, fully nonparametric methods. An interesting example in survival analysis is presented by Doss [53, 54]. Huber's pioneering work in classical statistics on a robust via media between these two extremes has been too technically demanding to yield a flexible set of methods that pass continuously from one extreme to the other. These ideas are discussed further in Chapter 3 on Dirichlet priors.

Similarly one can analyze generalized linear models with a nonparametric Bayesian choice of link functions. Bayesian nonparametrics is known to be a powerful, robust alternative to regression analysis based on probit or logit models. References are available in Chapter 7. There is some evidence of gaining an advantage in using Bayesian nonparametrics to model random effects in linear models for longitudinal data.

Sometimes things can go wrong if one uses a Dirichlet process prior inappropriately. Such a prior cannot be used for density estimation without some smoothing, but smoothing leads to formidable difficulties in calculating the posterior or the Bayes estimate of the density function. Solution of this computational problem by MCMC is fairly recent; see Chapter 5 for references and discussion. A major advantage of the Bayesian method is that choice of the smoothing parameter h, which is still a hard problem in classical density estimation, is relatively automatic. The Bayesian version of varying the smoothing parameter over different parts of the data is also relatively easy to implement. These are some of the major advances in Bayesian nonparametrics in recent years.

A major theoretical advance has occurred recently in Bayesian semiparametrics. One has the same advantages of flexibility here as discussed earlier, but unfortunately this is also an area where the Dirichlet process is inappropriate without some smoothing. Instead one can use Polya tree priors that sit on densities and satisfy some extra conditions. For details and references see Chapter 6.

A difficulty in Bayesian nonparametrics is that not much was known until recently about the asymptotic behavior of the posterior and various forms of frequentist validation. One method of frequentist validation of Bayesian analysis is to see if one can learn about the unknown true P_0 with vanishingly small error by examining where the posterior puts most of its mass. This idea and the first result of this sort are due to Laplace. A precise statement of this property leads to the notion of consistency of the posterior at P_0, due to Freedman [69]. In the case of finite-dimensional parameters, the posterior is usually consistent, and the data wash away the prior. For an infinite-dimensional parameter, this is an exception rather than the rule; see, for instance, examples of Freedman [69] and his theorem: For a multinomial with infinitely many classes, the set of (P_0, Π) for which posterior for the prior Π is consistent at P_0, is topologically small, i.e., of the first category. Freedman had also introduced the notion of tail free priors for which there is posterior consistency at P_0. A striking example of inconsistency was shown by Diaconis and Freedman [46] when a Dirichlet process is used for estimating a location parameter. In his discussion of [46], Barron points out that the use of a Dirichlet process prior in a location problem leads to a pathological behavior of the posterior for the location parameter. It is clear that inconsistency is a consequence of this pathology. Diaconis and Freedman [46] also suggested that such examples would occur even if one uses a prior on densities, e.g., a Polya tree prior sitting on densities.

Chapter 4 is devoted to general questions of consistency of the posterior and positive results. Applications appear in many other chapters and in fact run through the whole book. These results, as well as somewhat stronger results, like rates of convergence, are fairly recent and due to many authors, including ourselves.

To sum up, Bayesian nonparametrics is sufficiently well developed to take care of many problems. Computation of the posterior is numerically feasible for several classes of priors. We now know a fair amount of asymptotic behavior of posteriors for different priors to ensure consistency at plausible P_0s. Most important, Bayesian nonparametrics provides more flexibility than classical nonparametrics and a more robust analysis than both classical and Bayesian parametric inference. It deserves to be an important part of the Bayesian paradigm.

This monograph provides a systematic, theoretical development of the subject. A chapterwise summary follows:

1. After introducing some preliminaries, Chapter 1 discusses some fundamental aspects of Bayesian analysis in the relatively simple context of finite dimensional parameter space with dimension fixed for all sample sizes. Because this subject is treated well in many standard textbooks, the focus is on aspects such as nonsubjective priors, also called objective priors, posterior consistency and exchangeability. These are topics that usually do not receive much coverage in textbooks but are important for our monograph,

Because elicitation of subjective priors or quantification of expert knowledge is still not easy, most priors used in practice, especially in nonparametrics, are nonsubjective. We discuss the standard ways of generating such priors and how to modify them when some subjective or expert judgment is available (Section 1.61.7). We also briefly discuss common criticisms of nonsubjective Bayesian analysis and answers 1.6.2

Posterior consistency is introduced, and the classical theorem of Doob is proved with all details. Then, in the spirit of classical maximum likelihood theory, posterior consistency is established under regularity conditions using the uniform strong law of large numbers. Posterior consistency provides a frequentist validation that is especially important for inference on infinite-or high dimensional parameters because even with a massive amount of data, any inadequacy in the prior can still influence the posterior a lot. Posterior normality (Section 1.4) is a sharpening of posterior consistency that is related to Laplace approximation and plays an important role in the construction of reference and probability matching priors. Convergence of posterior distributions is usually studied under regularity conditions. A general approach that also works for nonregular problems is presented in Section 1.5. Exchangeability appears in the last sections Chapter 1.

In Chapter 2 we examine basic measure-theoretic questions that arise when we try to check measurability of a set or function or put a prior on such a large space as the set P of all probability measures on $\mathbb{R}$. The Kolmogorov construction based on consistent finite-dimensional distributions does not meet this requirement because the Kolmogorov sigma-field is too small to ensure measurability of important subsets like the set of all discrete distributions on $\mathbb{R}$ or the set of all P with a density with respect to the Lebesgue measure. Questions of measurability and convergence are discussed in Section 2.2.

An interesting fact is a proof that the set of discrete measures and the set of absolutely continuous probability measures are measurable. The main results in the chapter are the basic construction theorems 2.3.2 through 2.3.4. Tail free priors, including the Dirichlet process prior, may be constructed this way. The most important type of convergence, namely, weak convergence is discussed is detail in Section 2.5. The main result is a characterization of tightness in the spirit of Sethuraman and Tiwari (1982). Section 2.4 contains 0-1 laws for tail free priors as well as a theorem due to Kraft that can be used to construct a tail free prior for densities.

De Finetti's theorem appears in the last section.

The reader not interested in measure-theoretic issues may read this chapter quickly to understand the main results and get a flavor of some of the proofs. A reader with more measure-theoretic interest will gain a solid theoretical framework for handling priors for nonparametric problems and will also be rewarded with several measure-theoretic subtleties that are interesting.

The most important prior in Bayesian nonparametrics is the Dirichlet process prior, which plays a central role here as the normal in finite-dimensional problems. Most of Chapter 3 is devoted to this prior. The last section is on Polya tree priors.

We introduce a Dirichlet prior (3.1) first in the case of a finite sample space $\mathcal{X}$ and then for $\mathcal{X} = \mathbb{R}$ to help develop intuition for the main results regarding the latter. The Dirichlet prior D for $\mathcal{X} = \mathbb{R}$ is usually called the Dirichlet process prior. Section 3.2 contains calculation and justification of a formula for posterior and special properties. It also contains Sethuraman's clever and elegant construction, which applies to all $\mathcal{X}$ and suggests how one can simulate from this prior. Other results of interest include a characterization of support and convergence properties (Section 3.2) and the question of singularity of two Dirichlet process priors with respect to each other. Part of the reason why Dirichlet process priors have been so popular is the multitude of interesting properties mentioned earlier, of which the most important are the ease in calculation of posterior and the fact that the support is as rich as it should be for a prior for nonparametric problems.

A second and equally important reason for popularity is the flexibility, at least for mixtures of Dirichlet, and the relative case with which one can elicit the hyperparameters. These issues are discussed in 3.2.7

The last section extends most of this discussion to Polya tree priors which form a much richer class. Though not as mathematically tractable as D, they are still relatively easy to handle and one can use convenient, partly elicited hyperparameters.

As we have argued before, posterior consistency is a useful validation for a particular prior, especially in nonparametric problems. Chapter 4 deals with essentially three approaches to posterior consistency for three kinds of problems, namely, purely nonparametric problems of estimating a distribution function or its weakly continuous functionals, semiparametrics, and density estimation. The Dirichlet and, more generally, tail free priors have good consistency properties for the first class of problems. Posterior consistency for tail free priors is discussed in the first few pages of the chapter.

In Bayesian semiparametrics, for example estimation of a location parameter (Chapter 6) or the regression coefficient (Chapter 7), addition of Euclidean parameters destroys the tail free property of common priors like Dirichlet process and Polya tree. Indeed, the use of Dirichlet leads to a pathological posterior. Posterior consistency in this case is based on a theorem of Schwartz for a prior on densities. The two crucial conditions are that the true probability measure lie in the Kullback-Leibler support of the prior and there has to be uniformly exponentially consistent tests for $H_0 : f = f_0$

VS $H_1 : f \in V^c$, where V is a neighborhood whose posterior probability is being claimed to converge to one. This is presented in Section 4.2.

The Schwartz theorem is well suited for semiparametrics but not for density estimation because the second condition in the theorem does not hold for a V equal to an L_1-neighborhood of f_0. Barron (unpublished) has suggested a weakening of one of these conditions, suitably compensated by a condition on the prior. His conditions are necessary and sufficient for a certain form of exponential convergence of the posterior probability of V to one. Ghosal, Ghosh and Ramamoorthi (1999) make use of this theorem and some ideas of Barron, Schervish and Wasserman (1999) to modify Schwartz's result to make it suitable for showing posterior consistency with L_1-neighborhoods for a prior sitting on densities. All these results appear in Section 4.2.

Finally, Section 4.3 is devoted to another approach based on an inequality of LeCam, which bypasses the verification of the first condition of Schwartz.

Applications of these results are made in Chapters 5 through 8. Somewhat different but direct calculations leading to posterior consistency appear in Chapters 9 and 10.

Chapter 5 focuses on three kinds of priors for density estimation: Dirichlet mixtures of uniform, Dirichlet mixtures of normal, and Gaussian process priors. Dirichlet mixtures of normal are the most popular and the most studied. The Gaussian process priors seem very promising but have not been studied well. Dirichlet mixtures of uniform are essentially Bayesian histograms and have a relatively simple theory.

The chapter begins with fairly general construction of priors on densities in sections 5.2 and 5.3 and then specializes to Bayesian histograms and their consistency in Sections 5.40, 5.4.1, and 5.4.2. Dirichlet mixtures of normals are studied in Sections 5.6 and 5.7. The L_1-consistency of the posterior applies to the prior of Escobar and West in [168]. The final section contains an introduction to what is known about Gaussian process priors.

Interesting issues that emerge from this rather technical chapter is that checking the Kullback-Leibler support condition is especially hard for densities with $\mathbb{R}$ as support, whereas densities with bounded support are much easier to handle. A second source of technical difficulty is the need for efficient calculation of packing or covering numbers, also called Kolmogorov's metric entropy. These numbers play a basic role in Chapters 4,5 and 8.

Chapter 6 begins with the famous Diaconis-Freedman (1986) example where a Dirichlet process prior and a euclidean location parameter lead to posterior inconsistency. Barron (1986) has pointed out that there is a pathology in this case which is even worse than inconsistency. We argue, as suggested in Chapter 4, that the main problem leading to posterior inconsistency is that the tail free property does not hold. It is good to have a density but that does not seem to be enough. However, no counter example is produced.

The main contribution of the chapter is to suggest in Section 6.3 a strategy for proving posterior consistency for the location parameter in semiparametric setting and to provide in Section 6.4 a class of Polya tree priors which satisfy the conditions of Section 6.3 for a rich class of true densities. A major assumption needed in Section 6.3 holds only for densities with $\mathbb{R}$ as support. Later in the section we show how to extend these results to densities with bounded support. Whereas in density estimation bounded support helps, the converse seems to be true when one has to estimate a location parameter.

The discussion of Bayesian semiparametrics is continued in Chapter 7 . We assume a standard regression model

$$Y = \alpha + \beta x + \epsilon$$

with the error ϵ having a nonparametric density f. The main object is to estimate the regression coefficient but one may also wish to estimate the intercept α as well as the true density of ϵ. The classical counterpart of this is Bickel [19].

Because Y's are no longer i.i.d, the Schwartz theorem of Chapter 6 does not apply. In Section 7.2 - we prove a generalization that is valid for n independent but not necessarily identically distributed random variables.

The theorem needs two conditions which are exact analogues of the two conditions in Schwartz's theorem and one additional condition on the second moment of a log likelihood ratio. Verification of these conditions is discussed in Section 7.4.

In Section 7.3 we discuss sufficient conditions for the existence of uniformly consistent tests for β alone or (α, β) or (α, β, f).

Finally in sections 7.6 we verify the remaining two conditions for Polya tree priors and Dirichlet mixtures of normals. Verification of conditions require methods that are substantially different from those in Chapter 5.

Chapter 8 deals with three different but related topics, namely, three methods of construction of nonsubjective priors in infinite dimensional problems involving densities, consistency proof for such priors using LeCam's inequality and rates convergence for such and other priors. They are discussed in sections 8.2,8.5 and 8.6 respectively. In several examples it is shown that the rates of convergence are the best possible. However, for most commonly used priors getting rates of convergence is still a very hard open problem.

Chapters 9 and 10 deal with right censored data. Here, the object of interest is the distribution of a positive random variable X, viewed as survival time. What we have are observations of is $Z = X \wedge Y, \Delta = I(X \leq Y)$, where Y is a censoring random variable, independent of X.

Chapter 9 begins with a model studied by Susarla and Van Ryzin [155] where the distribution of X is given a Dirichlet process prior. We give a representation of the posterior and establish its consistency. Section 2 is a quick review of the notion of cumulative hazard function and identifiability of the distribution of X from that of (Z, Δ). This is then used in the next section where we start with a Dirichlet prior for the distribution of (Z, Δ) and use the identifiability result to transfer it to a prior for the distribution of X. We expect that this method will be useful in constructing priors for other kind of censored data. Section 9.4 is a preliminary study of Dirichlet priors for interval censored data. We show that, unlike the right censored case, letting $\alpha(\mathbb{R}) \to 0$ does not give the nonparametric maximum likelihood estimate.

Chapter 10 deals with neutral to right priors. These priors were introduced by Doksum in 1974 [48] and after some initial work by Ferguson and Phadia [64] remained dormant. There has been renewed interest in these priors since the introduction of Beta processes by Hjort [100]. Neutral to right priors, via the cumulative hazard function, gives rise to independent increment processes which in turn are described by their Lévy representations. In Section 10.1 after giving the definition and basic properties of neutral to right priors we move onto Section 10.2 where we briefly review the connection to independent increment processes and Lévy representations. Section 10.3 describes some properties of the prior in terms of the Lévy measure and Section 10.4 is devoted to Beta processes. The remaining parts of the chapter is devoted to posterior consistency and is partly driven by a surprising example of inconsistency due to Kim and Lee [114].

Chapter 11 contains some exercises. These were not systematically developed. However we have included in the hope that going through them will give the reader some additional insight into the material.

Most work on Bayesian nonparametrics concentrates on estimation. This monograph is no exception. However there is interesting new work on Bayes Factors and their consistency [13], [37]. Even in the context of estimation, in the context of censored data, not much has been done beyond the independent right censored model. There certainly is lot more to be done.

1
Preliminaries and the Finite Dimensional Case

1.1 Introduction

The basic Bayesian model consists of a parameter θ and a prior distribution Π for θ that reflects the investigator's belief regarding θ. This prior is updated by observing $X_1, X_2, \ldots, X_n$, which are modeled as i.i.d. P_θ given θ. The updating mechanism is Bayes theorem, which results in changing Π to the posterior $\Pi(\cdot|X_1, X_2, \ldots, X_n)$. The posterior reflects the investigator's belief as revised in the light of the data $X_1, X_2, \ldots, X_n$. One may also report the predictive distribution of the future observations or summary measures like the posterior mean or variance. If there is a decision problem with a specified loss function, one can choose the decision that minimizes the expected loss, with the associated loss calculated under the posterior. This decision is the Bayes solution, or the Bayes rule. Ideally, a prior should be chosen subjectively to express personal or expert knowledge and belief. Such evaluations and quantifications are not easy, especially in high- or infinite-dimensional problems. In practice, mathematically tractable priors, for example, conjugate priors, are often used as convenient and partly nonsubjective models of knowledge and belief. Certain aspects of these priors are chosen subjectively.

Finally, there are completely nonsubjective priors, the choice of which also leads to useful posteriors. For the finite-dimensional case a brief account appears in Section 1.6. For a moderate amount of data, i.e., for a moderate n, the effect of prior on the posterior is often negligible. In such cases the posterior arising from a nonsubjective prior may be considered a good approximation for the posterior that one would have gotten from a subjective prior.

The posterior, like the prior, is a probability measure on the parameter space Θ, except that it depends on $X_1, X_2, \ldots, X_n$ and the study of the posterior as $n \to \infty$ is naturally connected to the theory of convergence of probability measures. In Section 1.2.1, we present a brief survey of weak convergence of probability measures as well as relations between various metrics and divergence measures.

A recurring theme throughout this monograph is posterior consistency, which helps validate Bayesian analysis. Section 1.3 contains a formalization and brief discussion of posterior consistency for separable metric space Θ. In Sections 1.3 and 1.4 we study in some detail the case when Θ is finite-dimensional and $\theta \mapsto P_\theta$ is smooth. This is the framework of conventional parametric theory. Most of the results and asymptotics are classical, but some are relatively new. While the main emphasis of this monograph is in the nonparametric, and hence infinite-dimensional situation, we hope that Sections 1.3 and 1.4 will serve to clarify the points of contact and points of difference with the finite-dimensional case.

1.2 Metric Spaces

1.2.1 *preliminaries*

Let $(\mathbf{S}, \rho)$ be a metric space so that ρ satisfies (i) $\rho(s_1, s_2) = \rho(s_2, s_1)$, (ii) $\rho(s_1, s_2) \geq 0$ and $\rho(s_1, s_2) = 0$ iff $s_1 = s_2$ and (iii) $\rho(s_1, s_3) \leq \rho(s_1, s_2) + \rho(s_2, s_3)$.

Some basic properties of metric spaces are summarized here.

A sequence s_n in $\mathbf{S}$ *converges* to s iff $\rho(s_n, s) \to 0$. The *ball* with center s_0 and radius δ is the set $B(s_0, \delta) = \{s : \rho(s_0, s) < \delta\}$. A set U is *open* if every s in U has a ball $B(s, \delta)$ contained in U. A set V is *closed* if its complement V^c is open. A useful characterization of a closed set is: V is closed iff $s_n \in V$ and $s_n \to s$ implies $s \in V$. The intersection of closed sets is a closed set. For any set $A \subset \mathbf{S}$, the smallest closed set containing A, which is the intersection of all closed sets containing A, is called the *closure* of A and will be denoted by $\bar{A}$. Similarly A^o, the union of all open sets contained in A is called the *interior* of A. The *boundary* ∂A of the set A is defined as $\partial A = \bar{A} \cap \overline{(A^c)}$.

A subset A of $\mathbf{S}$ is *compact* if every open cover of A has a finite subcover, i.e., if $\{U_\alpha : \alpha \in \Lambda\}$ are open sets and $A \subset \cup_{\alpha \in \Lambda} U_\alpha$, then there exists $\alpha_1, \alpha_2, \ldots, \alpha_n$

such that $A \subset \cup_1^n U_{\alpha_i}$. A set A is compact iff every sequence in A has a convergent subsequence with limit in A.

The metric space $\mathbf{S}$ is *separable* if it has a countable dense subset, i.e., if there is a countable set $\mathbf{S}_0$ with $\bar{\mathbf{S}}_0 = \mathbf{S}$. Most of the sets that we consider are separable. In particular, if $\mathbf{S}$ is compact metric it is separable. Let $\mathbf{S}$ be separable and let $\mathbf{S}_0$ be a countable dense set. Consider the countable collection $\{B(s_i, 1/n) : s_i \in \mathbf{S}_0; n = 1, 2, \ldots\}$. If U is an open set and if $s \in U$, then for some $n > 1$, there is a ball $B(s, 1/n) \subset U$. Let $s_i \in \mathbf{S}_0$ with $\rho(s_i, s) < 1/2n$. Then s is in $B(s_i, 1/2n)$ and $B(s_i, 1/2n) \subset B(s, 1/n) \subset U$. This shows that in a separable space every open set is a countable union of balls. This fact fails to hold when $\mathbf{S}$ is not separable.

The Borel σ-algebra on $\mathbf{S}$ is the σ-algebra generated by all open sets and will be denoted by $\mathcal{B}(\mathbf{S})$. The remarks in the last paragraph show that if $\mathbf{S}$ is separable then $\mathcal{B}(\mathbf{S})$ is the same as the σ-algebra generated by open balls. In the absence of separability these two σ-algebras will be different.

It would sometimes be necessary to check that a given class of sets $\mathcal{C}$ is the Borel σ-algebra. A useful device to do this is the π-λ theorem given below. See Pollard [[140], Section 2.10] for a proof and some discussion.

Theorem 1.2.1. *[π-λ theorem] A class $\mathcal{D}$ of subsets of $\mathbf{S}$ is a π-system if it is closed under finite intersection, i.e., if A, B are in $\mathcal{D}$ then $A \cap B \in \mathcal{D}$. A class $\mathcal{C}$ of subsets of $\mathbf{S}$ is a λ-system if*

(i) $\mathbf{S}$ is in $\mathcal{C}$;

(ii) $A_n \in \mathcal{C}$ and $A_n \uparrow A$, then $A \in \mathcal{C}$;

(iii) $A, B \in \mathcal{C}$ and $A \subset B$, then $B - A \in \mathcal{C}$.

If $\mathcal{C}$ is a λ-system that contains a π-system $\mathcal{D}$, then $\mathcal{C}$ contains the σ-algebra generated by $\mathcal{D}$.

Remark 1.2.1. An easy application of the π-λ theorem shows that if two probability measures on $\mathbf{S}$ agree on all closed sets then they agree on $\mathcal{B}(\mathbf{S})$.

Remark 1.2.2. If two probability measures on $\mathbb{R}^K$ agree on all sets of the form $(a_1, b_1] \times (a_2, b_2], \ldots \times (a_k, b_k]$ then they agree on all Borel sets in $\mathbb{R}^k$.

Definition 1.2.1. Let P be a probability measure on $(\mathbf{S},\mathcal{B}(\mathbf{S}))$.The smallest closed set of P-measure 1 is called the *support*, or more precisely the topological support, of P.

When $\mathbf{S}$ is separable the support of P always exists. To see this let $\mathcal{U}_0 = \{U : U \text{ open}, P(U) = 0\}$, then $U_0 = \cup_{U \in \mathcal{U}_0} U$ is open. Because U_0 is a countable union of balls in $\mathcal{U}_0$, $P(U_0) = 0$. It follows easily that $F = U_0^c$ is the support of P. The support can be equivalently defined as a closed set F with $P(F) = 1$ and such that if $s \in F$ then $P(U) > 0$ for every neighborhood U of s. If $\mathbf{S}$ is not separable then the support of P may not exist.

1.2.2 Weak Convergence

We need elements of the theory of weak convergence of probability measures. The details of the material discussed below can be found, for instance, in Billingsley [[21], Chapter 1].

Let $\mathbf{S}$ be a metric space and $\mathcal{B}(\mathbf{S})$ be the Borel σ-algebra on $\mathbf{S}$. Denote by C($\mathbf{S}$) the set of all bounded continuous functions on $\mathbf{S}$. Note that every function in C($\mathbf{S}$) is $\mathcal{B}(\mathbf{S})$ measurable.

Definition 1.2.2. A sequence $\{P_n\}$ of probability measures on $\mathbf{S}$ is said to converge weakly to a probability measure P, written as $\{P_n\} \to P$ weakly, if

$$\int f \, dP_n \to \int f \, dP \qquad \text{for all} \quad f \in C(\mathbf{S})$$

The following "Portmanteau" theorem gives most of what we need.

Theorem 1.2.2. *The following are equivalent:*

1. $\{P_n\} \to P$ *weakly;*
2. $\int f \, dP_n \to \int f \, dP$ *for all* f *bounded and uniformly continuous;*
3. $\limsup P_n(F) \le P(F)$ *for all* F *closed;*
4. $\liminf P_n(U) \ge P(U)$ *for all* U *open;*
5. $\lim P_n(B) = P(B)$ *for all* $B \in \mathcal{B}(\mathbf{S})$*with* $P(\partial B) = 0$.

In applications, P_ns are often distributions on $\mathbf{S}$ induced by random variables X_ns taking values in $\mathbf{S}$. If $\mathbf{S}$ is not separable, then P_n is defined on a σ-algebra much smaller than $\mathcal{B}(\mathbf{S})$. In this case, to avoid measurability problems inner and outer probabilities have to be used. For a version of Theorem 1.2.2 in this more general setting see van der Vaart and Wellner [[161], 1.3.4]. The other useful result is Prohorov's theorem.

Theorem 1.2.3. *[Prohorov] If* $\mathbf{S}$ *is a complete separable metric space, then every subsequence of* P_n *has a weakly convergent subsequence iff* P_n *is tight, i.e., for every* $\epsilon > 0$*, there exists a compact set* K *with* $P_n(K) > 1 - \epsilon$ *for all* n.

When $\mathbf{S}$ is a complete separable metric space the space $\mathbf{M}(\mathbf{S})$-the space of probability measures on $\mathbf{S}$-is also metrizable, complete, and separable under weak convergence. In this case if $\int f\, dP_n \to \int f\, dP$ for f in a countable dense set in $C(\mathbf{S})$, then $P_n \to P$ weakly. We note that sets in $\mathbf{M}(\mathbf{S})$ of the form

$$\left\{Q : \left|\int f_i\, dP - \int f_i\, dQ\right| < \delta, i = 1, 2, \ldots, k;\ f_i \in C(\mathbf{S})\right\}$$

constitute a base for the neighborhoods at P, i.e., any open set is a union of family of sets of the form displayed above. The space $\mathbf{M}(\mathbf{S})$ and the space of probability measures on $\mathbf{M}(\mathbf{S})$ are of considerable interest to us. We will return to a detailed analysis of these spaces later; here are a few preliminary facts used later in this chapter.

The space $\mathbf{M}(\mathbf{S})$ has many natural metrics.

Weak convergence. As discussed earlier $\mathbf{M}(\mathbf{S})$ is metrizable, i.e., there is a metric ρ on $\mathbf{M}(\mathbf{S})$ such that $\rho(P_n, P) \to 0$ iff $P_n \to P$ weakly [see section 6 in Billingsley [21]]. The exact form of this metric is not of interest to us.

Total variation of L_1. The total variation distance between P and Q is given by $\|P - Q\|_1 = 2 \sup_B |P(B) - Q(B)|$. If p and q are densities of P and Q with respect to some measure μ, then $\|P - Q\|_1$ is the L_1-distance $\int |p - q|\, d\mu$ between p and q. Sometimes, when there can be no confusion with other metrics, we will omit the subscript 1 and denote the L_1 distance by just $\|P - Q\|$ or in terms of densities as $\|p - q\|$.

Hellinger metric. If p and q are densities of P and Q with respect to some σ-finite measure μ, the Hellinger distance between P and Q is defined by $H(P, Q) = \left[\int (\sqrt{p} - \sqrt{q})^2\, d\mu\right]^{1/2}$. This distance is convenient in the i.i.d. context because $A(P^n, Q^n) = A^n(P, Q)$, where $A(P, Q) = \int \sqrt{p}\sqrt{q}\, d\mu$, is called the *affinity* between P and Q and

$$H^2(P^n, Q^n) = 2(1 - (A(P, Q))^n)$$

The Hellinger metric is equivalent to the L_1-metric. The next proposition shows this.

Proposition 1.2.1.

$$\|P-Q\|_1^2 \le H^2(P,Q)\; 2(1+A(P,Q)) \le \|P-Q\|_1\; 2(1+A(P,Q))$$

Proof. Let μ dominate P and Q and let p, q, be densities of P and Q with respect to μ. Then

$$\begin{aligned}\left[\int |p-q|\, d\mu\right]^2 &= \left[\int |\sqrt{p}-\sqrt{q}||\sqrt{p}+\sqrt{q}|\, d\mu\right]^2 \\ &\le \int (\sqrt{p}-\sqrt{q})^2\, d\mu \int (\sqrt{p}+\sqrt{q})^2\, d\mu\end{aligned}$$

which is the first inequality. Also $H^2(P,Q) \le \|P-Q\|_1$ because

$$(\sqrt{p}-\sqrt{q})^2 \le p+q-\min(p,q) = |p-q|$$

□

As a corollary to the above proposition, we have the following.

Corollary 1.2.1. *Replacing $A(P,Q)$ by its upper bound 1 gives*

$$\|P-Q\|_1^2 \le 4H^2(P,Q) \le 4\|P-Q\|_1$$

Writing $H^2(P,Q) = 2(1-A(P,Q))$ in the first inequality, a bit of algebra gives

$$A(P,Q) \le \sqrt{1-\frac{\|P-Q\|_1^2}{4}}$$

Note that none of the three quantities discussed-the L_1 metric, the Hellinger metric, or the affinity $A(P,Q)$-depends on the dominating measure μ. The same holds for the Kullback- Leibler divergence(K-L divergence) which is considered next.

Kullback-Leibler divergence. The Kullback-Leibler divergence between two probability measures, though not a metric, has played a central role in the classical theory of testing and estimation and will play an important role in the later chapters of this text. Let P and Q be two probability measures and let p, q be their densities with respect to some measure μ. Then

$$K(P,Q) = \int p \log\frac{p}{q}\, d\mu \ge \int (1-\frac{q}{p})dP \ge 0$$

and $K(P,Q) = 0$ iff $P = Q$. Here is a useful refinement due to Hannan [92].

Proposition 1.2.2.

$$K(P, Q) \geq \frac{\|P - Q\|_1^2}{4}$$

Proof.

$$\begin{aligned}
\int p \log(p/q)\, d\mu &= \int 2(-\log \frac{\sqrt{q}}{\sqrt{p}}) p d\mu \\
&\geq \int 2\left(1 - (\sqrt{q}/\sqrt{p})\right) p d\mu \\
&= 2\left(1 - A(P, Q)\right) = H^2(P, Q)
\end{aligned}$$

The corollary to the previous proposition yields the conclusion. □

Kemperman [112] has shown that $K(P, Q) \geq \|P - Q\|_1^2/2$ and that this inequality is sharp.

Much of our study involves the convergence of sequences of functions of the form $T_n(X_1, X_2, \ldots, X_n) : \Omega \mapsto M(\Theta)$ where $\Omega = (\mathbf{X}^\infty, \mathcal{A}^\infty)$ with a measure P_0^∞. The different metrics on $M(\Theta)$ provide ways of formalizing the convergence of T_n to T. Thus

(i) $T_n \overset{weakly}{\to} T$ almost surely P_0 if

$$P_0^\infty \left\{\omega : T_n(\omega) \overset{weakly}{\to} T(\omega)\right\} = 1$$

(ii) $T_n \overset{weakly}{\to} T$ in P_0 probability if

$$P_0^\infty \left\{\omega : \rho(T_n(\omega), T(\omega)) > \epsilon\right\} \to 0$$

where ρ is a metric that generates weak convergence.

$T_n \overset{L_1}{\to} T$ almost surely P_0 or in P_0-probability can be defined similarly.

1.3 Posterior Distribution and Consistency

1.3.1 Preliminaries

We begin by formalizing the setup. Let Θ be the parameter space. We assume that Θ is a complete separable metric space endowed with its Borel σ-algebra $\mathcal{B}(\Theta)$. For each $\theta \in \Theta$, P_θ is a probability measure on a measurable space $(\mathbf{X}, \mathcal{A})$ such that, for each $A \in \mathcal{A}, \theta \mapsto P_\theta(A)$ is $\mathcal{B}(\Theta)$ measurable.

$X_1, X_2, \ldots$ is a sequence of $\mathbf{X}$-valued random variables that are, for each $\theta \in \Theta$, independent and identically distributed as P_θ . It is convenient to think of $X_1, X_2, \ldots$ as the coordinate random variables defined on $\Omega = (\mathbf{X}^\infty, \mathcal{A}^\infty)$ and P_θ^∞ as the i.i.d. product measure defined on Ω. We will denote by Ω_n the space $(\mathbf{X}^n, \mathcal{A}^n)$ and by P_θ^n the n-fold product of P_θ. When convenient we will also abbreviate $X_1, X_2, \ldots, X_n$ by $\mathbf{X_n}$.

Suppose that Π is a *prior*, i.e., a probability measure on $(\Theta, \mathcal{B}(\Theta))$. For each n, Π and the P_θs together define a *joint distribution* of θ and $\mathbf{X_n}$ namely, the probability measure $\lambda_{n,\Pi}$ on Ω_n by

$$\lambda_{n,\Pi}\ (B \times A) = \int_B P_\theta^n(A)\ d\Pi(\theta)$$

The marginal distribution λ_n of $X_1, X_2, \ldots, X_n$ is

$$\lambda_n(A) = \lambda_{n,\Pi}\ (\Theta \times A)$$

These notions also extend to the infinite sequence $X_1, X_2, \ldots$. We denote by λ_Π the joint distribution of θ,$X_1, X_2, \ldots$ and by λ the marginal distribution on Ω.

Any version of the conditional distribution of θ given $X_1, X_2, \ldots, X_n$ is called a *posterior distribution* given $X_1, X_2, \ldots, X_n$. Formally, a function $\Pi(\cdot\,|\cdot\,) : \mathcal{B}(\Theta) \times \Omega_n \mapsto [0,1]$ is called a posterior given $X_1, X_2, \ldots, X_n$ if

(a) for each $\omega \in \Omega_n, \Pi(\cdot\ |\omega)$ is a probability measure on $\mathcal{B}(\Theta)$;

(b) for each $B \in \mathcal{B}(\Theta)$, $\Pi(B|\cdot\ )$ is $\mathcal{A}^n$ measurable; and

(c) for each $B \in \mathcal{B}(\Theta)$ and $A \in \mathcal{A}$,

$$\lambda_{n,\Pi}\ (B \times A) = \int_A \Pi(B|\omega)\ d\lambda_n(\omega)$$

In the case that we consider, namely, when the underlying spaces are complete and separable, a version of the posterior always exists [Dudley [58], 10.2]. By condition (b), $\Pi(\cdot\ |\omega)$ is a function of $X_1, X_2, \ldots, X_n$ and hence we will write the postrior conveniently as $\Pi(\cdot|X_1, X_2, \ldots, X_n)$ or as $\Pi(\cdot|\mathbf{X_n})$.

Typically, a candidate for the posterior can be guessed or computed heuristically from the context. What is then required is to verify that it satisfies the three conditions listed earlier. When the P_θs are all dominated by a σ- finite measure μ, it is easy to see that, if $p_\theta = dP_\theta/d\mu$, then

$$\Pi(A|\mathbf{X_n}) = \frac{\int_A \prod_1^n p_\theta(X_i)\, d\Pi(\theta)}{\int_\Theta \prod_1^n p_\theta(X_i)\, d\Pi(\theta)}$$

Thus in the dominated case, $\prod_1^n p_\theta(X_i)/\int \prod_1^n p_\theta(X_i) d\Pi(\theta)$ is a version of the density with respect to Π of $\Pi(\cdot|\mathbf{X_n})$.

In the last expression the posterior given $X_1, X_2, \ldots, X_n$ is the same as that given a permutation $X_{\pi(1)}, X_{\pi(2)}, \ldots, X_{\pi(n)}$. Said differently, the posterior depends only on the empirical measure $(1/n)\sum_1^n \delta_{X_i}$, where for any x, δ_x denotes the measure degenerate at x. This property holds also in the undominated case. A simple sufficiency argument shows that there is a version of the posterior given $X_1, X_2, \ldots, X_n$ that is a function of the empirical measure.

Definition 1.3.1. For each n, let $\Pi(\cdot|\mathbf{X_n})$ be a posterior given $X_1, X_2, \ldots, X_n$. The sequence $\{\Pi(\cdot|\mathbf{X_n})\}$ is said to be consistent at θ_0 if there is a $\Omega_0 \subset \Omega$ with $P_{\theta_0}^\infty(\Omega_0) = 1$ such that if ω is in Ω_0, then for every neighborhood U of θ_0,

$$\Pi(U|\mathbf{X_n}(\omega)) \to 1$$

Remark 1.3.1. When Θ is a metric space $\{\theta : \rho(\theta, \theta_0) < 1/n : n \geq 1\}$ forms a base for the neighborhoods of θ_0, and hence one can allow the set of measure 1 to depend on U. In other words, it is enough to show that for each neighborhood U of θ_0,

$$\Pi(U|\mathbf{X_n}(\omega)) \to 1 \text{ a.e. } P_{\theta_0}^\infty$$

Further, when Θ is a separable metric space it follows from the Portmanteau theorem that consistency of the sequence $\{\Pi(\cdot|\mathbf{X_n})\}$ at θ_0 is equivalent to requiring that $\{\Pi(\cdot|\mathbf{X_n})\} \overset{weakly}{\to} \delta_{\theta_0}$ a.e.P_{θ_0}.

Thus the posterior is consistent at θ_0, if with P_{θ_0} probability 1, as n gets large, the posterior concentrates around θ_0.

Why should one require consistency at a particular θ_0? A Bayesian may think of θ_0 as a plausible value and question what would happen if θ_0 were indeed the true value and the sample size n increases. Ideally the posterior would learn from the data and put more and more mass near θ_0. The definition of consistency captures this requirement.

The idea goes back to Laplace, who had shown the following. If $X_1, X_2, \ldots, X_n$ are i.i.d. Bernoulli with $P_\theta(X = 1) = \theta$ and $\pi(\theta)$ is a prior density that is continuous and positive on $(0,1)$, then the posterior is consistent at all θ_0 in $(0,1)$. Von Mises [162] calls this the second fundamental law of large numbers; the first being Bernoulli's weak law of large numbers.

An elementary proof of Lapalace's result for a beta prior may be of some interest. Let the prior density with respect to Lebesgue measure on $(0,1)$ be

$$\Pi(\theta) = \frac{\Gamma(\alpha+\beta)}{\Gamma(\alpha)\,\Gamma(\beta)}\theta^{\alpha-1}(1-\theta)^{\beta-1}$$

Then the posterior density given $X_1, X_2, \ldots, X_n$ is

$$\frac{\Gamma(\alpha+\beta+n)}{\Gamma(\alpha+r)\,\Gamma(\beta+(n-r))}\theta^{\alpha+r-1}(1-\theta)^{\beta+(n-r)-1}$$

where r is the number of X_is equal to 1. An easy calculation shows that the posterior mean is

$$E(\theta|X_1, X_2, \ldots, X_n) = \left(\frac{\alpha+\beta}{\alpha+\beta+n}\right)\frac{\alpha}{\alpha+\beta} + \left(\frac{n}{\alpha+\beta+n}\right)\frac{r}{n}$$

which is a weighted combination of the consistent estimate r/n of the true value θ_0 and the prior mean $\alpha/(\alpha+\beta)$. Because the weight of r/n goes to 1,

$$E(\theta|X_1, X_2, \ldots, X_n) \to \theta_0 \text{ a.e. } P_{\theta_0}$$

A similar easy calculation shows that the posterior variance

$$Var(\theta|X_1, X_2, \ldots, X_n) = \frac{(\alpha+r)(\beta+(n-r))}{(\alpha+\beta+n)^2(\alpha+\beta+n+1)}$$

goes to 0 with probability 1 under θ_0. An application of Chebyshev's inequality completes the proof.

1.3.2 Posterior Consistency and Posterior Robustness

Posterior consistency is also connected with posterior robustness. A simple result is presented next [84].

Theorem 1.3.1. *Assume that the family $\{P_\theta : \theta \in \Theta\}$ is dominated by a σ-finite measure μ and let p_θ denote the density of P_θ. Let θ_0 be an interior point of Θ and π_1, π_2 be two prior densities with respect to a measure ν, which are positive and* *continuous at* θ_0. *Let* $\pi_i(\theta|\mathbf{X_n}), i = 1,2$ *denote the posterior densities of* θ *given* $\mathbf{X_n}$. *If* $\pi_i(\cdot|\mathbf{X_n}), i = 1,2$ *are both consistent at* θ_0 *then*

$$\lim_{n\to\infty} \int |\pi_1(\theta|\mathbf{X_n}) - \pi_2(\theta|\mathbf{X_n})| \; d\nu(\theta) = 0 \quad a.s\ P_{\theta_0}$$

Proof. We will show that with $P_{\theta_0}^\infty$ probability 1,

$$\int_\Theta \pi_2(\theta|\mathbf{X_n}) \left| 1 - \frac{\pi_1(\theta|\mathbf{X_n})}{\pi_2(\theta|\mathbf{X_n})} \right| d\nu(\theta) \to 0$$

Fix $\delta > 0, \eta > 0$, and $\epsilon > 0$ and use the continuity at θ_0 to obtain a neighborhood U of θ_0 such that for all $\theta \in U$

$$\left| \frac{\pi_1(\theta)}{\pi_2(\theta)} - \frac{\pi_1(\theta_0)}{\pi_2(\theta_0)} \right| < \delta \text{ and } |\pi_j(\theta_0) - \pi_j(\theta)| < \delta \text{ for } j = 1,2.$$

By consistency there exists $\Omega_0, P_{\theta_0}^\infty(\Omega_0) = 1$, such that for $\omega \in \Omega_0$,

$$\Pi_j(U|\mathbf{X_n}(\omega)) = \frac{\int_U \prod_1^n p_\theta(X_i(\omega))\ \pi_j(\theta)\ d\nu(\theta)}{\int_\Theta \prod_1^n p_\theta(X_i(\omega))\ \pi_j(\theta)\ d\nu(\theta)} \to 1$$

Fix $\omega \in \Omega_0$ and choose n_0 such that, for $n > n_0$,

$$\Pi_j(U|\mathbf{X_n}(\omega)) \geq 1 - \eta \text{ for } j = 1,2$$

Note that

$$\frac{\pi_1(\theta|\mathbf{X_n})}{\pi_2(\theta|\mathbf{X_n})} = \frac{\pi_1(\theta) \int_\Theta \prod_1^n p_\theta(X_i)\ \pi_2(\theta)\ d\nu(\theta)}{\pi_2(\theta) \int_\Theta \prod_1^n p_\theta(X_i)\ \pi_1(\theta)\ d\nu(\theta)}$$

Hence for $n > n_0$ and $\theta \in U$, after some easy manipulation, we have

$$\begin{aligned}
\left(\frac{\pi_1(\theta_0)}{\pi_2(\theta_0)} - \delta\right)&(1-\eta)\frac{\int_U \prod_1^n p_\theta(X_i(\omega))\ \pi_2(\theta)\ d\nu(\theta)}{\int_U \prod_1^n p_\theta(X_i(\omega))\ \pi_1(\theta)\ d\nu(\theta)} \\
&\leq \frac{\pi_1(\theta|\mathbf{X_n}(\omega))}{\pi_2(\theta|\mathbf{X_n}(\omega))} \\
&\leq \left(\frac{\pi_1(\theta_0)}{\pi_2(\theta_0)} + \delta\right)(1-\eta)^{-1}\frac{\int_U \prod_1^n p_\theta(X_i(\omega))\ \pi_2(\theta)\ d\nu(\theta)}{\int_U \prod_1^n p_\theta(X_i(\omega))\ \pi_1(\theta)\ d\nu(\theta)}
\end{aligned}$$

and by the choice of U,

$$(\pi_j(\theta_0) - \delta) \int_U \prod_1^n p_\theta(X_i(\omega))\, d\nu(\theta) \leq \int_U \prod_1^n p_\theta(X_i(\omega))\pi_j(\theta) d\nu(\theta)$$
$$\leq (\pi_j(\theta_0) + \delta) \int_U \prod_1^n p_\theta(X_i(\omega))\, d\nu(\theta) \tag{1.1}$$

Using (1.1) we have, again for $\theta \in U$,

$$\left(\frac{\pi_1(\theta_0)}{\pi_2(\theta_0)} - \delta\right)(1-\eta)\left(\frac{\pi_2(\theta_0) - \delta}{\pi_1(\theta_0) + \delta}\right) \leq \frac{\pi_1(\theta|\mathbf{X_n}(\omega))}{\pi_2(\theta|\mathbf{X_n}(\omega))}$$
$$\leq \left(\frac{\pi_1(\theta_0)}{\pi_2(\theta_0)} + \delta\right)(1-\eta)^{-1}\left(\frac{\pi_2(\theta_0) + \delta}{\pi_1(\theta_0) - \delta}\right)$$

so that for δ, η small

$$\left|\frac{\pi_1(\theta|\mathbf{X_n}(\omega))}{\pi_2(\theta|\mathbf{X_n}(\omega))} - 1\right| < \epsilon$$

Hence, for $n > n_0$,

$$\int |\pi_1(\theta|\mathbf{X_n}(\omega)) - \pi_2(\theta|\mathbf{X_n}(\omega))|\, d\nu(\theta)$$
$$\leq \int_U \pi_2(\theta|\mathbf{X_n}(\omega)) \left|1 - \frac{\pi_1(\theta|\mathbf{X_n}(\omega))}{\pi_2(\theta|\mathbf{X_n}(\omega))}\right| d\nu(\theta) + 2\eta$$
$$\leq \epsilon(1-\eta) + 2\eta$$

This completes the proof. □

Another notion related to Theorem 1.3.1 is that of merging where, instead of the posterior, one looks at the predictive distribution of $X_{n+1}, X_{n+2}, \ldots$ given $X_1 \ldots, X_n$. Here the attempt is to formalize the idea that two Bayesians starting with different priors Π_1 and Π_2 would eventually agree in their prediction of the distribution of future observations.

For a prior Π if we define, for any measurable subset C of Ω

$$\lambda_\Pi (C|\mathbf{X_n}) = \int_\Theta P_\theta^\infty(C)\Pi(d\theta|\mathbf{X_n})$$

then, $\lambda_{\Pi}(\cdot|\mathbf{X_n})$ is a version of the predictive distribution of $X_{n+1}, X_{n+2}, \ldots$ given $X_1, X_2, \ldots, X_n$. Note that given $\mathbf{X_n}$, the predictive distribution is a probability measure on $\Omega = \mathbb{R}^\infty$.

Let $\lambda_{\Pi_1}(\cdot|\mathbf{X_n})$ and $\lambda_{\Pi_2}(\cdot|\mathbf{X_n})$ be two predictive distributions, corresponding to priors Π_1 and Π_2.

An early result in merging is due to Blackwell and Dubins [24]. They showed that if Π_2 is absolutely continuous with respect to Π_1, then for θ in a set of Π_2 probability 1, the total variation distance between $\lambda_{\Pi_1}(\cdot|\mathbf{X_n})$ and $\lambda_{\Pi_2}(\cdot|\mathbf{X_n})$ goes to 0 almost surely P_θ^∞.

The connection with consistency was observed by Diaconis and Freedman [46]. Towards this, say that the predictive distributions *merge weakly with respect to* P_{θ_0} if there exists $\Omega_0 \subset \Omega$ with $P_\theta^\infty(\Omega_0) = 1$, such that for each $\omega \in \Omega_0$,

$$\left|\int \phi(\omega')\lambda_{\Pi_1}(d\omega'|\mathbf{X_n}(\omega)) - \int \phi(\omega')\lambda_{\Pi_2}(d\omega'|\mathbf{X_n}(\omega))\right| \to 0$$

for all bounded continuous functions ϕ on Ω.

Proposition 1.3.1. *Assume that $\theta \mapsto P_\theta$ is 1-1 and continuous with respect to weak convergence. Also assume that there is a compact set K such that $P_\theta(K) = 1$ for all θ.*

If Π_1 and Π_2 are two priors such that the posteriors $\Pi_1(\cdot|\mathbf{X_n})$ and $\Pi_2(\cdot|\mathbf{X_n})$ are consistent at θ_0, then the predictive distributions $\lambda_{\Pi_1}(\cdot|\mathbf{X_n})$ and $\lambda_{\Pi_2}(\cdot|\mathbf{X_n})$, merge weakly with respect to P_{θ_0}.

Proof. Let $\mathcal{G}$ be the class of all functions on Ω that are finite linear combinations of functions of the form

$$\phi(\omega) = \prod_1^k f_i(\omega_i)$$

where $f_1, f_2, \ldots, f_k$ are continuous functions on K. It is easy to see that if $\phi \in \mathcal{G}$ then $\theta \mapsto \int \phi(\omega')\, dP_\theta^\infty(\omega')$ is continuous. Further, by the Stone-Weirstrass theorem $\mathcal{G}$ is dense in the space of all continuous functions on K^∞.

From the definition of $\lambda_{\Pi_1}(\cdot|\mathbf{X_n})$ and $\lambda_{\Pi_2}(\cdot|\mathbf{X_n})$, if Ω_0 is the set where the posterior converges to δ_{θ_0}, then for $\omega \in \Omega_0$, for $\phi \in \mathcal{G}$,

$$\int \phi(\omega')\lambda_{\Pi_i}(d\omega'|\mathbf{X_n}(\omega)) = \int_\Theta \int_\Omega \phi(\omega')\, dP_\theta^\infty(\omega')\, \Pi_i(d\theta|(\mathbf{X_n}(\omega))$$

The inside integral gives rise to a bounded continuous function of θ. Hence by weak consistency at θ_0, for both $i = 1, 2$ the right-hand side converges to $\int_\Omega \phi(\omega')\, dP_{\theta_0}^\infty(\omega')$. This yields the conclusion. □

Further connections between merging and posterior consistency is explored in Diaconis and Freedman[46].

Note a few technical remarks: According to the definition, posterior consistency is a property that is specific to the fixed version $\Pi(\cdot|\mathbf{X_n})$. Measure theoretically, the posterior is unique only up to λ_n null sets. So the posterior is uniquely defined up to P_{θ_0} if $P_{\theta_0}^n$ is dominated by λ_n. Without this condition it is easy to construct examples of two versions $\{\Pi_1(\cdot|\mathbf{X_n})\}$ and $\{\Pi_2(\cdot|\mathbf{X_n})\}$ such that one is consistent and the other is not. It is easy to show that if $\{P_\theta \in \Theta\}$ are all mutually absolutely continuous and $\{\Pi_1(\cdot|\mathbf{X_n})\}$ and $\{\Pi_2(\cdot|\mathbf{X_n})\}$ are two versions of the posterior, then $\{\Pi_1(\cdot|\mathbf{X_n})\}$ is consistent iff $\{\Pi_2(\cdot|\mathbf{X_n}\}$ is.

1.3.3 Doob's Theorem

An early result on consistency is the following theorem of Doob [49].

Theorem 1.3.2. *Suppose that Θ and $\mathbf{X}$ are both complete separable metric spaces endowed with their respective Borel σ-algebras $\mathcal{B}(\Theta)$ and $\mathcal{A}$ and let $\theta \mapsto P_\theta$ be 1-1. Let Π be a prior and $\{\Pi(\cdot|\mathbf{X_n})\}$ be a posterior. Then there exists a $\Theta_0 \subset \Theta$, with $\Pi(\Theta_0) = 1$ such that $\{\Pi(\cdot|\mathbf{X_n})\}_{n\geq 1}$ is consistent at every $\theta \in \Theta_0$.*

Proof. The basic idea of the proof is simple. On the one hand, because for each θ the empirical distribution converges a.s. P_θ^∞ to P_θ, given any sequence of x_i's we can pinpoint the true θ. On the other hand, any version of the posterior distributions $\Pi(\cdot|\mathbf{X_n})$, via the martingale convergence theorem, converge a.s. with respect to the marginal λ_Π, to the posterior given the entire sequence. One then equates these two versions to get the result. A formal proof of these observations needs subtle measure theory.

As before let, $\Omega = \mathbf{X}^\mathcal{N}$, $\mathcal{B}$ be the product σ-algebra on Ω, λ_Π denote both the joint distribution of θ and $X_1, X_2, \ldots$ and the marginal distribution of $X_1, X_2, \ldots$. Let C be a subset of Θ, then by the martingale convergence theorem, as $n \to \infty$,

$$\Pi(C|X_1, X_2, \ldots, X_n) \to E(I_C|X_1, X_2, \ldots) \doteq f \text{ a.e. } \lambda_\Pi$$

We point out that the functions considered above are, formally, functions of two variables (θ, ω). I_C, is to be interpreted as $I_{C\times\Omega}$ and f is to be thought of as $f(\theta, \omega) = f(\omega)$ and so on.

We shall show that there exists a set Θ_0 with $\Pi(\Theta_0) = 1$ such that

$$\text{for } \theta \in \Theta_0 \cap C, \quad f = 1 \text{ a.e. } P_\theta^\infty \tag{1.2}$$

This would establish the theorem. To see this, take $\mathcal{U} = \{U_1, U_2, \ldots, \}$ a base for the open sets of Θ. Take $C = U_i$ in the above step and obtain the corresponding $\Theta_{0i} \subset \Theta$ satisfying (1.2). If we set $\Theta_0 = \cap_i \Theta_{0i}$ then (1.2) translates into " the posterior is consistent at all $\theta \in \Theta_0$".

To establish (1.2), let $\mathcal{A}_0$ be a countable algebra generating $\mathcal{A}$. Let

$$E = \{(\theta, \omega) : \lim_{n\to\infty} \frac{1}{n} \sum_1^n \delta_{X_i(\omega)}(A) = P_\theta(A) \text{ for all } A \in \mathcal{A}_0\}$$

The set E, since it arises from the limit of a sequence of measurable functions, is a measurable set and further by the law of large numbers for each θ the sections E_θ satisfy

(i) for all $\theta, P_\theta^\infty(E_\theta) = 1$

(ii) if $\theta \neq \theta', E_\theta \cap E_{\theta'} = \emptyset$

Define

$$f^*(\omega) = \begin{cases} 1 & \text{if}, \omega \in \cup_{\theta \in C} E_\theta \\ 0 & \text{otherwise.} \end{cases}$$

It is a consequence of a deep result in set theory that $\cup_{\theta \in C} E_\theta$ is measurable, from which it follows that f^* is measurable.

From its definition, f^* satisfies:

1. for all $\theta \in C$, $f^* = 1$ a.e. P_θ^∞

2. for all θ not in C, $f^* = 0$ a.e. P_θ^∞

   In other words for all θ, $f^* = I_C(\theta) f^*$ a.e. P_θ^∞

We claim that f^* is a version of $E(I_C | X_1, X_2, \ldots)$. For any measurable set $B \in \mathcal{B}$,

$$\int I_B f^* d\lambda_\Pi = \int I_B I_C(\theta) f^* dP_\theta^\infty d\Pi(\theta) = \int I_C(\theta) P_\theta^\infty(B) d\Pi(\theta) = \lambda_\Pi(C \times B)$$

Since f and f^* are both versions of $E(I_C | X_1, X_2, \ldots)$, we have

$$f = f^* \text{ a.e. } \lambda_\Pi$$

By Fubini's theorem, there exists a set Θ_0 with $\Pi(\Theta_0) = 1$, such that for θ in Θ_0

$$f = f^* \text{ a.e.} P_\theta^\infty$$

(1.2) follows easily from the properties 1 and 2 of f^* mentioned earlier.
This completes the proof. □

Remark 1.3.2. A well known result in set theory, the Borel Isomorphism theorem, states that any two uncountable Borel sets of complete separable metric spaces are isomorphic [[153],Theorem 3.3.13]. The result that we used from set theory is a version of this theorem which states that if **S** and **T** are Borel subsets of complete metric spaces and if ϕ is a 1-1 measurable function from **S** into **T**, then, the range of ϕ is a measurable set and ϕ^{-1} is also measurable. To get the result that we used, just set $\mathbf{S} = E, \mathbf{T} = \Omega$ and $\phi(\theta, \omega) = \omega$.

Remark 1.3.3. Another consequence of the Borel Isomorphism theorem is that Doob's theorem holds even when Θ and $\mathcal{X}$ are just Borel subsets of a complete separable metric space.

Many Bayesians are satisfied with Doob's theorem, which provides a sort of internal consistency but fails to answer the question of consistency at a specific θ_0 of interest to a Bayesian. Moreover in the infinite-dimensional case, the set of θ_0 values where consistency holds may be a very small set topologically [70] and may exclude infinitely many θ_0s of interest. Disturbing examples and general results of this kind appear in Freedman [69] in the context of an infinite-cell multinomial.

If θ_0 is not in the support of the prior Π then there exists an open set U such that $\Pi(U) = 0$. This implies that $\Pi(U|\mathbf{X_n})$ =0 a.s λ^n. Hence,it is not reasonable to expect consistency outside the support of Π. Ideally, one might hope for consistency at all θ_0 in the support of Π. This is often true for a finite-dimensional Θ. However, for an infinite-dimensional Θ this turns out to be too strong a requirement. We will often prove consistency for a large set of θ_0s . A Bayesian can then decide whether it includes all or most of the θ_0s of interest.

1.3.4 Wald-Type Conditions

We begin with a uniform strong law.

Theorem 1.3.3. *Suppose that K is a compact subset of a separable metric space. Let $T(\cdot,\cdot)$ be a real-valued function on $\theta \times \mathbb{R}$ such that*

(i) for each $x, T(\cdot, x)$ is continuous in θ, and

(ii) for each $\theta, T(\theta, \cdot)$ is measurable.

Let $X_1, X_2, \ldots$ i.i.d. random variables defined on $(\Omega, \mathcal{A}, P)$ with $E(T(\theta, X_1)) = \mu(\theta)$ and assume further that

$$E\left(\sup_{\theta\in K} |T(\theta, X_i)|\right) < \infty$$

Then, as $n \to \infty$,

$$\sup_{\theta\in K} \left|\frac{1}{n}\sum_1^n T(\theta, X_i) - \mu(\theta)\right| \to 0 \quad a.s.\ P$$

Proof. Continuity of $T(., x)$ and separability ensures that $\sup_{\theta\in K} |T(\theta, X_i)|$ is measurable. It follows from the dominated convergence theorem that $\theta \mapsto \mu(\theta)$ is continuous. Another application of the dominated convergence theorem shows that for any $\theta_0 \in K$,

$$\lim_{\delta\to 0} E\left[\sup_{|\theta-\theta_0|<\delta} |T(\theta, X_1) - \mu(\theta) - T(\theta_0, X_1) - \mu(\theta_0)|\right] = 0$$

Let $Z_{ji} = \sup_{\rho(\theta,\theta_i)<\delta_i} |T(\theta, X_j) - \mu(\theta) - T(\theta_i, X_j) - \mu(\theta_i)|$. By compactness of K, there exist $\theta_1, \theta_2, \ldots, \theta_k$ and $\delta_1, \delta_2, \ldots, \delta_k$ such that $K = \cup_1^k \{\theta : \rho(\theta, \theta_i) < \delta_i\}$, and $EZ_{1i} < \epsilon$ for $i = 1, 2, \ldots, k$.

By the strong law of large numbers, since $E(Z_{1,i}) < \epsilon$ for $i = 1, 2, \ldots, k$, there is a Ω_0 with $P(\Omega_0) = 1$ such that for $\omega \in \Omega_0$, $n > n(\omega)$, for $i = 1, 2, \ldots, k$,

$$\frac{1}{n}\sum_1^n Z_{j,i} < 2\epsilon$$

and

$$\left|\frac{1}{n}\sum_{j=1}^n T(\theta_i, X_j) - \mu(\theta_i)\right| < \epsilon$$

Now if $\theta \in \{\theta : \rho(\theta, \theta_i) < \delta_i\}$,

$$\begin{aligned}&\frac{1}{n}\left|\sum T(\theta, X_j(\omega)) - \mu(\theta)\right| \\ &\leq \frac{1}{n}\sum Z_{j,i}(\omega) + \left|\frac{1}{n}\sum T(\theta_i, X_j(\omega)) - \mu(\theta_i)\right| \leq 3\epsilon\end{aligned}$$

Hence $\sup_{\theta\in K} |\frac{1}{n}\sum T(\theta, X_j(\omega)) - \mu(\theta)| < 3k\epsilon$. □

Remark 1.3.4. A very powerful approach to uniform strong laws is through empirical processes. One considers a sequence of i.i.d. random variables X_i and studies uniformity over a family of functions $\mathcal{F}$ with an integrable envelope function ϕ, i.e., $E(\phi) < \infty$, and $|f(x)| \le |\phi(x)|, f \in \mathcal{F}$. Good references are Pollard [[139], II.2] and Van der Vaart and Wellner [[161], 2.4].

Here is an easy consequence of the last theorem. First a definition: Let Θ be a space endowed with a σ-algebra and $\theta \mapsto P_\theta$ be 1-1. For each θ in Θ, let $X_1, X_2, \ldots$ be i.i.d. P_θ . Assume that P_θ s are dominated by a σ-finite measure μ and $p_\theta = dP_\theta/d\mu$.

Definition 1.3.2. A measurable function $\hat{\theta}_n(X_1, X_2, \ldots, X_n)$ taking values in Θ is called a maximum likelihood estimate (MLE) if the likelihood function at $X_1, X_2, \ldots, X_n$ attains its maximum at $\hat{\theta}_n(X_1, X_2, \ldots, X_n)$ or formally,

$$\prod_1^n p_{\hat{\theta}_n(X_1,X_2,\ldots,X_n)}(X_i) = \sup_\theta \prod_1^n p_\theta(X_i)$$

Theorem 1.3.4. *Let Θ be compact metric. For a fixed θ_0, let*

$$T(\theta, x) = \log\left(p_\theta(x)/p_{\theta_0}(x)\right)$$

If $T(\theta, X_i)$ satisfy the assumptions of Theorem 1.3.3 with P=P_{θ_0}, then

1. *any MLE $\hat{\theta}_n$ is consistent at θ_0 ;*
2. *if Π is a prior on Θ and if θ_0 is in the support of Π then the posterior defined by the density (with respect to Π) $\prod_1^n p_\theta(X_i)/\int \prod_1^n p_\theta(X_i)\, d\Pi(\theta)$ is consistent at θ_0.*

Proof. (i) Take any open neighborhood U of θ_0 and let $K = U^c$. Note that $\mu(\theta) = E_{\theta_0}(T(\theta, X_i)) = -K(\theta_0, \theta) < 0$ for all θ and hence by the continuity of $\mu(\cdot)$, $\sup_{\theta \in K} \mu(\theta) < 0$.

On the one hand, by Theorem 1.3.3, given $0 < \epsilon < |\sup_{\theta \in K} \mu(\theta)|$, there exists $n(\omega)$, such that for $n > n(\omega)$,

$$\sup_{\theta \in K} \left| \frac{1}{n} \sum T(\theta, X_i) - \mu(\theta) \right| < \epsilon$$

On the other hand, $(1/n) \sum T(\hat{\theta}_n, X_i) \ge 0$. So $\hat{\theta}_n \notin K$ and hence $\hat{\theta}_n \in U$.

As a curiosity, we note that we have not used the measurability assumption on $\hat{\theta}_n$. We have shown that the samples where the MLE is consistent contain a measurable set of $P_{\theta_0}^\infty$ measure 1.

(ii) Let U be a neighborhood of θ_0 . We shall show that $\Pi(U|X_1, X_2, \ldots, X_n) \to 1$ a.s P_{θ_0}. As before, let $K = U^c$ and $T(\theta, X_i) = \log(p_\theta(X_i)/p_{\theta_0}(X_i))$ and $U_\delta = \{\theta : \rho(\theta, \theta_0) < \delta\}$. Let

$$A_1 = \inf_{\theta \in \bar{U}_\delta} \mu(\theta) \text{ and } A_2 = \sup_{\theta \in K} \mu(\theta)$$

Clearly $A_1 < 0, A_2 < 0$. Choose δ small enough so that $U_\delta \subset U$ and $|A_1| < |A_2|$. This can be done because $\mu(\theta)$ is continuous and as $\delta \downarrow 0$, $\inf_{\theta \in \bar{U}_\delta} \mu(\theta) \uparrow 0$.

Choose $\epsilon > 0$ such that $A_1 - \epsilon > A_2 + \epsilon$. By applying the uniform strong law of large numbers to K and $\bar{U}_\delta$, for ω in a set of P_{θ_0}-measure 1, there exists $n(\omega)$ such that for $n > n(\omega)$,

$$\left| \frac{1}{n} \sum T(\theta, X_i) - \mu(\theta) \right| < \epsilon \quad \forall \theta \in K \cup \bar{U}_\delta$$

Now

$$\begin{aligned}
\Pi(U|X_1, X_2, \ldots, X_n) &= \frac{\int_U e^{\sum_1^n T(\theta, X_i)} d\Pi(\theta)}{\int_U e^{\sum_1^n T(\theta, X_i)} d\Pi(\theta) + \int_{U^c} e^{\sum_1^n T(\theta, X_i)} d\Pi(\theta)} \\
&\geq 1 \Big/ \left(1 + \frac{\int_K e^{\sum_1^n T(\theta, X_i)} d\Pi(\theta)}{\int_{U_\delta} e^{\sum_1^n T(\theta, X_i)} d\Pi(\theta)} \right) \\
&\geq 1 \Big/ \left(1 + \frac{\Pi(K) e^{n(A_2 + \epsilon)}}{\Pi(U_\delta) e^{n(A_1 - \epsilon)}} \right)
\end{aligned}$$

Since $A_2 - A_1 + 2\epsilon < 0$ and $\Pi(U_\delta) > 0$, the last term converges to 1 as $n \to \infty$. $\square$

Remark 1.3.5. Theorem 1.3.4 is related to Wald's paper [163]. His conditions and proofs are similar but he handles the noncompact case by assumptions of the kind given next which ensure that the MLE $\hat{\theta}_n$ is inside a compact set eventually, almost surely. Here are two assumptions; we will refer to them as Wald's conditions:

1. Let $\Theta = \cup K_i$ where the K_is are compact and $K_1 \subset K_2 \subset \ldots$. For any sequence $\theta_i \in K_{(i-1)}^c \cap K_i, \qquad \lim_i p(x, \theta_i) = 0.$
2. Let $\phi_i(x) = \sup_{\theta \in K_{(i-1)}^c} (\log p(x, \theta)/p(x, \theta_0))$. Then $E_{\theta_0} \phi_i^+(X_1) < \infty$ for some i.

Assumption (1) implies that $\lim_{i\to\infty} \phi_i(x) = -\infty$. Using Assumption (2), the monotone convergence theorem and the dominated convergence theorem one can show

$$\lim_{i\to\infty} E_{\theta_0}\phi_i(X_1) = -\infty$$

Thus, given any $A_3 < 0$, we can choose a compact set K_j such that

$$E_\theta \phi_j = E_{\theta_0} \sup_{\theta \in K^c_{(j-1)}} \log p(X_i, \theta) - E_{\theta_0} p(X_i, \theta_0) < A_3$$

Using

$$\frac{1}{n} \sup_{\theta\in K_j^c} \sum_1^n \log p(X_i,\theta)/p(X_i,\theta_0) \le \frac{1}{n}\sum_1^n \sup_{\theta\in K_j^c} \log p(X_i,\theta)/p(X_i,\theta_0)$$

and applying the usual SLLN to $1/n \sum_{i=1}^n \phi_j(X_i)$, it can be concluded that eventually it is ≤ 0 a.s. P_{θ_0}. This implies that eventually, $\hat\theta_n \in K_j$ a.s P_{θ_0}. This result for the compact case can now be used to establish consistency of $\hat\theta_n$.

Remark 1.3.6. Suppose Θ is a convex open subset of $\mathbb{R}^p$ and for $\theta \in \Theta$,

$$\log f_\theta(x_i) = A(\theta) + \sum_1^p \theta_j x_i + \psi(x_i)$$

and $\left(\frac{\partial \log f_\theta}{\partial\theta}\right), \left(\frac{\partial^2 \log f_\theta}{\partial\theta^2}\right)$ exist. Then by Lehman[123]

$$I(\theta) = E_\theta\left(\frac{\partial \log f_\theta}{\partial\theta}\right)^2 = -E_\theta\left(\frac{\partial^2 \log f_\theta}{\partial\theta^2}\right) = \frac{d^2 A(\theta)}{d\theta^2} > 0$$

Thus the likelihood is log concave. In this case also the MLE is consistent without compactness by a simple direct argument using Theorem 1.3.4. Start with a bounded open rectangle around θ_0 and let K be its closure. Because K is compact, the MLE $\hat\theta_K$, with K as the parameter space exists and given any open neighborhood $V \subset K$ of θ_0, $\hat\theta_K$ lies in V with probability tending to 1. If $\hat\theta_K \in V$ it must be a local maximum and hence a global maximum because of log concavity. This completes the proof. In the log concave situation more detailed and general results are available in Hjort and Pollard [101]

Remark 1.3.7. Under the assumptions of either of the last remarks it can be verified that the posterior is consistent.

The next two examples show that even in the finite-dimensional case consistency of the MLE and the posterior do not always occur together.

Example. This example is due to Bahadur. Our presentation follows Lehman [124]. Here $\Theta = \{1, 2, \ldots, \}$. For each θ, we define a density f_θ on $[0,1]$ as follows:
Let $h(x) = e^{1/x^2}$. Define $a_0 = 1$ and a_n by $\int_{a_n}^{a_{n-1}} (h(x) - C)\, dx = 1 - C$ where $0 < C < 1$. Because $\int_0^1 e^{1/x^2} dx = \infty$ it is easy to show that a_ns are unique and tend to 0 as $n \to \infty$.

Define $f_k(x)$ on $[0,1]$ by

$$f_k(x) = \begin{cases} h(x) & \text{if } a_k < x < a_{k-1} \\ C & \text{otherwise} \end{cases}$$

For each k, let $X_1, X_2, \ldots, X_n$ be i.i.d. f_k. Denoting $\min(X_1, X_2, \ldots, X_n)$ by $X_1^{(n)}$, we can write the likelihood function as

$$L_{X_1, X_2, \ldots, X_n}(k) = \begin{cases} C^n & \text{if } a_{k-1} < X_1^{(n)} \\ \prod d_i & \text{if } a_{k-1} > X_1^{(n)} \end{cases}$$

where $d_i = I_{A_i}(X_i) h(X_i) + I_{A_i^c}(X_i) C$ and $A_i = (a_i, a_{i-1}]$.

Because $h(x) > 1$, the likelihood function attains its maximum in the finite set $\{k : a_k > X_1^{(n)}\}$, and hence an MLE exists.

Fix $j \in \Theta$. We shall show that any MLE $\hat{\theta}_n$ fails to be consistent at j by showing

$$P_j \left\{ \sum_1^n \log \frac{f_{\hat{\theta}_n}(X_i)}{f_j(X_i)} > 1 \right\} \to 1$$

Actually, we show more, namely, for each j, $\hat{\theta}_n$ converges in P_j probability to ∞. Fix m and consider the set $\Theta_1 = \{1, 2, \ldots, m\} \subset \Theta$. It is enough to show as $n \to \infty$,

$$P_j\{\hat{\theta}_n \notin \Theta_1\} \to 1$$

Define $k^*(X_1, X_2, \ldots, X_n)$ to be k if $X_1^{(n)} \in (a_k, a_{k-1})$. Because the likelihood function at $\hat{\theta}_n$ is larger than that at k^* it suffices to show that

$$\sum_1^n \log \frac{f_{K_n^*}(X_i)}{f_j(X_i)} \to \infty \text{ in } P_j \text{ probability}$$

Towards this first note that for any k and j,

$$\sum_1^n \log \frac{f_k(X_i)}{f_j(X_i)} = \sum^{(k)} \log \frac{h(X_i)}{C} - \sum^{(j)} \log \frac{h(X_i)}{C}$$

where $\sum^{(k)}$ is the sum over all i such that $X_i \in (a_k, a_{k-1})$. With k_n^* in place of k, we have

$$\sum_1^n \log \frac{f_{k_n^*}(X_i)}{f_j(X_i)} = \sum^{(*)} \log \frac{h(X_i)}{C} - \sum^{(j)} \log \frac{h(X_i)}{C}$$

where $\sum^{(*)}$ is the sum over all i such that $X_i \in (a_{k_n^*}, a_{k_n^*-1})$.

Because for each x, $h(x)/C > 1$, the first sum on the right-hand side is larger than $\log(h(X_{(1)}^{(n)})/C)$, one of the terms appearing in the sum. Formally,

$$\sum^{(*)} \log \frac{h(X_i)}{C} \geq \log \frac{h(X_{(1)}^{(n)})}{C}$$

On the other hand, because h is decreasing

$$\sum^{(j)} \log \frac{h(X_i)}{C} \leq \nu_{k,n} \log \frac{h(a_j)}{C}$$

where $\nu_{k,n}$ is the number of X_is in (a_j, a_{j-1}).

Thus

$$\sum_1^n \frac{1}{n} \log \frac{f_{k_n^*}(X_i)}{f_j(X_i)} \geq \frac{1}{n} \log \frac{h(X_1^{(n)})}{C} - \frac{1}{n} \nu_{k,n} \frac{\log h(a_j)}{C}$$

Because $(1/n)\nu_{k,n} \to P_j(a_j, a_{j-1})$, the second term converges to a finite constant. We complete the proof by showing

$$\frac{1}{n} \log h(X_1^{(n)}) = \frac{1}{n(X_1^{(n)})^2} \to \infty$$

in P_j probability.

Toward this, consider $X \sim P_j$ and $Y \sim U(0, 1/C)$. Then for all x,

$$P(X > x) \leq P(Y > x)$$

To see this, $P(Y > x) = 1 - Cx$ and for $P(X > x)$ note that

$$\text{if } x > a_{j-1} \text{ then } P(X > x) = C(1 - x) < 1 - Cx$$

$$\text{If } x \in (a_j, a_{j-1}) \text{ then } P(X > x) \leq 1 - a_j C \leq 1 - Cx$$

and

$$\text{if } x < a_j, \text{ then } P(X > x) = 1 - Cx$$

Consequently $X_{(1)}^{(n)}$ is stochastically smaller than $Y_{(1)}^{(n)}$ and because h is decreasing

$$P\{h(X_{(1)}^{(n)}) > x\} \geq P\{h(Y_{(1)}^{(n)}) > x\}.$$

Therefore to show that $(1/n)\log h(X_{(1)}^{(n)}) \to \infty$ in P_j probability, it is enough to show that $(1/n)\log h(Y_{(1)}^{(n)}) \to \infty$ in $U(0, 1/C)$ probability. This follows because

$$\frac{1}{n}\log h(Y_{(1)}^{(n)}) = \frac{1}{n(Y_{(1)}^{(n)})^2}$$

and easy computation shows that $nY_{(1)}^{(n)}$ has a limiting distribution and is hence bounded in probability and $Y_{(1)}^{(n)} \to 0$ a.s.

On the other hand, Θ being countable, Doob's theorem assures consistency of the posterior at all $j \in \Theta$. This result also follows from Schwartz's theorem which provides more insight on the behavior of the posterior.

Intuitively, a Bayesian with a proper posterior is better off in such situations because a proper prior assigns a small probability to large values of K, which cause problems for $\hat{\theta}_n$. For an illuminating discussion of integrating rather than maximizing the likelihood, see the discussion of a counterexample due to Stein in [9].

Example. This is an example where the posterior fails to be consistent at θ_0 in the support of Π. This example is modeled after an example of Schwartz [145], but is much simpler. In the next example Θ is finite-dimensional. In the infinite-dimensional case there are many such examples due to Freedman [69] and Diaconis and Freedman [46], [45].

Let $\Theta = (0, 1) \cup (2, 3)$ and $X_1, X_2, \ldots, X_n$ be i.i.d $U(0, \theta)$. Let θ_0 =1. Π is a prior with density π, which is positive and continuous on Θ with $\pi(\theta) = e^{-1/(\theta-\theta_0)^2}$ on $(0, 1)$.Because $\int_0^1 \pi(\theta)\, d\theta < 1$, there exists such a prior density π, which is also positive on $(2, 3)$.

We will argue that the posterior density fails to be consistent at θ_0 by showing that the posterior probability of $(2, 3)$ goes to one in P_{θ_0} probability. The proof rests on the following facts both of which are easy to verify:
Let $X_{(n)}$ denote the maximum of $X_1, X_2, \ldots, X_n$. Then under P_{θ_0}, i.e., under $U(0,1)$, $n(X_{(n)} - \theta_0) = O_P(1)$. In fact, $n(X_{(n)} - \theta_0)$ converges to an exponential distribution. The second fact is $(1/n)\log(1 - X_{(n)}^{n-1}) \to 0$ in P_{θ_0} probability, because by direct calculation the distribution of $(1 - X_{(n)}^{n-1}) \overset{w}{\to} U(0,1)$.
Now the posterior probability of $(2, 3)$ is given by

$$\frac{\int_2^3 \frac{1}{\theta^n} I_{(0,\theta)}(X_{(n)})\ \pi(\theta)\ d\theta}{\int_0^1 \frac{1}{\theta^n} I_{(0,\theta)}(X_{(n)})\ \pi(\theta)\ d\theta + \int_2^3 \frac{1}{\theta^n} I_{(0,\theta)}(X_{(n)})\ \pi(\theta)\ d\theta}$$

Because $0 \leq X_{(n)} \leq 1$ a.e. P_{θ_0}, the numerator is equal to $\int_2^3 (1/\theta^n)\ \pi(\theta)\ d\theta$ and the first integral in the denominator is $\int_{X_{(n)}}^1 \frac{1}{\theta^n}\ \pi(\theta)\ d\theta$. So the posterior probability of $(2, 3)$ reduces to

$$\frac{1}{\left(1 + \frac{\int_{X_{(n)}}^1 \theta^{-n}\ \pi(\theta)\ d\theta}{\int_2^3 \theta^{-n}\ \pi(\theta)\ d\theta}\right)} = \frac{1}{\left(1 + \frac{I_1}{I_2}\right)}$$

Now

$$I_1 \leq \pi(X_{(n)}) \int_{X_{(n)}}^1 \theta^{-n}\ d\theta = \frac{\pi(X_{(n)})}{n-1} \frac{(1 - X_{(n)}^{n-1})}{X_{(n)}^{n-1}}$$

and $(1/n)\log I_1$ is less than

$$-\frac{n-1}{n}\log X_{(n)} - \frac{\log(n-1)}{n} + \frac{1}{n}\log(1 - X_{(n)}^{n-1}) + \frac{1}{n}\log \pi(X_{(n)})$$

As $n \to \infty$ the first two terms on the right side go to 0. The third goes to 0 by the second remark. The last term, using the explicit form of π on $(0, 1)$, goes to $-\infty$ in P_{θ_0} probability. Thus $(1/n)\log I_1 \to -\infty$ in P_{θ_0} probability.
On the other hand

$$\frac{1}{3^n}\Pi(2,3) < \int_2^3 \frac{1}{\theta^n}\ \pi(\theta)\ d\theta < \frac{1}{2^n}\Pi(2,3)$$

Hence

$$-(\log 3)\Pi(2,3) \leq \frac{1}{n}\log I_2 \leq -(\log 2)\Pi(2,3)$$

and thus $\log(I_1/I_2) \to -\infty$ in P_{θ_0} probability. Equivalently, $I_1/I_2 \to 0$ in P_{θ_0} probability.

In this example, the MLE is consistent. We could have taken the parameter space to be $[\epsilon, 1] \cup [2, 3]$ and ensured compactness. What goes wrong here, as we shall later recognize, is the lack of continuity of the Kullback-Leibler information and, of course, the behavior of Π in the neighborhood of θ_0. If a prior Π satisfies $\Pi(\theta_0, \theta_0 + h) > 0$, for all $h > 0$, then similar calculations or an application of the Schwartz theorem, to be proved later, show that the posterior is consistent.

Remark 1.3.8. We have seen that consistency of MLE neither implies nor is implied by consistency of the posterior. The following condition implies both. Let V be any open set containing θ_0. Then the condition is

$$\sup_{\theta \in V^c} \prod_1^n f_\theta(X_i)/f_{\theta_0}(X_i) \to 0 \text{ a.s } \theta_0$$

Theorem 1.3.4 implies this stronger condition.

1.4 Asymptotic Normality of MLE and Bernstein–von Mises Theorem

A standard result in the asymptotic theory of maximum likelihood estimates is its asymptotic normality. In this section we briefly review this and its Bayesian parallel-the Bernstein–von Mises theorem-on the asymptotic normality of the posterior distribution. A word about the asymptotic normality of the MLE: This is really a result about the consistent roots of the likelihood equation $\partial \log f_\theta/\partial\theta = 0$. If a global MLE $\hat{\theta}_n$ exists and is consistent, then under a differentiability assumption it is easy to see that for each P_{θ_0}, $\hat{\theta}_n$ is a consistent solution of the likelihood equation almost surely P_{θ_0}. On the other hand, if f_θ is differentiable in θ, then for each θ_0 it is possible to construct [Serfling [147] 33.3; Cramér [35]] a sequence T_n that is a solution of the likelihood equation and that converges to θ_0. The problem, of course, is that T_n depends on θ_0 and so will not qualify as an estimator. If there exists a consistent estimate θ'_n, then a consistent sequence that is also a solution of the likelihood equation can be constructed by picking $\hat{\theta}_n$ to be the solution closest to θ'_n. For a sketch of this argument, see Ghosh [89].

As before, let $X_1, X_2, \ldots, X_n$ be i.i.d. f_θ, where f_θ is a density with respect to some dominating measure μ and $\theta \in \Theta$, and Θ is an open subset of $\mathbb{R}$. We make the following regularity assumptions on f_θ:

(i) $\{x : f_\theta(x) > 0\}$ is the same for all $\theta \in \Theta$

(ii) $L(\theta, x) = \log f_\theta(x)$ is thrice differentiable with respect to θ in a neighborhood $(\theta_0 - \delta, \theta_0 + \delta)$. If $\dot{L}, \ddot{L}$,and $\dddot{L}$ stand for the first, second, and third derivatives, then $E_{\theta_0}\dot{L}(\theta_0)$ and $E_{\theta_0}\ddot{L}(\theta_0)$ are both finite and

$$\sup_{\theta \in (\theta_0 - \delta, \theta_0 + \delta)} |\dddot{L}(\theta, x)| < M(x) \quad \text{and } E_{\theta_0} M < \infty$$

(iii) Interchange of the order of expectation with respect to θ_0 and differentiation at θ_0 are justified, so that

$$E_{\theta_0}\dot{L}(\theta_0) = 0, E_{\theta_0}\ddot{L}(\theta_0) = -E_{\theta_0}(\dot{L}(\theta_0))^2$$

(iv) $I(\theta_0) \doteq E_{\theta_0}(\dot{L}(\theta_0))^2 > 0$.

Theorem 1.4.1. *If $\{f_\theta : \theta \in \Theta\}$ satisfies conditions (i)–(iv) and if $\hat{\theta}_n$ is a consistent solution of the likelihood equation then $\sqrt{n}(\hat{\theta}_n - \theta_0) \xrightarrow{\mathcal{D}} N(0, 1/I(\theta_0))$.*

Proof. Let $L_n(\theta) = \sum_1^n L(\theta, X_i)$. By Taylor expansion

$$0 = \dot{L}_n(\hat{\theta}_n) = \dot{L}_n(\theta_0) + (\hat{\theta}_n - \theta_0)\ddot{L}_n(\theta_0) + \frac{(\hat{\theta}_n - \theta_0)^2}{2}\dddot{L}_n(\theta')$$

where $\theta_0 < \theta' < \hat{\theta}_n$. Thus,

$$\sqrt{n}(\hat{\theta}_n - \theta_0) = \frac{\frac{1}{\sqrt{n}}\dot{L}_n(\theta_0)}{-\frac{1}{n}\ddot{L}_n(\theta_0) - \frac{1}{2}(\hat{\theta}_n - \theta_0)\frac{1}{n}\dddot{L}_n(\theta')}$$

By the central limit theorem, the numerator converges in distribution to $N(0, I(\theta_0))$; the first term in the denominator goes to $I(\theta_0)$ by SLLN; the second term is $o_P(1)$ by the assumptions on $\hat{\theta}_n$ and $\dddot{L}$. □

We next turn to asymptotic normality of the posterior. We wish to prove that if $\hat{\theta}_n$ is a consistent solution of the likelihood equation, then the posterior distribution of $\sqrt{n}(\theta - \hat{\theta}_n)$ is approximately $N(0, 1/I(\theta_0))$. Early forms of this theorem go back to Laplace, Bernstein, and von Mises [see [46] for references]. A version of this theorem appears in Lehmann [124]. Condition (v) in Theorem 1.4.2 is taken from there. Other related references are Bickel and Yahav [20], Walker [164], LeCam [121], [120] and

Borwanker et al. [27]. Ghosal [75, 76, 77] has developed posterior normality results in cases where the dimension of the parameter space is increasing. Further refinements developing asymptotic expansions appear in Johnson [107],[108] , Kadane and Tierney [158] and Woodroofe [173]. Lindley [129], Johnson [108] and Ghosh et al. [82], provide expansions of the posterior that refine posterior normality. See the next section for an alternative unified treatment of regular and nonregular cases.

Theorem 1.4.2. *Suppose $\{f_\theta : \theta \in \Theta\}$ satisfies assumptions (i)–(iv) of the Theorem 1.4.1 and $\hat{\theta}_n$ is a consistent solution of the likelihood equation. Further, suppose*
(v) for any $\delta > 0$, there exists an $\epsilon > 0$ such that

$$P_{\theta_0}\left\{\sup_{|\theta-\theta_0|>\delta} \frac{1}{n}\left(L_n(\theta) - L_n(\theta_0)\right) \le -\epsilon\right\} \to 1$$

(vi) The prior has a density $\pi(\theta)$ with respect to Lebesgue measure, which is continuous and positive at θ_0.

Let $\mathbf{X_n}$ stand for $X_1, X_2, \ldots, X_n$ and $f_\theta(\mathbf{X_n})$ for its joint density. Denote by $\pi^*(s|\mathbf{X_n})$ the posterior density of $s = \sqrt{n}(\theta - \hat{\theta}_n(\mathbf{X_n}))$. Then as $n \to \infty$,

$$\int_{\mathbb{R}} \left| \pi^*(s|\mathbf{X_n}) - \frac{\sqrt{I(\theta_0)}}{\sqrt{2\pi}} e^{-\frac{s^2 I(\theta_0)}{2}} \right| ds \quad \stackrel{P_{\theta_0}}{\to} 0 \tag{1.3}$$

Proof. Because $s = \sqrt{n}(\theta - \hat{\theta}_n)$,

$$\pi^*(s|\mathbf{X_n}) = \frac{\pi(\hat{\theta}_n + \frac{s}{\sqrt{n}}) f_{\hat{\theta}_n + s/\sqrt{n}}(\mathbf{X_n})}{\int_{\mathbb{R}} \pi(\hat{\theta}_n + \frac{t}{\sqrt{n}}) f_{\hat{\theta}_n + t/\sqrt{n}}(\mathbf{X_n})\, dt}$$

To avoid notational mess, we suppress the $\mathbf{X_n}$ and rewrite the last line as

$$\frac{\pi(\hat{\theta}_n + \frac{s}{\sqrt{n}}) e^{L_n(\hat{\theta}_n + s/\sqrt{n}) - L_n(\hat{\theta}_n)}}{\int_{\mathbb{R}} \pi(\hat{\theta}_n + t/\sqrt{n}) e^{L_n(\hat{\theta}_n + \frac{t}{\sqrt{n}}) - L_n(\hat{\theta}_n)}\, dt}$$

Thus we need to show

$$\int_{\mathbb{R}} \left| \frac{\pi(\hat{\theta}_n + s/\sqrt{n}) e^{L_n(\hat{\theta}_n + s/\sqrt{n}) - L_n(\hat{\theta}_n)}}{\int_{\mathbb{R}} \pi(\hat{\theta}_n + t/\sqrt{n}) e^{L_n(\hat{\theta}_n + t/\sqrt{n}) - L_n(\hat{\theta}_n)}\, dt} - \sqrt{\frac{I(\theta_0)}{2\pi}} e^{-\frac{s^2 I(\theta_0)}{2}} \right| ds \quad \stackrel{P_{\theta_0}}{\to} 0 \tag{1.4}$$

It is enough to show that

$$\int_{\mathbb{R}} \left| \pi(\hat{\theta}_n + \frac{t}{\sqrt{n}}) e^{L_n(\hat{\theta}_n + t/\sqrt{n}) - L_n(\hat{\theta}_n)} - \pi(\theta_0) e^{-\frac{t^2 I(\theta_0)}{2}} \right| dt \stackrel{P_{\theta_0}}{\to} 0 \tag{1.5}$$

To see this, note that writing C_n for $\int_{\mathbb{R}} \pi(\hat{\theta}_n + t/\sqrt{n}) e^{L_n(\hat{\theta}_n + t/\sqrt{n}) - L_n(\hat{\theta}_n)} dt$, (1.4) is

$$C_n^{-1} \left[\int_{\mathbb{R}} \left| \pi(\hat{\theta}_n + \frac{s}{\sqrt{n}}) e^{L_n(\hat{\theta}_n + s/\sqrt{n}) - L_n(\hat{\theta}_n)} - C_n \sqrt{\frac{I(\theta_0)}{2\pi}} e^{-\frac{s^2 I(\theta_0)}{2}} \right| ds \right] \stackrel{P_{\theta_0}}{\to} 0$$

Because (1.5) implies that $C_n \to \pi(\theta_0)\sqrt{2\pi / I(\theta_0)}$ it is enough to show that the integral inside the brackets goes to 0 in probability, and this term is less than $I_1 + I_2$, where

$$I_1 = \int_{\mathbb{R}} \left| \pi(\hat{\theta}_n + \frac{s}{\sqrt{n}}) e^{L_n(\hat{\theta}_n + s/\sqrt{n}) - L_n(\hat{\theta}_n)} - \pi(\theta_0) e^{-\frac{s^2 I(\theta_0)}{2}} \right| ds$$

and

$$I_2 = \int_{\mathbb{R}} \left| \pi(\theta_0) e^{-\frac{s^2 I(\theta_0)}{2}} - C_n \sqrt{\frac{I(\theta_0)}{2\pi}} e^{-\frac{s^2 I(\theta_0)}{2}} \right| ds$$

Now I_1 goes to 0 by (1.5) and I_2 is equal to

$$\left| \pi(\theta_0) - C_n \sqrt{\frac{I(\theta_0)}{2\pi}} \right| \int_{\mathbb{R}} e^{-\frac{s^2 I(\theta_0)}{2}} ds$$

which goes to 0 because $C_n \to \pi(\theta_0)\sqrt{2\pi / I(\theta_0)}$.

To achieve a further reduction, set

$$h_n = -\frac{1}{n} \sum_1^n \ddot{L}(\hat{\theta}_n, X_i) = -\frac{1}{n} \ddot{L}_n(\hat{\theta}_n, X_i)$$

Because as $n \to \infty$, $h_n \to I(\theta_0)$ a.s. P_{θ_0}, to verify (1.5) it is enough if we show that

$$\int_{\mathbb{R}} \left| \pi(\hat{\theta}_n + \frac{t}{\sqrt{n}}) e^{L_n(\hat{\theta}_n + t/\sqrt{n}) - L_n(\hat{\theta}_n)} - \pi(\hat{\theta}_n) e^{-\frac{t^2 h_n}{2}} \right| dt \stackrel{P_{\theta_0}}{\to} 0 \tag{1.6}$$

To show (1.6), given any δ, $c > 0$, we break $\mathbb{R}$ into three regions:
$A_1 = \{t : |t| < c \log \sqrt{n}\}$,

$A_2 = \{t : c\log\sqrt{n} < |t| < \delta\sqrt{n}\}$, and
$A_3 = \{t : |t| > \delta\sqrt{n}\}$.

We begin with A_3.

$$\int_{A_3}\left|\pi(\hat{\theta}_n + \frac{t}{\sqrt{n}})e^{L_n(\hat{\theta}_n + t/\sqrt{n}) - L_n(\hat{\theta}_n)} - \pi(\hat{\theta}_n)e^{-\frac{t^2 h_n}{2}}\right| dt$$
$$\leq \int_{A_3}\pi(\hat{\theta}_n + \frac{t}{\sqrt{n}})e^{L_n(\hat{\theta}_n + t/\sqrt{n}) - L_n(\hat{\theta}_n)}\, dt + \int_{A_3}\pi(\hat{\theta}_n)e^{-\frac{t^2 h_n}{2}}\, dt$$

The first integral goes to 0 by assumption (v). The second is seen to go to 0 by the usual tail estimates for a normal.

Because $\hat{\theta}_n \to \theta_0$, by Taylor expansion, for large n,

$$L_n(\hat{\theta}_n + \frac{t}{\sqrt{n}}) - L_n(\hat{\theta}_n) = \frac{t^2}{2n}\ddot{L}_n(\hat{\theta}_n) + \frac{1}{6}(\frac{t}{\sqrt{n}})^3\dddot{L}_n(\theta') = -\frac{t^2 h_n}{2} + R_n$$

for some $\theta' \in (\theta_0, \hat{\theta}_n)$. Now consider

$$\int_{A_1}\left|\pi(\hat{\theta}_n + \frac{t}{\sqrt{n}})e^{-\frac{t^2 h_n}{2} + R_n} - \pi(\hat{\theta}_n)e^{-\frac{t^2 h_n}{2}}\right| dt$$
$$\leq \int_{A_1}\pi(\hat{\theta}_n + \frac{t}{\sqrt{n}})\left|e^{-\frac{t^2 h_n}{2} + R_n} - e^{-\frac{t^2 h_n}{2}}\right| dt + \int_{A_1}\left|\pi(\hat{\theta}_n + \frac{t}{\sqrt{n}}) - \pi(\hat{\theta}_n)\right| e^{-\frac{t^2 h_n}{2}}\, dt$$

Because π is continuous at θ_0, the second integral goes to 0 in P_{θ_0} probability. The first integral equals

$$\int_{A_1}\pi(\hat{\theta}_n + \frac{t}{\sqrt{n}})e^{-\frac{t^2 h_n}{2}}\left|e^{R_n} - 1\right| dt$$
$$\leq \int_{A_1}\pi(\hat{\theta}_n + \frac{t}{\sqrt{n}})e^{-\frac{t^2 h_n}{2}}e^{|R_n|}\,|R_n|\, dt \tag{1.7}$$

Now,

$$\sup_{t\in A_1} R_n = \sup_{t\in A_1}(\frac{t}{\sqrt{n}})^3\dddot{L}_n(\theta') \leq c^3\frac{(\log\sqrt{n})^3}{n}O_P(1) = o_P(1)$$

and hence (1.7) is

$$\leq \sup_{t\in A_1}\pi(\hat{\theta}_n + \frac{t}{\sqrt{n}})\int_{A_1}e^{-\frac{t^2 h_n}{2}}e^{|R_n|}\,|R_n|\, dt = o_P(1)$$

Next consider

$$\begin{aligned}
&\int_{A_2} \left| \pi(\hat{\theta}_n + \frac{t}{\sqrt{n}}) e^{\frac{t^2 h_n}{2} + R_n} - \pi(\hat{\theta}_n) e^{-\frac{t^2 h_n}{2}} \right| \, dt \\
&\leq \int_{A_2} \pi(\hat{\theta}_n + \frac{t}{\sqrt{n}}) e^{\frac{t^2 h_n}{2} + R_n} \, dt + \int_{A_2} \pi(\hat{\theta}_n) e^{-\frac{t^2 h_n}{2}} \, dt
\end{aligned}$$

The second integral is

$$\begin{aligned}
&\leq 2\pi(\hat{\theta}_n) e^{-\frac{h_n c \log \sqrt{n}}{2}} [\delta\sqrt{n} - c \log \sqrt{n}] \\
&\leq K\pi(\hat{\theta}_n) \frac{\sqrt{n}}{n^{c h_n/4}}
\end{aligned}$$

so that by choosing c large, the integral goes to 0 in P_{θ_0} probability.

For the first integral, because $t \in A_2$, and $c \log \sqrt{n} < |t| < \delta\sqrt{n}$, we have $|t|/\sqrt{n} < \delta$. Thus $|R_n| = (\frac{|t|}{\sqrt{n}})^3 \frac{1}{6} \dddot{L}_n(\theta') \leq \delta \frac{t^2}{6} \frac{1}{n} \dddot{L}_n(\theta')$

Because $\sup_{\theta' \in (\theta_0 - \delta, \theta_0 + \delta)} (1/n) \left| \dddot{L}_n(\theta') \right|$ is $O_P(1)$, by choosing δ small we can ensure that

$$P_{\theta_0} \left\{ |R_n| < \frac{t^2}{4} h_n \text{ for all } t \in A_2 \right\} > 1 - \epsilon \text{ for } n > n_0 \tag{1.8}$$

or

$$P_{\theta_0} \left\{ -\frac{t^2 h_n}{2} + R_n < -\frac{t^2 h_n}{4} \text{ for all } t \in A_2 \right\} > 1 - \epsilon \tag{1.9}$$

Hence, with probability greater than $1 - \epsilon$,

$$\begin{aligned}
&\int_{A_2} \pi(\hat{\theta}_n + \frac{t}{\sqrt{n}}) e^{-\frac{t^2 h_n}{2} + R_n} \, dt \\
&\leq \sup_{\theta \in A_2} \pi(\hat{\theta}_n + t/\frac{t}{\sqrt{n}}) \int_{A_2} e^{-t^2 h_n/4} \, dt \\
&\to 0 \text{ as } n \to \infty
\end{aligned}$$

Finally, the three steps can be put together, first by choosing a δ to ensure (1.8) and then by working with this δ in steps 1 and 3. □

An asymptotic normality result also holds for Bayes estimates.

Theorem 1.4.3. *In addition to the assumptions of Theorem 1.4.2 assume that* $\int |\theta| \pi(\theta) \, d\theta < \infty$. *Let* $\theta_n^* = \int_{\mathbb{R}} \theta \, \Pi(d\theta | X_1, X_2, \ldots, X_n)$ *be the Bayes estimate with respect to squared error loss. Then* *(i)* $\sqrt{n}(\hat{\theta}_n - \theta_n^*) \to 0$ *in* P_{θ_0} *probability*

(ii) $\sqrt{n}(\theta_n^* - \theta_0)$ *converges in distribution to* $N(0, 1/I(\theta_0))$.

Proof. The assumption of finite moment for π and a slight refinement of detail in the proof of Theorem 1.4.2 strengthens the assertion to

$$\int_{\mathbb{R}} (1+|t|)\pi(\hat{\theta}_n + \frac{t}{\sqrt{n}}) \left| e^{L_n(\hat{\theta}_n + t/\sqrt{n}) - L_n(\hat{\theta}_n)} - e^{-\frac{t^2 h_n}{2}} \right| dt \stackrel{P_{\theta_0}}{\to} 0 \tag{1.10}$$

Consequently

$$\int_{\mathbb{R}} (1+|t|) \left| \pi^*(t|\mathbf{X_n}) - \frac{\sqrt{I(\theta_0)}}{\sqrt{2\pi}} e^{-\frac{t^2 I(\theta_0)}{2}} \right| \, dt \quad \stackrel{P_{\theta_0}}{\to} \; 0$$

and hence $\left| \int_{\mathbb{R}} t \left| \pi^*(t|\mathbf{X_n}) - (\sqrt{I(\theta_0)/2\pi}) e^{-\frac{t^2 I(\theta_0)}{2}} \right| \, dt \right| \stackrel{P_{\theta_0}}{\to} 0$. Note that because

$$\sqrt{\frac{I(\theta_0)}{2\pi}} \int_{\mathbb{R}} t e^{-\frac{t^2 I(\theta_0)}{2}} \, dt = 0$$

we have $\int_{\mathbb{R}} t \; \pi^*(dt|\mathbf{X_n}) \to 0$.

To relate these observations to the theorem, note that

$$\theta_n^* = \int_{\mathbb{R}} \theta \; \Pi(d\theta|X_1, X_2, \ldots, X_n) = \int_{\mathbb{R}} (\hat{\theta}_n + \frac{t}{\sqrt{n}}) \; \pi^*(dt|\mathbf{X_n})$$

and hence $\sqrt{n}(\theta_n^* - \hat{\theta}_n) = \int_{\mathbb{R}} t \; \pi^*(dt|\mathbf{X_n})$.

Assertion (ii) follows from (i) and the asymptotic normality of $\hat{\theta}_n$ discussed earlier. □

Remark 1.4.1. This theorem shows that the posterior mean of θ can be approximated by $\hat{\theta}_n$ up to an error of $o_P(n^{-1/2})$. Actually, under stronger assumptions one can show [82] that the error is of the order of n^{-1} . A result of this type also holds for the posterior variance.

Remark 1.4.2. With a stronger version of assumption (v), namely, for any δ,

$$\sup_{|\theta - \theta_0| > \delta} \frac{1}{n} [L_n(\theta) - L_n(\theta_0)] \leq -\epsilon \qquad \text{eventually a.e. } P_{\theta_0}$$

and $\hat{\theta}_n \to \theta_0$ a.s., we can have the L_1-distance in (1.3) go to 0 a.s. P_{θ_0}.

Remark 1.4.3. If we have almost sure convergence at each θ_0, then by Fubini, the L_1-distance evaluated with respect to the joint distribution of $\theta, X_1, X_2, \ldots, X_n$ goes to 0. For refinements of such results see [82].

Remark 1.4.4. Multiparameter extensions follow in a similar way.

Remark 1.4.5. It follows immediately from (1.5) that

$$\log \int_{\mathbb{R}} \prod_1^n f_\theta(X_i)\pi(\theta)d\theta = L_n(\hat{\theta}_n) + \log C_n - \frac{1}{2}\log n$$

$$= L_n(\hat{\theta}_n) - \frac{1}{2}\log n + \frac{1}{2}\log 2\pi - \frac{1}{2}\log I(\theta_0) + \log \pi(\theta_0) + o_P(1)$$

In the multiparameter case with a p dimensional parameter, this would become

$$\log \int_{\mathbb{R}} \prod_1^n f_\theta(X_i)\pi(\theta)d\theta = L_n(\hat{\theta}_n) - \frac{p}{2}\log n + \frac{p}{2}\log 2\pi - \frac{1}{2}\log ||I(\theta_0)|| + \log \pi(\theta_0) + o_P(1)$$

where $||I(\theta_0)||$ stands for the determinant of the Fisher information matrix.

This is identical to the approximation of Schwarz [146] needed for developing his BIC (Bayes information criteria) for selecting from K given models. Schwarz recommends the use of the penalized likelihood under model j with a p_j-dimensional parameter, namely,

$$L_n(\hat{\theta}_n) - \frac{p_j}{2}\log n$$

to evaluate the jth model. One chooses the model with highest value of this criterion.

The proof suggested here does not assume exponential families as in Schwarz[146] but assumes that the true density f_0 is in the model being considered. To have a similar approximation when f_0 is not in the model, one assumes

$$\inf_\theta \int f_0 \log \frac{f_0}{f_\theta}$$

is attained at θ_0. We use this θ_0 in the assumptions of this section.

Remark 1.4.6. The main theorem in this section remains true if we replace the normal distribution $N(0, 1/I(\theta_0))$ by $N(0, 1/a)$ where $a = -(1/n)(d^2 \log L/d\theta^2)|_{\hat{\theta}_n}$ is the observed Fisher information per unit observation. To a Bayesian, this form of the theorem is more appealing because it does not involve a true (but unknown) value θ_0. The proof requires very little change.

1.5 Ibragimov and Hasminskiĭ Conditions

Ibragimov and Hasminskiĭ, henceforth referred to as IH, in their text [102] used a very general framework for parametric models that includes both the regular model treated in the last section and nonregular problems like $U(0,\theta)$. In fact, IH verify their conditions for various classes of nonregular problems and some stochastic processes. Within their framework we will provide a necessary and sufficient condition for a suitably normed posterior to have a limit in probability. This theorem includes Theorem 1.4.2 on posterior normality under slightly different conditions and with results on nonregular cases. It also answers some questions on nonregular problems raised by Smith [152].

We begin with notations and conditions appropriate for this section. Let Θ be an open set in $\mathbb{R}^k$. For simplicity we take k to be 1.

The joint probability distribution of $X_1, X_2, \ldots, X_n$ is denoted by P_θ^n and its density with respect to Lebesgue measure (or any other σ- finite measure) by $p(\mathbf{X_n}, \theta)$. Let ϕ_n be a sequence of positive constants converging to 0. If $k > 1$ then ϕ_n would be a k-dimensional vector of such constants. In the so-called regular case treated in the last section, $\phi_n = 1/\sqrt{n}$. In the nonregular cases, typically $\phi_n \to 0$ at a faster rate. Consider the map U defined by $U(\theta) = \phi_n^{-1}(\theta - \theta_0)$, where θ_0 is the true value. Let $\mathcal{U}_n$ be the range of this map, i.e., $\mathcal{U}_n = \{U(\theta) : \theta \in \Theta\}$. The variable u is a suitably scaled deviation of θ from θ_0. The likelihood ratio process is defined as

$$Z_n(u, \underline{X}_n) = \frac{p(\underline{X}_n, \theta_0 + \phi_n u)}{p(\underline{X}_n, \theta_0)}$$

The IH conditions can be thought of as two conditions on the Hellinger distance and one on weak convergence of finite-dimensional distributions of Z_n.

IH conditions

1. For some $M > 0$, $m_1 \geq 0$, $\alpha > 0$, $n_0 \geq 1$,

$$E_{\theta_0} \left| Z_n^{\frac{1}{2}}(u_1) - Z_n^{\frac{1}{2}}(u_2) \right|^2 \leq M(1 + A^{m_1})|u_1 - u_2|^\alpha$$
$$\forall u_1, u_2 \in \mathcal{U}_n \text{ with } |u_1| \leq A, |u_2| \leq A$$

   for all $n \geq n_0$.

   Note that the left-hand side is the square of the Hellinger distance between $p(\underline{X}_n, \theta_0 + \phi_n u_1)$ and $p(\underline{X}_n, \theta_0 + \phi_n u_2)$. The condition is like a Lipschitz condition in the rescaled parameter space but uniformly in n.

2. For all $u \in \mathcal{U}_n$ and $n \geq n_0$,

$$E_{\theta_0} \left| Z_n^{\frac{1}{2}}(u) \right| \leq e^{-g_n(|u|)}$$

where g_n is a sequence of real-valued functions satisfying the following conditions:

   (a) for each $n \geq 1$, $g_n(y) \uparrow \infty$ as $y \to \infty$,

   (b) for any $N > 0$,

$$\lim_{\substack{y\to\infty \\ n\to\infty}} y^N e^{-g_n(y)} = 0$$

3. The finite-dimensional distributions of $\{Z_n(u) : u \in \mathcal{U}_n\}$ converge to those of a stochastic process $\{Z(u) : u \in \mathbf{R}\}$.

For i.i.d. $X_1, X_2, \ldots, X_n$ with compact Θ, condition 2 will hold if ϕ_n^{-1} is bounded by a power of n, as is usually the case. This may be seen as follows: Note that

$$E_{\theta_0} Z_n^{\frac{1}{2}}(u) = [A(\theta_0, \theta_0 + \phi_n u)]^n$$

where $[A(\theta_0, \theta_0 + \phi_n u)]^n$ is the affinity between p_{θ_0} and $p_{(\theta_0+\phi_n u)}$ given by $\int \sqrt{p_{\theta_0} p_{(\theta_0+\phi_n u)}}dx$. Define

$$g_n(y) = \begin{cases} -n \log A(\theta_0, \theta_0 + \phi_n y) & \text{if } y \in \mathcal{U}_n \\ \infty & \text{otherwise} \end{cases}$$

Condition 2(a) and 2(b) follow trivially. For non compact cases the condition is similar to the Wald conditions. The following result appears in IH (theorem I.10.2).

Theorem 1.5.1. *Let Π be a prior with continuous positive density at θ_0 with respect to the Lebesgue measure. Under the IH conditions and with squared error loss, the normalized Bayes estimate $\phi_n^{-1}(\tilde{\theta}_n - \theta_0)$ converges in distribution to $\int uZ(u)\,du / \int Z(u)\,du$.*

A similar result holds for other loss functions. This result of IH is similar to the result that was derived as a corollary to the Bernstein–von Mises theorem on posterior normality. So it is natural to ask if such a limit, not necessarily normal, exists for the posterior under conditions of IH.

We begin with a fact that immediately follows from the Hewitt-Savage 0-1 law.

Proposition 1.5.1. *Suppose $X_1, X_2, \ldots, X_n$ are i.i.d. and Π is a prior. Let $\hat{\theta}(X_1, X_2, \ldots, X_n)$ be a symmetric function of $X_1, X_2, \ldots, X_n$. Let*

$$t = \phi_n^{-1}\left(\theta - \hat{\theta}(X_1, X_2, \ldots, X_n)\right)$$

and let A be a Borel set. Suppose

$$\Pi(t \in A | X_1, X_2, \ldots, X_n) \stackrel{P_{\theta_0}}{\to} Y_A$$

Then Y_A is constant a.e. P_{θ_0}.

In view of this, the following definition of convergence of posterior seems appropriate, at least in the i.i.d. case.

Definition 1.5.1. For some symmetric function $\hat{\theta}(X_1, X_2, \ldots, X_n)$ the posterior distribution of $t = \phi_n^{-1}\left(\theta - \hat{\theta}(X_1, X_2, \ldots, X_n)\right)$ has a limit Q if

$$\sup_A \{|\Pi(t \in A | X_1, X_2, \ldots, X_n) - Q(A)|\} \stackrel{P_{\theta_0}}{\to} 0$$

In this case, $\hat{\theta}(X_1, X_2, \ldots, X_n)$ is called a *proper centering.*

We now state our main result.

Theorem 1.5.2. *Suppose the IH conditions hold and Π is a prior with continuous positive density at θ_0 with respect to the Lebesgue measure. If a proper centering $\hat{\theta}(X_1, X_2, \ldots, X_n)$ exists, then there exists a random variable W such that*

(a) $\phi_n^{-1}(\theta_0 - \hat{\theta}(X_1, X_2, \ldots, X_n))$ converges in distribution to W.

(b) For almost all $\eta \in \mathbb{R}$, with respect to the Lebesgue measure $\xi(\eta - W) = q(\eta)$ is nonrandom, where $\xi(u) = Z(u)/\int_{\mathbb{R}} Z(u)\, du$, $u \in \mathbb{R}$.

Conversely if b holds for some random variable W, then the posterior mean given $X_1, X_2, \ldots, X_n$, is a proper centering with $Q(A) = \int_A q(t)\, dt$.

Remark 1.5.1. Under the IH conditions it can be shown that the posterior mean given $X_1, X_2, \ldots, X_n$ exists. (See the proof of IH theorem 10.2)

Remark 1.5.2. It is proved in Ghosal et al. [79] that under IH conditions the posterior with centering at θ_0 converges weakly to $\xi(.)$ a.s. P_{θ_0}. Theorem 1.5.2 shows that if weak convergence is to be strengthened to convergence in probability by centering at a suitable $\hat{\theta}(X_1, X_2, \ldots, X_n)$, then conditions (a) and (b) are needed.

Example 1.5.1. We sketch how the current theorem leads to (a version of) the Bernstein–von Mises theorem. Assume that the X_is are i.i.d. and conditions 1 and 2 of IH hold and that the following stochastic expansion used earlier in this chapter is valid.

$$\log Z_n(u) = \frac{u}{\sqrt{n}} \sum_1^n \frac{\partial \log p(X_i, \theta)}{\partial \theta}|_{\theta_0} - \frac{u^2}{2} I(\theta_0) + o_P(1).$$

Then

$$\log Z_n(u) \xrightarrow{D} uV - \frac{u^2}{2} I(\theta_0) \text{ where } V \text{ is a } N(0, I(\theta_0)) \text{ random variable.}$$

Let $\log Z(u) = uV - (u^2/2)I(\theta_0)$. This implies that

$$(\log Z_n(u_1), \log Z_n(u_2), \dots \log Z_n(u_m))$$

converges in distribution to

$$(\log Z(u_1), \log Z(u_2), \dots \log Z(u_m))$$

i.e., condition 3 of IH holds. An elementary calculation now shows that $W = V/I(\theta_0)$ and $q(\eta)$ is the normal density at η with mean 0 and variance $I^{-1}(\theta_0)$.

Some feeling about condition 1 in the regular case may be obtained as follows: Easy calculation shows

$$E_{\theta_0}\left(Z_n^{\frac{1}{2}}(u_1)(Z_n^{\frac{1}{2}}(u_2)\right) = A(u_1, u_2)^n$$

If we expand $(p_{\theta_0 + (u/\sqrt{n})}) =^{\frac{1}{2}}$ up to the quadratic term and integrate, we get the following approximation.

$$\{1 + C\frac{(u_1 - u_2)^2}{n} + 3R_n\}$$

Because

$$E_{\theta_0}\left|(Z_n^{\frac{1}{2}}(u_1) - Z_n^{\frac{1}{2}}(u_2)\right|^2 = 2 - 2A(u_1, u_2)^n$$

it can be bounded as required in condition 2 under appropriate conditions on the negligibility of the remainder term R_n. A useful sufficient condition is provided in lemma 1.1 of IH.

Example. The following is a nonregular case where the posterior converges to a limit:

$$p(x, \theta) = \begin{cases} e^{-(x-\theta)} & x > \theta \\ 0 & \text{otherwise} \end{cases}$$

The norming constant ϕ_n is n^{-1} and a convenient centering is $\hat{\theta}(X_1, X_2, \ldots, X_n) = \min(X_1, X_2, \ldots, X_n)$. Conditions 1 and 2 of IH are verified in chapter 5 of IH under very general assumptions that cover the current example. We shall verify the easy condition 3 and the necessary and sufficient condition of Theorem 1.5.2. Let $V_n = n(\hat{\theta}(X_1, X_2, \ldots, X_n) - \theta)$ and W be a random variable exponentially distributed on $(-\infty, 0)$ with mean -1. Then V_n and W have the same distribution for all n. Also

$$Z_n(u) = \begin{cases} e^u & \text{if } u - V_n < 0 \\ 0 & \text{otherwise} \end{cases}$$

Define $Z(u)$ similarly with W replacing V_n. Because W and V_n have the same distribution, the finite-dimensional distributions of Z_n and Z are the same. Moreover

$$\xi(u) = \begin{cases} e^{u+W} & \text{if } u + W < 0 \\ 0 & \text{otherwise} \end{cases}$$

and so $q(\eta) = e^{\eta}$ if $\eta < 0$ and 0 otherwise. The case when P_θ is uniform can be reduced to this case by a suitable transformation of X and θ.

Example. This example deals with the hazard rate change point problem. Consider $X_1, X_2, \ldots, X_n$ i.i.d. with hazard rate

$$\frac{f_\theta(x)}{1 - F_\theta(x)} = \begin{cases} a & \text{if } 0 < x < \theta \\ b & \text{if } x > \theta \end{cases}$$

Typically a is much bigger than b. This density has been used to model electronic components with initial high hazard rate and cancer relapse times. For details see Ghosh et al.[85].

Let $\hat{\theta}(X_1, X_2, \ldots, X_n)$ be the MLE of θ. It can be shown that $\phi_n = n^{-1}$ is the right norming constant and that the IH conditions hold. But the necessary condition that $\xi(u - W)$ is nonrandom fails. On the other hand, if a, b are also unknown, it can be shown that the posterior distribution of $\left(\sqrt{n}(a - \hat{a}), \sqrt{n}(b - \hat{b})\right)$ has a limit in the sense of theorem 1.5.2. For details see [85] and [79]

Remark 1.5.3. Ghosal et al. [84] show that typically in non-regular examples the necessary condition of Theorem 1.5.2 fails.

Remark 1.5.4. Theorems 2.2 and 2.3 of [84] imply consistency of the posterior under conditions of IH.

Remark 1.5.5. If $\sum \phi_n^s < \infty$ for some $s > 0$, then posterior consistency holds in the a.s sense.

1.6 Nonsubjective Priors

This section contains a brief discussion of nonsubjective priors. This term has been generally used in the literature for the so-called noninformative priors. In this section we use it as a generic description of all priors that are not elicited in a fully subjective manner.

1.6.1 *Fully Specified*

Fully specified nonsubjective priors try to quantify low information in one sense or another. Because there is no completely satisfactory definition of information, many choices are available. Only the most common are discussed. A comprehensive survey is by Kass and Wasserman [111]. A quick overview is available in Ghosh and Mukherjee [86] and Ghosh [83]. In particular, we use this term to describe conjugate priors and their mixtures.

For convenience we take $\Theta = \mathbb{R}^p$. The use of uniform distribution, namely, the Lebesgue measure, as a prior goes back to Bayes and Laplace. It has been criticized as being *improper* (i.e., total measure is not finite), a property that applies to all the priors considered in this section, and is a consequence of Θ being unbounded. An improper prior may be used only if it leads to a proper posterior for all samples. This posterior may then be used to calculate Bayes estimates and so on. However, even then there arise problems with testing hypotheses and model selection. Because we will not consider testing for infinite-dimensional Θ we will not pursue this. For finite-dimensional Θ, attractive possibilities are available. See, for example, Berger and Pericchi [16] and Ghosh and Samanta [88]

As pointed out by Fisher, choice of uniform distribution is not invariant in the following sense. Take a smooth 1-1 function $\eta(\theta)$ of θ. Argue that if one has no information about θ then the same is true of $\eta(\theta)$, and hence one can quantify this belief by a uniform distribution for η. Going back to θ one gets a nonuniform prior π for θ satisfying

$$\pi(\theta) = |\frac{d\eta}{d\theta}|$$

where $|d\eta/d\theta|$ is the Jacobian, i.e., the determinant of the $p \times p$ matrix $[\partial\eta_i/\partial\theta_j]$.

It appears that Fisher's criticism led to the decline of Bayesian methods based on uniform priors. This also helped the growth of methods based on maximizing the likelihood. However, Basu [9] makes a strong case for a uniform distribution after a suitable finite discrete approximation to Θ. This idea will be taken up in Chapter 8.

A natural Bayesian answer to Fisher's criticism is to look for a method that produces priors $\pi_1(\theta)$, $\pi_2(\eta)$ for θ and η such that one can pass from one to the other by the usual Jacobian formula

$$\pi_1(\theta) = \pi_2(\eta(\theta))|\frac{d\eta}{d\theta}| \tag{1.11}$$

Suppose the likelihood satisfies regularity conditions and the $p \times p$ Fisher's information matrix

$$I(\theta) = \left[E_\theta\left(\frac{\partial \log f_\theta}{\partial \theta_i} \cdot \frac{\partial \log f_\theta}{\partial \theta_j}\right)\right]$$

is positive definite. Then Jeffreys suggested the use of

$$\pi_1(\theta) = \{\det I(\theta)\}^{1/2}$$

This is known as the Jeffreys prior. It is easily verified that (1.11) is satisfied if we set

$$\pi_2(\eta) = \left\{\det\left[E_\theta\left(\frac{\partial \log f_\theta}{\partial \eta_i} \cdot \frac{\partial \log f_\theta}{\partial \eta_j}\right)\right]\right\}^{1/2}$$

using the Fisher information matrix in the η-space. One apparently unpleasant aspect is the dependence of the prior on the experiment. This is examined in the next subsection.

The Jeffreys prior was the most popular nonsubjective prior until the introduction of reference priors by Bernardo [18]. The algorithm described next is due to Berger and Bernardo [14], [15]. We follow the treatment given in Ghosh [83].

For a discrete random variable or vector W with probability function $p(w)$, the Shannon entropy is

$$S(p) = S(W) = -E_p(\log p(W))$$

It can be axiomatically developed and is a basic quantity in information and communication theory. Maximization of entropy, which is equivalent to minimizing information, leads to a discrete uniform distribution, provided W assumes only finitely many values.

Unfortunately, no such universally accepted measure exists if W is not discrete. In the general case we may still define

$$S(p) = S(W) = -E_p(\log p(W))$$

where p is the density with respect to some σ-finite measure μ. Unfortunately, this $S(p)$ depends on μ and is rarely used directly in information or communication theory. Further, if one maximizes $S(p)$ one gets p =constant, i.e. one gets essentially μ.

A different measure, also due to Shannon, was used by Lindley [128] and Bernardo [18]. Consider two random vectors V, W with joint density p. Then

$$S(p) \equiv S(V,W) = S(V) + S_V(W)$$

where

$$S_V(W) = E(I(W|V))$$
$$I(W|V) = -E\{\log p(W|V)|V\}$$

Here $S_V(W)$ is the part of the entropy of W that can be explained by its dependence on V. The residual entropy is

$$S(W) - S_V(W) = E\left\{E\left(\log \frac{p(W|V)}{p(W)}|V\right)\right\} \geq 0$$

Because

$$\int p(w|v) \log\left(p(w|v)/p(w)\right) \mu(dw) \geq 0$$

this quantity is taken as a measure of entropy in the construction of reference priors.

Let $X = (X_1, X_2, \ldots, X_n)$ have density $p(x|\theta)$ and let the prior be $p(\theta)$ and posterior density be $p(\theta|x)$. Lindley's measure of information in X is

$$S(X, p(\theta)) = E\left\{E\left(\log \frac{p(\theta|x)}{p(\theta)}|X\right)\right\} \tag{1.12}$$

So it is a measure of how close the prior is to the posterior. If the prior is most informative, i.e., degenerate at a point, then the quantity is 0. Maximizing the quantity should therefore make the prior as noninformative as possible provided $S(X, p(\theta))$ is the correct measure of entropy.

Bernardo[18] recommended taking a limit first as $n \to \infty$ and then maximizing. Taking a limit seems to introduce some stability and removes dependence on n. Subsequent research has shown that maximizing for a fixed n may lead to discrete priors, which are unacceptable as noninformative.

To ensure that a limit exists, one assumes i.i.d. observations with enough regularity conditions for posterior normality in a sufficiently strong sense. Details are available in Clarke and Barron [33].

Suppose K_i is an increasing sequence of compact sets whose union is the whole parameter space Θ. To avoid confusion with the density p the dimension of θ is taken as d. Then using the posterior normality

$$\begin{aligned} S(x,p) &= -E\left(\log p(\theta)\right) + E\left(\log p(\theta|X)\right) \\ &= -E\left(\log p(\theta)\right) + E \log N(\theta) + o(1) \end{aligned}$$

where N is the normal density with mean $\hat{\theta}$ and dispersion matrix $I^{-1}(\hat{\theta})/n$.

The second term on the right equals

$$-nE\left(\sum \frac{(\theta_i - \hat{\theta}_i)(\theta_j - \hat{\theta}_j)I_{ij}(\hat{\theta})}{2}\right) + E\log\left\{\det I(\hat{\theta})\right\}^{1/2} + \frac{d}{2}\log\frac{n}{2\pi}$$

If we approximate $I_0(\hat{\theta})$ by $I_0(\theta)$ and $E(\theta_i - \hat{\theta}_i)(\theta_j - \hat{\theta}_j)$ by $I_{ij}(\theta)/n$, $S(x,p)$ simplifies to

$$\frac{d}{2}\log\frac{n}{2\pi e} + \int_{K_i} p(\theta)\log\{\det I(\theta)\}^{1/2} - \int_{K_i} p(\theta)\log p(\theta) + o(1) \tag{1.13}$$

Thus as $n \to \infty$, $S(X,p)$ is decomposed into a term that does not depend on $p(\theta)$ and

$$J(p, K_i) = \int_{K_i} p(\theta)\log\frac{\{\det I(\theta)\}^{1/2}}{p(\theta}) \, d\theta$$

which is maximized at

$$p_i(\theta) = \begin{cases} const. \ \{\det I(\theta)\}^{1/2} & \text{if } \theta \in K_1 \\ = 0 & \text{otherwise} \end{cases}$$

If one lets $i \to \infty$, p_is may be regarded as converging to the Jeffreys prior. This is a rederivation of the Jeffreys prior from an information theoretic point of view by Bernardo [18]. To get a reference prior, one writes $\theta = (\theta_1, \theta_2)$, where θ_1 is the parameter of interest and θ_2 is a nuisance parameter. Let d_i be the dimension of θ_i, and for convenience take $\Theta = \Theta_1 \times \Theta_2$.

For a fixed θ_1, let $p(\theta_2|\theta_1)$ be a conditional prior for θ_2 given θ_1. By integrating out θ_2, one is left with θ_1 and X. Then one finds the marginal prior $p(\theta_1)$ as described earlier. This depends on the choice $p(\theta_2|\theta_1)$. Bernardo [18] recommended use of the Jeffreys prior $const \cdot \det\{I_{22}(\theta)\}^{1/2}$, treating θ_2 as variable and with θ_1 held constant. Here $I_{22}(\theta) = [I_{ij}(\theta), i, j, = d_1 + 1, \ldots, d_1 + d_2]$.

Fix compact sets K_{i1}, K_{i2} of Θ_1 and Θ_2. Consider priors concentrating on $K_{i1} \times K_{i2}$. Let $p_i(\theta_2|\theta_1)$ be a given conditional prior. Our first object is to maximize the entropy in θ_1 and find the marginal $p(\theta_1)$.

Let

$$\begin{aligned} S(X, p_i(\theta_1)) &= E\left\{\log\frac{p_i(\theta_1|X)}{p_i(\theta_1)}\right\} \\ &= S(X, p_i(\theta_1, \theta_2)) - \int_{K_{i1}} p_i(\theta_1) S(X, p_i(\theta_2|\theta_1)) \, d\theta_1 \end{aligned} \tag{1.14}$$

Assuming that one can interchange integration with respect to θ_1, using the asymptotic form of (1.13) of $S(X, p(\theta_1, \theta_2)$,

$$S(X, p_i(\theta_1)) = \frac{d_1}{2} \log \frac{n}{2\pi e} + \int_{K_{i1}} p_{i1}(\theta_1) \left\{ \log \frac{\psi_i(\theta_1)}{p_i(\theta_1)} \right\} d\theta_1 + o(1)$$

where

$$\psi_i(\theta_1) = \exp \left\{ \int_{K_{i1}} p_i(\theta_2|\theta_1) \log \left(\frac{\det I(\theta)}{\det I_{22}(\theta)} \right)^{1/2} d\theta_2 \right\}$$

Maximizing $S(X, p_i(\theta_1))$ asymptotically,

$$p_i(\theta_1) = const\ \psi_i(\theta_1) \text{ on } K_{i1}$$

where the constant is for normalization.

Then

$$p_i(\theta_1, \theta_2) = \begin{cases} constant\ \psi_i(\theta_1) p_i(\theta_2|\theta_1) & \text{on } K_{i1} \times K_{i2} \\ 0 & \text{elsewhere} \end{cases}$$

Finally take

$$p(\theta_2|\theta_1) = \begin{cases} c_i(\theta_1) \left\{ \det I_{22}(\theta) \right\}^{1/2} & \text{on } K_{i2} \\ 0 & \text{otherwise} \end{cases}$$

To choose a limit in some sense, fix $\theta_0 = (\theta_{10}, \theta_{20})$ and assume

$$\lim p_i(\theta_1, \theta_2)/p_i(\theta_{10}, \theta_{20}) = p(\theta_1, \theta_2)$$

exists for all $\theta \in \Theta$. Then $p(\theta_1, \theta_2)$ is the reference prior when θ_1 is more important than θ_2. If the convergence to $p(\theta_1, \theta_2)$ is uniform on compacts, then for any pair of sets B_1, B_2 contained in a fixed $K_{i_0 1} \times K_{i_0 2}$

$$\lim \frac{\int_{B_1} p_i(\theta_1, \theta_2)\ d\theta}{\int_{B_2} p_i(\theta_1, \theta_2)\ d\theta} = \frac{\int_{B_1} p(\theta_1, \theta_2)\ d\theta}{\int_{B_2} p(\theta_1, \theta_2)\ d\theta}$$

Berger and Bernardo [15] recommend a d-dimensional break up of θ as $(\theta_1, \theta_2, \ldots, \theta_d)$ and a d-step algorithm starting with

$$p(\theta_d|\theta_1, \ldots, \theta_{d-1}) = c(\theta_1, \theta_2, \ldots, \theta_{d-1}) \sqrt{I_{dd}(\theta)} \text{ on } K_{id}$$

Some justification for this is provided in Datta and Ghosh [38].

There is still another class of nonsubjective priors obtained by matching what a frequentist might do (because, presumably, that is how a Bayesian without prior information would act). Technically, this amounts to matching posterior and frequentist probabilities up to a certain order of approximation. This leads to a differential equation involving the prior. For $d = 1$ the Jeffreys prior is the unique solution. For $d > 1$, reference priors are often a solution of the matching equation. More details are given in Ghosh [83].

Finally, there is one class of problems in which there is some sort of consensus on what nonsubjective prior to use. These are problems where a nice group G of transformations leaves the problem invariant and either acts transitively on Θ, i.e., $\{g(\theta_0); g \in G\} = \Theta$, or reduces Θ to a one-dimensional maximal invariant parameter. See, for example, Berger [13]. In the next example G acts transitively. In such problems the right invariant Haar measure is a common choice and is a reference prior. The Jeffreys prior is a left invariant Haar measure which causes problems [see, e.g., Dawid, Stone, and Zidek [39]). For examples involving one-dimensional maximal invariants, see Datta and Ghosh [38]. Here also reference priors do well.

Example 1.6.1. X_is are i.i.d. normal with mean θ_2 and variance θ_1; θ_1 is the parameter of importance. The information matrix is

$$I(\theta) = \begin{pmatrix} \frac{1}{2\theta_1^2} & 0 \\ 0 & \frac{1}{\theta_1} \end{pmatrix}$$

and so the reference prior may be obtained through the following steps:

$$p_i(\theta_2|\theta_1) = d_i \text{ on } K_{i2}$$

$$\psi_i(\theta_1) = \exp[\int_{K_{i2}} d_i \log \frac{1}{\sqrt{2}\theta_1}] \, d\theta_2$$

$$p_i(\theta_1, \theta_2) = c_i \frac{1}{\theta_1} \text{ on } K_{i2} \times K_{i2}$$

$$p_i(\theta_1, \theta_2) = \theta_{10}/\theta_1$$

which is also known to arise from the right invariant Haar measure for (μ, σ). The Jeffreys prior is proportional to θ_1^{-3}, which corresponds to the left invariant Haar measure.

If the mean is taken to be θ_1 and variance θ_2, then the reference prior is proportional to θ_1^{-1}. But, in general, a reference prior depends on how the components are ordered.

1.6.2 *Discussion*

Nonsubjective priors are best thought of as providing a tool for calculating posteriors. Theorems like posterior normality indicate that the effect of the prior washes away as the sample size increases. Hence a posterior obtained from a nonsubjective prior may be thought of as an approximation to a posterior obtained from a subjective prior.

Though there is no unique choice for a nonsubjective prior, the posterior obtained from different nonsubjective priors will usually be close to each other, even for moderate values of n. Thus lack of uniqueness may not matter very much.

It is true that a nonsubjective prior usually depends on the experiment, e.g., through the information matrix $I(\theta)$. This would not seem paradoxical if one remembers that nonsubjective priors have low information, and it seems that information cannot be defined except in the context of an experiment. The measure of information used by Bernardo [18] clarifies this.

Nonsubjective priors are typically improper but some justification comes from the work of Heath and Sudderth [97], [96]. They show that, at least for amenable groups, the posterior obtained from a right invariant measure can be obtained from a proper finitely additive prior.

For improper priors one has to verify that the posteriors are proper. In many cases this is not easy. Some Bayesians use improper priors and restrict it to a large compact set. In general, this is not advisable. It is a remarkable fact that for the Jeffreys or reference priors, the posteriors are often proper, but there exist simple counterexamples; see for example, [38]. If the likelihood shows marked inhomogeneities asymptotically, as in the so-called nonergodic cases, one must take these into account through suitable conditioning.

1.7 Conjugate and Hierarchical Priors

Let X_is be i.i.d. Consider exponential densities with a special parametrization

$$f_\theta(x) = \exp\{A(\theta) + \sum_1^p \theta_j T_j(x) + \psi(x)\}$$

Given $X_1, X_2, \ldots, X_n$, the sufficient statistic is $(\sum_1^n T_1(x_i), \ldots, \sum_1^n T_p(x_i))$. Assume Θ is an open p- dimensional rectangle. Because

$$E_\theta\left(\frac{\partial \log f_\theta}{\partial \theta_j}\right) = 0$$

one has

$$\frac{\partial A(\theta)}{\partial \theta_j} = E_\theta(T_j) = \eta_j(\theta)$$

$\eta = (\eta_1, \eta_2, \ldots, \eta_p)$ provides another natural parametrization. Note that the MLE $\hat{\eta} = T/n$.

A class of priors C is said to be a conjugate family if given $p \in C$ the posterior for all n belongs to C. One can generate such families by choosing a σ-finite measure ν on Θ and defining elements of C by

$$p(\theta|m, t_1, t_2, \ldots, t_p) = const.\ \exp\{mA(\theta) + \sum_1^p \theta_j t_j\} \tag{1.15}$$

where m is a positive integer and $t_1, t_2, \ldots, t_p$ are elements in the sample space of $T_1, T_2 \ldots, T_p$. The constants $m, t_1, t_2, \ldots, t_p$ are parameters of the prior distribution chosen such that the prior is proper.

Usually, ν is a nonsubjective prior. Then the prior displayed in (1.15) can be interpreted as a posterior when the prior is ν and one has a conceptual sample of size m yielding values of sufficient statistics $T = (t_1, t_2, \ldots, t_p)$, i.e., compared with ν it represents prior information equivalent to a sample of size m.

The case when ν is the Lebesgue measure deserves special attention. Under certain conditions, one can prove the following by an argument involving integration by parts,

$$E(\eta|X_1, X_2, \ldots, X_n) = \frac{mE(\eta) + n\hat{\eta}}{m+n} \tag{1.16}$$

which shows that the posterior mean is a convex combination of the prior mean and a suitable frequentist estimate. The relation strengthens the interpretation of m as a measure of information in the prior. The elements of C corresponding to the Lebesgue measure are usually called *conjugate* priors. Diaconis and Ylvisaker [47] have shown that these are the only priors that satisfy (1.16). One can elicit the values of $t_1, t_2, \ldots, t_p$ by eliciting the prior mean and m by comparing prior information with information from a sample. This makes these priors relatively easy to elicit, but because one is only eliciting some aspects of the prior, a conjugate prior is a nonsubjective prior with some parameters reflecting prior belief.

Example. f_θ is normal density with mean μ and standard deviation σ. Here $\theta_1 = \mu/2\sigma^2, \theta_2 = -1/\sigma^2$, $A(\theta) = -(\mu^2/2\sigma^2) - \log\sigma$, and $T_1(x) = x, T_2(x) = x^2$. A conjugate prior is of the form

$$p(\theta) = Const.\ e^{mA(\theta)+t_1\theta_1+t_2\theta_2}$$

which can be displayed as the product of a normal and inverse gamma.

Example. f_θ is Bernoulli with parameter θ. Conjugate priors are beta distributions.

Example. f_θ is multinomial with parameters $\theta_1, \theta_2, \ldots, \theta_p$, where $\theta_i \geq 0, \sum \theta_i = 1$. Conjugate priors are Dirichlet distributions discussed in the next chapter.

Conjugate priors have been criticized on two grounds. The relation (1.16) may not be reasonable if there is conflict between the prior and the data. For example, if $p = 1$ and the prior mean is 0 and $\hat{\eta}$ is 20, should one believe the data or the prior? A convex combination of two incompatible estimates is unreasonable.

For $N(\mu, \sigma^2)$, a t-prior for μ and a nonsubjective prior for σ ensures that in cases like this the posterior mean shifts more toward the data, i.e., a choice of such a prior means that, in cases of conflict, one trusts the data. The t-prior is a scale mixture of normal. In general, it seems that mixtures on conjugate priors will possess this kind of property, but we have not seen any general investigation in the literature.

The other criticism of conjugate priors is that only one parameter m is left to model the prior belief on uncertainty. Once again, a mixture of conjugate priors offers more flexibility.

These mixtures may be thought of as modeling prior belief in a hierarchy of stages called hierarchical priors. The reason for their current popularity in Bayesian analysis is that they are flexible and posterior quantities can be calculated by Markov chain Monte Carlo. A good source is Schervish [144].

1.8 Exchangeability, De Finetti's Theorem, Exponential Families

Subjective priors can be elicited in special simple cases, a relatively recent treatment is Kadane et al. [109]. However there is one class of problems where subjective judgments can be made relatively easily and can lead to both a model and a prior.

Suppose $\{X_i\}$ is a sequence of random variables. This sequence is said to be *exchangeable* if for any n distinct $i_1, i_2, \ldots, i_n$,

$$P\{X_{i_1} \in B_1, X_{i_2} \in B_2, \ldots, X_{i_n} \in B_n\} \\ = P\{X_1 \in B_1, X_2 \in B_2, \ldots, X_n \in B_n \quad (1.17)$$

Suppose $\{X_i\}$ take values in $\{0, 1\}$. One may be able to judge if the $\{X_i\}$s are exchangeable. In some sense, such judgments are fundamental to science when one makes inductions about future based on past experience. The next theorem of De

Finetti shows that this subjective judgment leads to a model and affirms the existence of a prior.

Theorem 1.8.1. *If a sequence of random variables $\{X_i\}$ is exchangeable and if each X_i takes values in $\{0,1\}$ then there exists a distribution Π such that*

$$P\{X_1 = x_1, X_2 = x_2, \ldots, X_n = x_{n1}\} = \int_0^1 \theta^r (1-\theta)^{n-r} d\Pi(\theta)$$

with $r = \sum_1^n x_i$

The theorem implies that one has a Bernoulli model and a prior Π. To specify a prior, one needs additional subjective judgments. For example, if given $X_1, X_2, \ldots, X_n$ one predicts $X_{n+1} = (\alpha + \sum x_i)/(\alpha + \beta + n)$, Π then must be a beta prior.

Regazzini [67] has shown that judgments on Exchangeability, along with certain judgments on predictive distributions of X_{n+1} given $X_1, X_2, \ldots, X_n$ lead to a similar representation theorem, which leads to an exponential model along with a mixing distribution, which may be interpreted as a prior. Earlier Bayesian derivations of exponential families is due to Lauritzen [117] and Diaconis and Freedman [44]. A good treatment is in Schervish [144] where partial exchangeability and its modeling through hierarchical priors is also discussed.

2
$M(\mathcal{X})$ and Priors on $M(\mathcal{X})$

2.1 Introduction

As mentioned in Chapter 1, in the nonparametric case the parameter space Θ is typically the set of all probability measures on $\mathcal{X}$. We denote the set of all probability measures on $\mathcal{X}$ by $M(\mathcal{X})$. The cases of interest to us are when $\mathcal{X}$ is a finite set and when $\mathcal{X} = \mathbb{R}$. The Bayesian aspect requires prior distributions on $M(\mathcal{X})$, in other words, probabilities on the space of probabilities. In this chapter we develop some measure-theoretic and topological features of the space $M(\mathcal{X})$ and discuss various notions of convergence on the space of prior distributions.

The results in this chapter, except for the last section, are mainly used to assert the existence of the priors discussed later. Thus, for a reader who is prepared to accept the existence theorems mentioned later, a cursory reading of this chapter would be adequate. On the other hand, for those who are interested in measure-theoretic aspects, a careful reading of this chapter will provide a working familiarity with the measure-theoretic subtleties involved. The last section where formal definitions of consistency are discussed, can be read independently. While we generally consider the case $\mathcal{X} = \mathbb{R}$, most of the arguments would go through when $\mathcal{X}$ is a complete separable metric space.

2.2 The Space $M(\mathcal{X})$

As before, let $\mathcal{X}$ be a complete separable metric space with $\mathcal{B}$ the corresponding Borel σ-algebra on $\mathcal{X}$. Denote by $M(\mathcal{X})$ the space of all probability measures on $(\mathcal{X}, \mathcal{B})$.

As seen in the chapter 1 there are many reasonable notions of convergence on the space $M(\mathcal{X})$, but they are not all equally convenient for our purpose. We begin with a brief discussion of these.

Total Variation Metric. Recall that the total variation metric was defined by

$$\|P - Q\| = 2 \sup_B |P(B) - Q(B)|$$

If p and q are densities of P and Q with respect to some σ-finite measure μ, then $\|P - Q\|$ is just the L_1-distance $\int |p - q| \, d\mu$ between p and q. The total variation metric is a strong metric. If $x \in \mathcal{X}$ and δ_x is the probability degenerate at x, then $U_x = \{P : \|P - \delta_x\| < \epsilon\} = \{P : P(x) > 1 - \epsilon\}$ is a neighborhood of δ_x. Further if $x \neq x'$ then $U_x \cap U_{x'} = \emptyset$. Thus, when $\mathcal{X}$ is uncountable, $\{U_x : x \in \mathcal{X}\}$ is an uncountable collection of disjoint open sets, the existence of which renders $M(\mathcal{X})$ nonseparable. Further, no sequence of discrete measures can converge to a continuous measure and vice versa. These properties make the total variation metric uninteresting when considered on all of $M(\mathcal{X})$.

The total variation metric when restricted to sets of the form L_μ—all probability measures dominated by a σ-finite measure μ—is extremely useful and interesting. In this context we will refer to the total variation as the L_1-metric. It is a standard result that L_μ with the L_1-metric is complete and separable.

Hellinger Metric. This metric was also discussed in Chapter 1. Briefly the Hellinger distance between P and Q is defined by

$$H(P, Q) = \left[\int (\sqrt{p} - \sqrt{q})^2 \, d\mu \right]^{1/2}$$

where p and q are densities with respect to μ. The Hellinger metric is equivalent to the L_1 metric. Associated with the Hellinger metric is a useful quantity $A(P, Q)$ called *affinity*, defined as $A(P, Q) = \int \sqrt{p}\sqrt{q} \, d\mu$. The relation $H^2(P^n, Q^n) = 2 - 2(A(P, Q))^n$, where P^n, Q^n are n-fold product measures, makes the Hellinger metric convenient in the i.i.d. context.

Setwise convergence. The metrics defined in the last section provide corresponding notions of convergence. Another natural way of saying P_n converges to P is to require that $P_n(B) \to P(B)$ for all Borel sets B. A way of formalizing this topology is as follows. Let $\mathcal{F}$ be the class of functions $\{P \mapsto P(B) : B \in \mathcal{B}\}$. On $M(\mathcal{X})$ give the smallest topology that makes the functions in $\mathcal{F}$ continuous. It is easy to see that under this topology, if f is a bounded measurable function, then $P \mapsto \int f\, dP$ is continuous. Sets of the form $\{P : |P(B_i) - P_0(B_i)| < \epsilon_i,\ B_1, B_2, \ldots, B_k \in \mathcal{B}\}$ give a neighborhood base at P_0.

Setwise convergence is an intuitively appealing notion, but it has awkward topological properties that stem from the fact that convergence of $P_n(B)$ to $P(B)$ for sets in an algebra does not ensure the convergence for all Borel sets. We summarize some additional facts as a proposition.

Proposition 2.2.1. *Under setwise convergence:*

(i) $M(\mathcal{X})$ is not separable,

(ii) If P_0 is a continuous measure then P_0 does not have a countable neighborhood base, and hence the topology of setwise convergence is not metrizable.

Proof. (i) $U_x = \{P : P\{x\} > 1 - \epsilon\}$ is a neighborhood of δ_x, and as x varies form an uncountable collection of disjoint open sets.

(ii) Suppose that there is a countable base for the neighborhoods at P_0. Let $\mathcal{B}_0$ be a countable family of sets such that sets of the type

$$\mathcal{U} = \{P : |P(B_i) - P_0(B_i)| < \epsilon_i,\ B_1, B_2, \ldots, B_k \in \mathcal{B}_0\}$$

form a neighborhood base at P_0. It then follows that $P_n(B) \to P(B)$ for all Borel sets B iff $P_n(B) \to P(B)$ for all sets in $\mathcal{B}_0$.

Let $\mathcal{B}_n = \sigma(B_1, B_2, \ldots, B_n)$ where $B_1, B_2, \ldots$ is an enumeration of $\mathcal{B}_0$. Denote by $B_{n1}, B_{n2}, \ldots B_{nk(n)}$ the atoms of $\mathcal{B}_n$. Define P_n to be the discrete measure that gives mass $P_0(B_{ni})$ to x_{ni} where x_{ni} is a point in B_{ni}. Clearly $P_n(B_{mj}) \to P_0(B_{mj})$ for all m_j. On the other hand $P_n(\cup_{i,m}\{x_{mi}\}) = 1$ for all n but $P_0((\cup_{i,m}\{x_{mi}\}) = 0$.

□

These shortcomings persist even when we restrict attention to subsets $M(\mathcal{X})$ of the form L_μ.

Supremum Metric. When $\mathcal{X}$ is $\mathbb{R}$, the Glivenko-Cantelli theorem on convergence of empirical distribution suggests another useful metric, which we call the supremum metric. This metric is defined by

$$d_K(P,Q) = \sup_t |P(-\infty,t] - Q(-\infty,t]|$$

Under this metric $M(\mathcal{X})$ is complete but not separable.

Weak Convergence. In many ways weak convergence is the most natural and useful topology on $M(\mathcal{X})$. Say that

$$P_n \to P \text{ weakly} \quad \text{or } P_n \overset{\text{weakly}}{\to} P \text{ if}$$

$$\int f\, dP_n \to \int f\, dP$$

for all bounded continuous functions f on $\mathcal{X}$. For any P_0 a neighborhood base consists of sets of the form $\cap_1^k\{P : \left|\int f_i\, dP_0 - \int f_i\, dP\right| < \epsilon\}$ where $f_i, i = 1,2,\ldots,k$ are bounded continuous functions on $\mathcal{X}$. One of the things that makes the weak topology so convenient is that under weak convergence $M(\mathcal{X})$ is a complete separable metric space.

The main results that we need with regard to weak convergence are the Portmanteau theorem and Prohorov's theorem given in Chapter 1.

Because $M(\mathcal{X})$ is a complete separable metric space under weak convergence, we define the *Borel σ-algebra* $\mathcal{B}_M$ on $M(\mathcal{X})$ to be the smallest σ-algebra generated by all weakly open sets, equivalently all weakly closed sets. This σ-algebra has a more convenient description as the smallest σ-algebra that makes the functions $\{P \mapsto P(B) : B \in \mathcal{B}\}$ measurable. Let $\mathcal{B}_0$ be the σ-algebra generated by all weakly open sets. Consider all B such that $P \mapsto P(B)$ is $\mathcal{B}_0$-measurable. This class contains all closed sets, and from the π-λ theorem (Theorem 1.2.1) it follows easily that $\mathcal{B}_M$ is the σ-algebra generated by all weakly open sets.

We have discussed two other modes of convergence on $M(\mathcal{X})$: the total variation and setwise convergence. It is instructive to pause and investigate the σ-algebras corresponding to these and their relationship with $\mathcal{B}_M$.

Because these are nonseparable spaces, there is no good acceptable notion of a Borel σ-algebra. In the case of total variation metric, the two common σ-algebras considered are

(i) $\mathcal{B}_o$—the σ-algebra generated by open sets and

(ii) $\mathcal{B}_b$—the σ-algebra generated by open balls.

The σ-algebra $\mathcal{B}_o$ generated by open sets is much larger than $\mathcal{B}_M$. To see this, restrict the σ-algebra to the space of degenerate measures $D_{\mathcal{X}} = \{\delta_x : x \in \mathcal{X}\}$. Then each δ_x is relatively open, and this will force the restriction of $\mathcal{B}_o$ to $D_{\mathcal{X}}$ to be the power set. On the other hand, $\mathcal{B}_M$ restricted to $D_{\mathcal{X}}$ is just the inverse of the Borel σ-algebra on $\mathcal{X}$ under the map $\delta_x \mapsto x$.

Because every open ball is in $\mathcal{B}_M$, so is every set in the σ-algebra generated by these balls. It can be shown that $\mathcal{B}_b$ is properly contained in $\mathcal{B}_M$.

Similar statements hold when we consider the σ-algebras for setwise convergence. The corresponding σ-algebras here would be those generated by open sets and those generated by basic neighborhoods at a point. A discussion of these different σ-algebras can be found in [71].

We next discuss measurability issues on $M(\mathcal{X})$. Following are a few of elementary propositions.

Proposition 2.2.2. *(i) If $\mathcal{B}_0$ is an algebra generating $\mathcal{B}$ then*

$$\sigma\{P \mapsto P(B) : B \in \mathcal{B}_0\} = \mathcal{B}_M$$

(ii) $\sigma\left\{P \mapsto \int f\, dP : f \text{ bounded measurable}\right\} = \mathcal{B}_M$

Proof. (i) Let $\tilde{\mathcal{B}} = \{B : P \mapsto P(B) \text{ is } \mathcal{B}_M \text{ measurable}\}$. Then $\tilde{\mathcal{B}}$ is a σ-algebra and contains $\mathcal{B}_0$. The result now follows from Theorem 1.2.1.

(ii) It is enough to show that $P \mapsto \int f\, dP$ is $\mathcal{B}_M$ measurable. This is immediate for f simple, and any bounded measurable f is a limit of simple functions.

□

Proposition 2.2.3. *Let $f_P(x)$ be a bounded jointly measurable function of (P, x). Then $P \mapsto \int f_P(x)\, dP(x)$ is $\mathcal{B}_M$ measurable.*

Proof. Consider

$$\mathcal{G} = \left\{F \subset M(\mathcal{X}) \times \mathcal{X} \text{ such that } P(F^P) \text{ is } \mathcal{B}_M \text{ measurable}\right\}$$

Here F^P is the P-section $\{x : (P, x) \in F\}$ of F. $\mathcal{G}$ is a λ-system that contains the π-class of all sets of the form $C \times B; C \in \mathcal{B}_M$, $B \in \mathcal{B}$, and by Theorem 1.2.1 is the product σ-algebra on $M(\mathcal{X}) \times \mathcal{X}$. This proves the proposition when $f_P(x) = I_F(P, x)$. The proof is completed by verifying when $f_P(x)$ is simple, and by passing to limits. □

Proposition 2.2.3 can be used to prove the measurability of the set of discrete probabilities.

Proposition 2.2.4. *The set of discrete probabilities is a measurable subset of* $M(\mathcal{X})$.

Proof. If $E = \{(P,x) : P\{x\} > 0\}$ is a measurable set, then setting $f_P(x) = I_E(P,x)$, the set of discrete measures is just $\{P : \int f_P(x)dP = 1\}$ and would be measurable by Proposition 2.2.3. To see that $E = \{(P,x) : P\{x\} > 0\}$ is measurable, we show that $(P,x) \mapsto P\{x\}$ is jointly measurable in (P,x). Consider the set of all a measurable subsets F of $\mathcal{X} \times \mathcal{X}$ such that $(P,x) \mapsto P(F^x)$ is measurable in (P,x). As before, $F^x = \{y : (x,y) \in F\}$. This class contains all Borel sets of the form $B_1 \times B_2$ and is a λ-system, and by Theorem 1.2.1 is the Borel σ-algebra on $\mathcal{X} \times \mathcal{X}$. In particular $(P,x) \mapsto P(F^x)$ is measurable when $F = \{(x,x) : x \in \mathcal{X}\}$ is the diagonal and $E = \{(P,x) : P(F^x > 0)\}$. ☐

Consider $f_P(x)$ used in Proposition 2.2.4. Then P is continuous iff $\int f_P(x)dP = 0$. It follows that the set of continuous measures is a measurable set.

If μ is a σ-finite measure on $\mathbb{R}$, then L_μ is a measurable subset of $M(\mathcal{X})$. To see this, assume without loss of generality that μ is a probability measure. Let $\mathcal{B}_n$ be an increasing sequence of algebras, with finitely many atoms, whose union generates $\mathcal{B}$. Denote the atoms of $\mathcal{B}_n$ by $B_{n1}, B_{n2}, \ldots B_{k(n)}$, and for any probability measure P, set $f_P(x) = \lim \sum_1^{k(n)} P(B_{ni})/\mu(B_{ni})$ when it exists and 0 otherwise. To complete the argument note that $L_\mu = \{P : \int f_P(x)d\mu = 1\}$.

2.3 (Prior) Probability Measures on $M(\mathcal{X})$

2.3.1 $\mathcal{X}$ Finite

Suppose $\mathcal{X} = \{1, 2, \ldots, k\}$. In this case $M(\mathcal{X})$ can be identified with the $(k-1)$ dimensional probability simplex $S_k = \{p_1, p_2, \ldots, p_k : 0 \le p_i \le 1, \sum p_i = 1\}$. One way of defining a prior on $M(\mathcal{X})$ is by defining a measure on S_k. Any such measure defines the joint distribution of $\{P(A) : A \subset \mathcal{X}\}$, because for any A, $P(A) = \sum_{i \in A} p_i$, where $p_k = 1 - \sum_1^{k-1} p_i$.

An example of a prior distribution on S_k is the uniform distribution—the normalized Lebesgue measure on $\{p_1, p_2, \ldots, p_{k-1} : 0 \le p_i \le 1, \sum p_i \le 1\}$. Another example is the Dirichlet density which is given by

$$\Pi(p_1, p_2, \ldots, p_{k-1}) = \frac{\Gamma(\sum_1^k \alpha_i)}{\prod \Gamma(\alpha_i)} p_1^{\alpha_1 - 1} p_2^{\alpha_2 - 1} \ldots p_{k-1}^{\alpha_{k-1} - 1} (1 - \sum_1^{k-1} p_i)^{\alpha_k - 1}$$

where $\alpha_1, \alpha_2, \dots, \alpha_k$ are positive real numbers. This density will be studied in greater detail later.

A different parametrization of $M(\mathcal{X})$ yields another method of constructing a prior on $M(\mathcal{X})$. Assume for ease of exposition that $\mathcal{X}$ contains 2^k elements $\{x_1, x_2, \dots, x_{2^k}\}$. Let

$$B_0 = \{x_1, x_2, \dots, x_{2^{k-1}}\} \text{ and } B_1 = \{x_{2^{k-1}+1}, x_{2^{k-1}+2}, \dots, x_{2^k}\}$$

be a partition of $\mathcal{X}$ into two sets. Let B_{00}, B_{01} be a partition of B_0 into two halves and B_{10}, B_{11} be a similar partition of B_1. Proceeding this way we can get partitions $B_{\epsilon_1\epsilon_2\dots\epsilon_i0}, B_{\epsilon_1\epsilon_2\dots\epsilon_i1}$ of $B_{\epsilon_1\epsilon_2\dots\epsilon_i}$ where each ϵ_i is 0 or 1 and $i < k$. Clearly, this partition stops at $i = k$.

We next note that the partitions can be used to identify $\mathcal{X}$ with $E_k = \{0,1\}^k$. Any $x \in \mathcal{X}$ corresponds to a sequence $\epsilon_1(x)\epsilon_2(x)\dots\epsilon_k(x)$ where $\epsilon_i(x) = 0$ if x is in $B_{\epsilon_1(x)\epsilon_2(x)\dots\epsilon_{i-1}(x)0}$ and 1 if x is in $B_{\epsilon_1(x)\epsilon_2(x)\dots\epsilon_{i-1}(x)1}$. Conversely, any sequence $\epsilon_1\epsilon_2\dots\epsilon_k$ corresponds to the point $\cap_1^k B_{\epsilon_1\epsilon_2\dots\epsilon_i}$. Thus there is a correspondence—depending on the partition—between the set $M(\mathcal{X})$ of probability measures on $\mathcal{X}$ and the set $M(E_k)$ of probability measures on E_k.

Any probability measure on E_k is determined by quantities like

$$y_{\epsilon_1\epsilon_2\dots\epsilon_k} = P\left(\epsilon_{i+1} = 0 \mid \epsilon_1, \epsilon_2, \dots, \epsilon_i\right)$$

Specifically, let E_k^* be the set of all sequences of 0 and 1 of length less than k, including the empty sequence $\emptyset$. If $0 \le y_{\underline{\epsilon}} \le 1$ is given for all $\underline{\epsilon} \in E_k^*$, then there is a probability on E_k by

$$P(\epsilon_1\epsilon_2\dots\epsilon_k) = \prod_{i=1,\epsilon_i=0}^{k} y_{\epsilon_1\epsilon_2\dots\epsilon_{i-1}} \prod_{i=1,\epsilon_i=1}^{k} (1 - y_{\epsilon_1\epsilon_2\dots\epsilon_{i-1}})$$

where $i = 1$ corresponds to the empty sequence $\emptyset$. Hence construction of a prior on E_k amounts to a specification of the joint distribution for $\{y_{\underline{\epsilon}} : \underline{\epsilon} \in E_k^*\}$.

A little reflection will show that all we have done is to reparametrize a probability P on $\mathcal{X}$ by

$$P(B_0), P(B_{00}|B_0), P(B_{10}|B_1), \dots, P(B_{\epsilon_1\epsilon_2\dots\epsilon_{k-1}0}|B_{\epsilon_1\epsilon_2\dots\epsilon_{k-1}0})$$

Of interest to us is the case where the $Y_{\underline{\epsilon}}$s, equivalently $P(B_{\underline{\epsilon}0}|B_{\underline{\epsilon}})$s, are all independent. The case when these are independent beta random variables—the Polya tree processes—will be studied in Chapter 3

Yet another method of obtaining a prior distribution on $M(\mathcal{X})$ is via De Finetti' theorem. De Finetti's theorem plays a fundamental role in Bayesian inference, and we refer the reader to [144] for an extensive discussion.

Let $X_1, X_2, \ldots, X_n$ be $\mathcal{X}$-valued random variables. $X_1, X_2, \ldots, X_n$ is said to be exchangeable if $X_1, X_2, \ldots, X_n$ and $X_{\pi(1)}, X_{\pi(2)}, \ldots, X_{\pi(n)}$ have the same distribution for every permutation π of $\{1, 2, \ldots, n\}$. A sequence $X_1, X_1, \ldots$ is said to be exchangeable if $X_1, X_2, \ldots, X_n$ is exchangeable for every n.

Theorem 2.3.1. *[De Finetti] A sequence of $\mathcal{X}$-valued random variables is exchangeable iff there is a unique measure Π on $M(\mathcal{X})$ such that for all n,*

$$\int_{M(\mathcal{X})} \prod_1^n p(x_i)\, d\Pi(p) = Pr\{X_1 = x_1, X_2 = x_2, \ldots, X_n = x_n\}$$

In general it is not easy to construct Π from the distribution of the X_is. Typically, we will have a natural candidate for Π. By uniqueness, it is enough to verify the preceding equation. On the other hand, given Π, the behavior of $X_1, X_1, \ldots$ often gives insight into the structure of Π.

As an example, let $\mathcal{X} = \{x_1, x_2, \ldots, x_k\}$. Let $\alpha_1, \alpha_2, \ldots, \alpha_k$ be positive integers. Let $\bar{\alpha}(i) = \alpha_i / \sum \alpha_j$. Consider the following urn scheme: Suppose a box contains balls of k- colors, with α_i balls of color i. Choose a ball at random, so that $P(X_1 = i) = \bar{\alpha}(i)$. Replace the ball and add one more of the same color. Clearly, $P(X_2 = j | X_1 = i) = (\alpha_j + \delta_i(j))/(\sum \alpha_i + 1)$ where $\delta_i(j) = 1$ if $i = j$ and 0 otherwise. Repeat this process to obtain $X_3, X_4, \ldots$ Then

(i) $X_1, X_2, \ldots$ are exchangeable; and

(ii) the prior Π for this case is the Dirichlet density on S_k.

2.3.2 $\mathcal{X} = \mathbb{R}$

We next turn to construction of measures on $M(\mathcal{X})$. Because the elements of $M(\mathcal{X})$ are functions on $\mathcal{B}$, $M(\mathcal{X})$ can be viewed as a subset of $[0,1]^{\mathcal{B}}$ where the product space $[0,1]^{\mathcal{B}}$ is equipped with the canonical product σ-algebra, which makes all the coordinate functions measurable. Note that the restriction of the product σ-algebra to $M(\mathcal{X})$ is just $\mathcal{B}_M$. A natural attempt to construct measures on $M(\mathcal{X})$ would be to use Kolomogorov's consistency theorem to construct a probability measure on $[0,1]^{\mathcal{B}}$, which could then be restricted to $M(\mathcal{X})$. However $M(\mathcal{X})$ is not measurable as a subset of $[0,1]^{\mathcal{B}}$, and that makes this approach somewhat inconvenient. To see that $M(\mathcal{X})$ is not measurable, note that singletons are measurable subsets of $M(\mathcal{X})$ but not so in the product space.

When $\mathcal{X} = \mathbb{R}$, distribution functions turn out to be a useful crutch to construct priors on $M(\mathbb{R})$. To elaborate:

(i) Let Q be a dense subset of $\mathbb{R}$ and let $\mathcal{F}^*$ be all real-valued functions on Q such that

(a) F is right-continuous on Q,

(b) F is nondecreasing, and

(c) $\lim_{t\to\infty} = F(t) = 1, \lim_{t\to-\infty} F(t) = 0$.

(ii) Let $\mathcal{F}$ be all real-valued functions on $\mathbb{R}$ such that

(a) F is right-continuous on $\mathbb{R}$,

(b) F is non decreasing, and

(c) $\lim_{t\to\infty} F(x) = 1, \lim_{t\to-\infty} F(x) = 0$.

(iii) $M(\mathbb{R}) = \{P : P \text{ is a probability measure on } \mathbb{R}\}$

There is a natural 1-1 correspondence between these three sets: Let $\phi_1 : M(\mathbb{R}) \mapsto \mathcal{F}$ be the function that takes a probability measure P to its distribution function $F_P(t) = P(-\infty, t]$ and let $\phi_2 : \mathcal{F} \to \mathcal{F}^*$ be the function that maps a distribution function to its restriction on Q. These maps are 1-1, onto, and bi-measurable. Thus any probability measure on $\mathcal{F}^*$ can be transferred to a probability on $\mathcal{F}$ and then to $M(\mathbb{R})$. A prior on $\mathcal{F}^*$ only involves the distributions of

$$(F(t_1), F(t_2) - F(t_1), \ldots, F(t_k) - F(t_{k-1}))$$

for t_is in Q. However, because any $F(t)$ is a limit of $F(t_n), t_n \in Q$, the distributions of quantities like $(F(t_1), F(t_2) - F(t_1), \ldots, F(t_k) - F(t_{k-1}))$ for t_i-real can be recovered, at least as limits. On the other hand since a general Borel set B has no simple description in terms of intervals, one can assert the existence of a distribution for $P(B)$ that is compatible with the prior on $\mathcal{F}^*$, but it may not be possible to arrive at anything resembling an explicit description of the distribution.

It is convenient to use the notation $\mathcal{L}(\cdot|\Pi)$ to stand for the distribution or law of a quantity under the distribution Π.

Theorem 2.3.2. *Let Q be a countable dense subset of $\mathbb{R}$. Suppose for every k and every collection $t_1 < t_2 < \ldots < t_k$ with $\{t_1, t_2, \ldots, t_k\} \subset Q$, $\Pi_{t_1,t_2,\ldots,t_k}$ is a probability measure on $[0,1]^k$ which is a specification of a distribution of $((F(t_1), F(t_2), \ldots, F(t_k))$ such that*

(i) if $\{t_1, t_2, \ldots, t_k\} \subset \{s_1, s_2, \ldots, s_l\}$ then the marginal distribution on $(t_1, t_2, \ldots, t_k)$ obtained from $\Pi_{s_1,s_2,\ldots,s_l}$ is $\Pi_{t_1,t_2,\ldots,t_k}$;

(ii) *if* $t_1 < t_2$ *then* $\Pi_{t_1,t_2}\{F(t_1) \le F(t_2)\} = 1$;

(iii) *if* $(t_{1n}, t_{2n}, \dots, t_{kn}) \downarrow (t_1, t_2, \dots, t_k)$ *then* $\Pi_{(t_{1n},t_{2n},\dots,t_{kn})}$ *converges in distribution to* $\Pi_{(t_1,t_2,\dots,t_k)}$; *and*

(iv) *if* $t_n \downarrow -\infty$ *then* $\Pi_{t_n} \to 0$ *in distribution and if* $t_n \uparrow \infty$ *then* $\Pi_{t_n} \to 1$ *in distribution.*

then there exists a probability measure Π *on* $M(\mathbb{R})$ *such that for every* $t_1 < t_2 < \dots < t_k$, *with* $\{t_1, t_2, \dots, t_k\} \subset Q$,

$$\mathcal{L}\left((F(t_1), F(t_2), , \dots, F(t_k))\,|\Pi\right) = \Pi_{t_1,t_2,\dots,t_k}.$$

Proof. By the Kolomogorov consistency theorem (i) ensures the existence of a probability measure Π on $[0,1]^Q$ with $\Pi_{(t_1,t_2,\dots,t_k)}$ as marginals. We will argue that $\Pi(\mathcal{F}^*) = 1$

Suppose $\mathcal{F}_1^* = \cap_{t_i<t_j}\left\{F \in [0,1]^Q : F(t_i) \le F(t_j)\right\}$. Because Q is countable by (ii), $\Pi(\mathcal{F}_1^*) = 1$.

Next, fix t and a sequence t_n in Q decreasing to t. On $\mathcal{F}_1^*$, $F(t_n)$ as a function of F is decreasing in n and hence has a limit. If $F^*(t) = \lim_n F(t_n)$ then $F^*(t) \ge F(t)$ and by assumption (iii) $E_\Pi F^*(t) = E_\Pi F(t)$, so that $F^*(t) = F(t)$ a.e. Π. Consequently

$$\Pi\{F \in \mathcal{F}_1^* : \text{ F is right-continuous at } t\} = 1$$

and the countability of Q yields

$$\Pi\{F : F \text{ is monotone and } F \text{ is right-continuous at all } t \in Q\} = 1$$

A similar argument shows that with Π probability 1, for F in $\mathcal{F}_1^*$, $\lim_{t\to\infty} = F(t) = 1$, and $\lim_{t\to-\infty} F(t) = 0$. This shows that $\Pi(\mathcal{F}^*) = 1$.

Thus we have established the existence of a probability measure on $\mathcal{F}^*$. Using the discussion preceding the theorem this prior can be lifted to all of $M(\mathbb{R})$. □

The assumptions of Theorem 2.3.2 require specification of finite-dimensional distribution only for t_is in Q and the conclusion also involves only the finite dimensional distributions for t_is in Q. It is easy to see that if one starts with $\Pi_{(t_1,t_2,\dots,t_k)}$ with t_i's real and satisfying the conditions of Theorem 2.3.2 then one would get a Π for which the marginals are $\Pi_{(t_1,t_2,\dots,t_k)}$ for t_is real.

A convenient way of specifying the distribution of $(F(t_1), F(t_2), \dots, F(t_k))$ for $t_1 < t_2 < \dots, t_k$, is by specifying the distribution, say $\Pi'_{t_1,t_2,\dots,t_k}$, of

$$(F(t_1), F(t_2) - F(t_1), \dots, F(t_k - F(t_{k-1}))$$

The convenience arises from the fact that $(-\infty, t_1], (t_1, t_2], \ldots, (t_k, \infty)$ can be thought of as $k+1$ cells and $(p_1, p_2, \ldots, p_{k+1})$ as the corresponding multinomial probabilities. Note that $\Pi'_{t_1,t_2,\ldots,t_k}$ is a probability measure on $S_k = \{(p_1, p_2, \ldots, p_k : p_i \geq 0, \sum_1^k p_i \leq 1\}$. If the specifications of the collection $\Pi'_{t_1,t_2,\ldots,t_k}$ satisfy assumptions (ii),(iii), and (iv) of Theorem 2.3.2, then so would the collection $\Pi_{t_1,t_2,\ldots,t_k} = \mathcal{L}((p_1, p_1 + p_2, \ldots, \sum_1^k p_i)|\Pi'_{t_1,t_2,\ldots,t_k})$. These observations give the following easy variant of Theorem 2.3.2.

Theorem 2.3.3. *Suppose that for every k and every collection $t_1 < t_2 < \ldots < t_k$ with $\{t_1, t_2, \ldots, t_k\} \subset \mathbb{R}$, $\Pi_{t_1,t_2,\ldots,t_k}$ is a probability measure on $S_k = \{(p_1, p_2, \ldots, p_k) : p_i \geq 0, \sum_1^k p_i \leq 1\}$ such that*

(i) if $\{t_1, t_2, \ldots, t_k\} \subset \{s_1, s_2, \ldots, s_l\}$ then the marginal distribution on $(t_1, t_2, \ldots, t_k)$ obtained from $\Pi_{s_1,s_2,\ldots,s_l}$ is $\Pi_{t_1,t_2,\ldots,t_k}$;

(ii) if $(t_{1n}, t_{2n}, \ldots, t_{kn}) \to (t_1, t_2, \ldots, t_k)$ then $\Pi_{(t_{1n},t_{2n},\ldots,t_{kn})}$ converges in distribution to $\Pi_{(t_1,t_2,\ldots,t_k)}$; and

(iii) if $t_n \downarrow -\infty$ then $\Pi_{t_n} \to 0$ in distribution and if $t_n \uparrow \infty$ then $\Pi_{t_n} \to 1$ in distribution.

then there exists a probability measure Π on $\mathcal{F}$ (equivalently on $M(\mathbb{R})$ such that for every $t_1 < t_2 < \ldots < t_k$, with $\{t_1, t_2, \ldots, t_k\} \subset \mathbb{R}$,

$$\mathcal{L}\left((F(t_1), F(t_2) - F(t_1), \ldots, F(t_k) - F(t_{k-1}))\,|\Pi\right) = \Pi_{t_1,t_2,\ldots,t_k}$$

Suppose $(B_1, B_2, \ldots, B_k)$ is a collection of disjoint subsets of $\mathbb{R}$; the next theorem shows that if the distribution of $P(B_1), P(B_2), \ldots, P(B_k)$ are themselves prescribed consistently then the prior Π would have the prescribed marginal distribution for $(P(B_1), P(B_2), \ldots, P(B_k))$.

Theorem 2.3.4. *Suppose for each collection of disjoint Borel sets $(B_1, B_2, \ldots, B_k)$ a distribution $\Pi_{B_1,B_2,\ldots,B_k}$ is assigned for $(P(B_1), P(B_2), \ldots, P(B_k))$ such that*

(i) $\Pi_{B_1,B_2,\ldots,B_k}$ is a probability measure on k-dimensional probability simplex S_k and if $A_1, A_2, \ldots, A_l$ is another collection of disjoint Borel sets whose elements are

union of sets from $(B_1, B_2, \ldots, B_k)$ *then*

$$\Pi_{A_1,A_2,\ldots,A_l} = \textit{distribution of} \left(\sum_{B_i \subset A_1} P(B_i), \sum_{B_i \subset A_2} P(B_i), \ldots, \sum_{B_i \subset A_l} P(B_i) \right)$$

(ii) if $B_n \downarrow \emptyset$; *and* $\Pi_{B_n} \to 0$ *in distribution,*

(iii) $P(\mathbb{R}) \equiv 1$.

Then there exists a probability measure Π *on* $M(\mathbb{R})$ *such that for any collection of disjoint Borel sets* $(B_1, B_2, \ldots, B_k)$, *the marginal distribution of* $(P(B_1), \ldots, P(B_k))$ *under* Π *is* $\Pi_{B_1,B_2,\ldots,B_k}$.

Remark 2.3.1. Given $\Pi_{B_1,B_2,\ldots,B_k}$ as earlier, we can extend the definition to obtain $\Pi_{A_1,A_2,\ldots,A_m}$ for any collection (not necessarily disjoint) of Borel sets $A_1, A_2, \ldots, A_m$. Toward this, let $B_1 = A_1, B_i = A_i - \cup_{j<i} A_j$, and define $\Pi_{A_1,A_2,\ldots,A_m}$ as the distribution of $(P(B_1, P(B_1) + P(B_2) + \ldots, \sum_1^m P(B_j))$ under $\Pi_{B_1,B_2,\ldots,B_m}$. The following proof shows that the marginal distribution under Π of $(P(A_1), P(A_2), \ldots, P(A_k))$ of any collection of Borel sets is $\Pi_{A_1,A_2,\ldots,A_k}$.

Proof. As in the Theorem 2.3.3 start with partitions of the form $B_i = (t_{i-1}, t_i]$ for $i = 1, 2, \ldots, k$; and let Π be the measure obtained on $\mathcal{F}$. Let ϕ_2 be the map from $\mathcal{F}$ to $M(\mathbb{R})$ defined by $\phi_2(F) = P_F$, where P_F is the probability measure corresponding to F. It is easy to see that this map is 1-1 and measurable. We will continue to denote by Π the induced measure on $M(\mathbb{R})$.

Π by construction sits on $M(\mathbb{R})$. What we then need to show is that the marginal distribution of $(P(B_1), P(B_2), \ldots, P(B_k))$ under Π is $\Pi_{B_1,B_2,\ldots,B_k}$.

Step 1 (ii) implies that

$$\text{if } (B_{1n}, B_{2n}, \ldots, B_{kn}) \downarrow (B_1, B_2, \ldots, B_{k1}) \text{ then}$$

$$(P(B_{1n}), P(B_{2n}), \ldots, P(B_{kn})) \to (P(B_1), P(B_2), \ldots, P(B_k)) \text{ in distribution.}$$

To see this,

$$\begin{aligned}((P(B_{1n}), &P(B_{2n}), \ldots, P(B_{kn})) \\ &= (P(B_1) + (P(B_{1n}) - P(B_1)), P(B_2) + (P(B_{2n}) - P(B_2)), \ldots, \\ &\qquad P(B_k) + (P(B_{kn}) - P(B_k)))\end{aligned}$$

and for each i, $(B_{in} - B_i) \downarrow \emptyset$ and hence $(P(B_{in}) - P(B_i))$ goes to 0 in distribution and hence in probability. As a result, the whole vector

$$((P(B_{1n}) - P(B_1)), (P(B_{2n}) - P(B_2)), \ldots, (P(B_{kn}) - P(B_k))) \downarrow 0 \text{ in probability}$$

Step 2 Denote by $\mathcal{B}_0$ the algebra generated by intervals of the form $(a, b]$. For any $B_1, B_2, \ldots, B_k$, let $\mathcal{L}(P(B_1), P(B_2), \ldots, P(B_k)|\Pi)$ denote the distribution of the vector $(P(B_1), P(B_2), \ldots, P(B_k))$ under Π. Fix k. Let $C_i = (a_i, b_i], i = 2, \ldots, k$. Consider

$$\hat{\mathcal{B}} = \{B_1 : \mathcal{L}(P(B_1), P(C_2), \ldots, P(C_k)|\Pi) = \Pi_{(B_1, C_2, \ldots, C_k)}\}$$

Then $\hat{\mathcal{B}}$ contains all sets of the form $(a, b]$, is closed under disjoint unions of such sets, and hence contains $\mathcal{B}_0$. In addition, by Step 1 this is a monotone class. So $\hat{\mathcal{B}}$ is $\mathcal{B}$.

Step 3 Now consider

$$\{B_2 : \mathcal{L}(P(B_1), P(B_2), P(C_3), \ldots, P(C_k)|\Pi) = \Pi_{(B_1, B_2, C_3, \ldots, C_k)}\}$$

From step 2, this class contains all sets of the form $(a, b]$, and their finite disjoint unions and hence contains $\mathcal{B}_0$. Further, it is a monotone class and so is $\mathcal{B}$. Continuing similarly, it follows that for any Borel sets $B_1, B_2, \ldots, B_k$,

$$\mathcal{L}(P(B_1), P(B_2), \ldots, P(B_k)|\Pi) = \Pi_{B_1, B_2, \ldots, B_k}$$

. □

Example 2.3.1. Let α be a finite measure on $\mathbb{R}$. For any partition $(B_1, B_2, \ldots, B_k)$, let $\Pi_{B_1, B_2, \ldots, B_k}$ on S_k be a Dirichlet $(\alpha(B_1), \alpha(B_2), \ldots, \alpha(B_k))$. We will show in Chapter 3 that this assignment satisfies the conditions of Theorem 2.3.4.

Remark 2.3.2. Theorem 2.3.4 on constructing a measure Π on $\mathcal{F}$ through finite-dimensional distribution can be viewed from a different angle. Toward this, for each n, divide the interval $[-2^n, 2^n]$ into intervals of length 2^{-n} and let $-2^n = t_{n1} < t_{n2} < \ldots < t_{nk(n)} = 2^n$ denote the endpoints of the intervals. These define a partition of $\mathbb{R}$ into $k(n) + 1$ cells in an obvious way. Any probability $(p_1, p_2, \ldots, p_{k(n)+1})$ on these $k(n) + 1$ cells corresponds to a distribution function on $\mathbb{R}$, which is constant on each interval and thus any probability $\Pi_{t_{n1}, t_{n2}, \ldots, t_{nk(n)}}$ on $S_{k(n)+1}$ defines a probability measure μ_n on $\mathcal{F}_n$ = all distribution functions, which are constant on the interval $(t_{ni}, t_{ni+1}]$. The consistency assumption on $\Pi_{t_{n1}, t_{n2}, \ldots, t_{nk(n)}}$ shows that the marginal distribution on $\mathcal{F}_n$ obtained from μ_{n+1} is just μ_n. Now it can be shown that

1. $\{\mu_n\}_{n\geq 1}$ is tight as a sequence of probability measures on $\mathcal{F}$. To see this, let $\varepsilon_i \downarrow 0$ and let K_i be a sequence of compact subsets of $\mathbb{R}$. Then

$$\{P : P(K_i) \geq 1 - \varepsilon_i \text{ for all } i\}$$

is a compact subset of $M(\mathbb{R})$. What is needed to show tightness is that given δ, there is a set of the form given earlier with μ_n measure greater than $1-\delta$ for all n. Use assumptions (i) and (iii) of Theorem 2.3.4 and show that for each i, you can get an n_i such that for all n, $\mu_n\{F : F(t_{n_i0}) > \varepsilon_i \text{ and } 1-F(t_{n_i,k(n_i)}) > \varepsilon_i\} < \delta/2^i$;

2. $\{\mu_n\}$ converges to a measure Π; and

3. Π satisfies the conclusions of Theorem 2.3.4.

2.3.3 Tail Free Priors

When $\mathcal{X}$ is finite, we have seen that by partitioning $\mathcal{X}$ into

$$\{B_0, B_1\}, \{B_{00}, B_{01}, B_{10}, B_{11}\}, \ldots$$

and reparametrizing a probability by $P(B_0), P(B_{00}|B_0)\ldots$, we can identify measures on $M(\mathcal{X})$ with E_k—the set of sequences of 0s and 1s of length k. Tail free priors arise when these conditional probabilities are independent. In this section we extend this method to the case $\mathcal{X} = \mathbb{R}$.

Let E be all infinite sequences of 0s and 1s, i.e., $E = \{0,1\}^{\mathbf{N}}$. Denote by E_k all sequences $\epsilon_1, \epsilon_2, \ldots, \epsilon_k$ of 0s and 1s of length k, and let $E^* = \cup_k E_k$ be all sequences of 0s and 1s of finite length. We will denote elements of E^* by $\underline{\epsilon}$.

Start with a partition

$$\mathcal{T}_0 = \{B_0, B_1\}$$

of $\mathcal{X}$ into two sets. Let

$$\mathcal{T}_1 = \{B_{00}, B_{01}, B_{10}, B_{11},\}$$

where B_{00}, B_{01} is a partition of B_0 and B_{10}, B_{11} is a partition of B_1. Proceeding this way,let $\mathcal{T}_n$ be a partition consisting of sets of the form $B_{\underline{\epsilon}}$, where $\underline{\epsilon} \in E_n$ and further $B_{\underline{\epsilon}1}$, $B_{\underline{\epsilon}0}$ is a partition of $B_{\underline{\epsilon}}$.

We assume that we are given a sequence of partitions $\mathcal{T} = \{\mathcal{T}_n\}_{n\geq 1}$ constructed as in the last paragraph such that the sets $\{B_{\underline{\epsilon}} : \underline{\epsilon} \in E^*\}$ generate the Borel σ-algebra.

Definition 2.3.1. A prior Π on $M(\mathbb{R})$ is said to be *tail free* with respect to $\underline{\mathcal{T}} = \{\mathcal{T}_n\}_{n\geq 1}$ if rows in

$$\{P(B_0)\}$$
$$\{P(B_{00}|B_0), P(B_{10}|B_1)\}$$
$$\{P(B_{000}|B_{00}), P(B_{000}|B_{00}), P(B_{010}|B_{01}), P(B_{100}|B_{10}), P(B_{110}|B_{11})\}$$
$$\cdots\cdots\cdots$$

are independent.

To turn to the construction of tail free priors on $M(\mathbb{R})$, start with a dense set of numbers Q, like the binary rationals in $(0,1)$, and write it as $Q = \{a_{\underline{\epsilon}} : \underline{\epsilon} \in E^*\}$ such that for any $\underline{\epsilon}$ $\underline{\epsilon}0 < \underline{\epsilon} < \underline{\epsilon}1$ and construct the following sequence of partitions of $\mathbb{R}$: $\underline{\mathcal{T}}_0 = \{B_0, B_1\}$ is a partition of $\mathbb{R}$ into two intervals, say

$$B_0 = (-\infty, a_0], B_1 = (a_0, \infty)$$

Let $\underline{\mathcal{T}}_1 = \{B_{00}, B_{01}, B_{10}, B_{11},\}$, where

$$B_{00} = (-\infty, a_{00}], B_{01} = (a_{00}, a_0]$$

and

$$B_{10} = (a_0, a_{01}], B_{11} = (a_{01}, \infty)$$

Proceeding this way, let $\underline{\mathcal{T}}_n$ be a partition consisting of sets of the form $B_{\epsilon_1,\epsilon_2,\ldots,\epsilon_n}$, where $\epsilon_1, \epsilon_2, \ldots, \epsilon_n$ are 0 or 1 and further $B_{\epsilon_1,\epsilon_2,\ldots,\epsilon_n 0}, B_{\epsilon_1,\epsilon_2,\ldots,\epsilon_n 1}$ is a partition of $B_{\epsilon_1,\epsilon_2,\ldots,\epsilon_n}$.

The assumption that Q is dense is equivalent to the statement that the sequence of partitions $\underline{\mathcal{T}} = \{\mathcal{T}_n\}_{n\geq 1}$ constructed as in the last paragraph are such that the sets $\{B_{\underline{\epsilon}} : \underline{\epsilon} \in E^*\}$ generate the Borel σ-algebra.

For each $\underline{\epsilon} \in E^*$, let $Y_{\underline{\epsilon}}$ be a random variable taking values in $[0,1]$. If we set $Y_{\underline{\epsilon}} = P(B_{\underline{\epsilon}0} \,|B_{\underline{\epsilon}})$, then for each k, $\{Y_{\underline{\epsilon}} : \underline{\epsilon} \in \cup_{i\leq k} E_i\}$ define a joint distribution for $P(B_{\underline{\epsilon}}) : \underline{\epsilon} \in E_k$. By construction, these are consistent. In order for these to define a prior on $M(\mathbb{R})$ we need to ensure that the continuity condition (ii) of Theorem 2.3.2 holds.

Theorem 2.3.5. *If* $Y_{\underline{\epsilon}} = P(B_{\underline{\epsilon}0} \,|B_{\underline{\epsilon}})$*, where* $Y_{\underline{\epsilon}} : \underline{\epsilon} \in E^*$ *is a family of* $[0,1]$ *valued random variables such that*

(i)

$$Y \perp \{Y_0, Y_1\} \perp \{Y_{00}, Y_{01}, Y_{10}, Y_{11}\} \perp \ldots$$

(ii) for each $\epsilon \in E^*$,

$$Y_{\epsilon 0} Y_{\epsilon 00} Y_{\epsilon 000} \ldots = 0 \text{ and } Y_1 Y_{11} \ldots = 0 \tag{2.1}$$

then there exists a tail free prior Π *on* $M(\mathbb{R})$ *(with respect to the partition under consideration) such that* $Y_\epsilon = P(B_{\epsilon 0}|B_\epsilon)$.

Proof. As noted earlier we need to verify condition (ii) of Theorem 2.3.2. In the current situation it amounts to showing that if $\underline{\epsilon}^o = \epsilon_1^0 \epsilon_2^0 \ldots \epsilon_k^0$ and as $n \to \infty$, $a_{\underline{\epsilon}_n}$ decreases to $a_{\underline{\epsilon}^0}$, then the distribution of $F\left(a_{\underline{\epsilon}_n}\right)$ converges to $F\left(a_{\underline{\epsilon}^0}\right)$. Because any sequence of $a_{\underline{\epsilon}}$ decreasing to $a_{\underline{\epsilon}^0}$ is a subsequence of $a_{\underline{\epsilon}^0 1}, a_{\underline{\epsilon}^0 10}, a_{\underline{\epsilon}^0 100}, \cdots$,

$$F\left(a_{\underline{\epsilon}^0 10\ldots 0}\right) = F\left(a_{\underline{\epsilon}^0}\right) + P(B_{\underline{\epsilon}^0 10\ldots 0})$$

and

$$P(B_{\underline{\epsilon}^0 1,0\ldots 0}) = P(B_{\underline{\epsilon}^0})(1 - Y_{\underline{\epsilon}^0}) Y_{\underline{\epsilon}^0 1} Y_{\underline{\epsilon}^0 10} \ldots$$

the result follows from (ii). □

These discussions can be usefully and elegantly viewed by identifying $\mathbb{R}$ with the space of sequences of 0s and 1s.

As before, let E be $\{0,1\}^{\mathbf{N}}$. Any probability on E gives rise to the collection of numbers $\{y_{\underline{\epsilon}} : \underline{\epsilon} \in E^*\}$, where $y_{\epsilon_1 \epsilon_2 \ldots \epsilon_n} = P(\epsilon_{n+1} = 0 | \epsilon_1 \epsilon_2 \ldots \epsilon_n)$. Conversely, setting $y_{\epsilon_1 \epsilon 2 \ldots \epsilon_n} = P(\epsilon_{n+1} = 0 | \epsilon_1 \epsilon_2 \ldots \epsilon_n)$, any set numbers $\{y_{\underline{\epsilon}} : \underline{\epsilon} \in E^*\}$, with $0 \le y_{\underline{\epsilon}} \le 1$ determines a probability on E. In other words,

$$P(\epsilon_1 \epsilon_2 \ldots \epsilon_k) = \prod_{i=1, \epsilon_i = 0}^{k} y_{\epsilon_1 \epsilon_2 \ldots \epsilon_{i-1}} \prod_{i=1, \epsilon_i = 1}^{k} (1 - y_{\epsilon_1 \epsilon_2 \ldots \epsilon_{i-1}}) \tag{2.2}$$

Hence, to define a prior on $M(E)$, we need to specify a joint distribution for $\{Y_{\underline{\epsilon}} : \underline{\epsilon} \in E^*\}$, where each $Y_{\underline{\epsilon}}$ is between 0 and 1.

As in the finite case, we want to use the partitions $\underline{\mathcal{T}} = \{\mathcal{T}_n\}_{n \ge 1}$ to identify $\mathbb{R}$ with sequences of 0s and 1s. and Let $x \in \mathbb{R}$. $\phi(x)$ is the function that sends x to the sequence $\underline{\epsilon}$ in E, where

$$\epsilon_1(x) = 0 \quad \text{if } x \in B_0 \qquad \epsilon_1(x) = 1 \quad \text{if } x \in B_1$$

$$\epsilon_i(x) = 0 \quad \text{if } x \in B_{\epsilon_1, \epsilon_2, \ldots, \epsilon_{i-1} 0} \qquad \epsilon_i(x) = 1 \quad \text{if } x \in B_{\epsilon_1 \epsilon_2 \ldots \epsilon_{i-1} 1}$$

Because each $\underline{\mathcal{T}}_n$ is a partition of $\mathbb{R}$, ϕ defines a function from $\mathbb{R}$ into E. ϕ is 1-1, measurable but not onto E. The range of ϕ will not contain sequences that are eventually 0. This is another way of saying that with binary expansions we consider the expansion with 1 in the tails rather than 0s. If $D = \{\underline{\epsilon} \in E : \epsilon_i = 0 \text{ for all } i \geq n \text{ for some } n\} \cup \{\underline{\epsilon} : \epsilon_i = 1 \text{ for all } i\}$, then ϕ is 1-1, measurable from $\mathbb{R}$ onto $D^c \cap E$. Further, ϕ^{-1} is measurable on $D^c \cap E$. Thus, as before, the set of probability measures $M(\mathbb{R})$ can be identified with $M^0(E)$—the set of probability measures on E that give mass 0 to D. This reduces the task of defining a prior on $M(\mathbb{R})$ to one of defining a prior on $M^0(E)$.

The condition $P(D) = 0$ gets translated to

$$y_{\underline{\epsilon}0}(y_{\underline{\epsilon}00})\ldots = 0 \text{ for all } \underline{\epsilon} \in E^* \text{ and } y_1 y_{11} \ldots = 0 \tag{2.3}$$

As before, defining a prior on $M(\mathbb{R})$, equivalently on $M^0(E)$, amounts to defining $\{Y_{\underline{\epsilon}} : \underline{\epsilon} \in E^*\}$ such that (2.3) is satisfied almost surely. Satisfying (2.3) almost surely corresponds to condition (ii) in Theorem 2.3.5.

A useful way to specify a prior on $M(E)$ is by having $Y_{\underline{\epsilon}}$ for $\underline{\epsilon}$ of different lengths be mutually independent, which yields tail free priors. In Chapter 3, we return to this construction, to develop Polya tree priors.

Tail free prior are conjugate in the sense that if the prior is tail free, then so is the posterior. To avoid getting lost in a notational mess we first state an easy lemma.

Lemma 2.3.1. *Let $\xi_1, \xi_2, \ldots, \xi_k$ be independent random vectors (not necessarily of the same dimension) with joint distribution $\mu = \prod_1^k \mu_i$. Let J be a subset of $\{1, 2, \ldots, k\}$ and let μ^* be the probability with*

$$\frac{d\mu^*}{d\mu} = C \prod_{j \in J} \xi_j$$

Then $\xi_1, \xi_2, \ldots, \xi_k$ are independent under μ^*.

Proof. Clearly $C = \prod_{j \in J} [\int \xi_j d\mu_j]^{-1}$. Further,

$$\text{Prob}(\xi_i \in B_i : 1 \leq i \leq k) = \int_{(\xi_i \in B_i : 1 \leq i \leq k)} C[\prod_{j \in J} \xi_j] d\mu$$

$$= \prod_{i \notin J} \mu_i(B_i) \prod_{j \in J} \frac{\int_{B_j} \xi_j d\mu_j}{\int \xi_j d\mu_j}$$

□

Theorem 2.3.6. *Suppose Π is a tail free prior on $M(\mathbb{R})$ with respect to the sequence of partitions $\{\mathcal{T}_k\}_{k\geq 1}$. Given P, let $X_1, X_2, \ldots, X_n$ be,i.i.d. P; then the posterior is also tail free with respect to $\{\mathcal{T}_k\}_{k\geq 1}$.*

Proof. We will prove the result for $n = 1$; the general case follows by iteration. Consider the posterior distribution given $\mathcal{T}_k$. Because $\{B_{\underline{\epsilon}} : \underline{\epsilon} \in E_k\}$ are the atoms of $\mathcal{T}_k$, it is enough to find the posterior distribution given $X \in B_{\underline{\epsilon}'}$ for each $\underline{\epsilon}' \in E_k$.

Let $\underline{\epsilon}' = \epsilon_1\epsilon_2 \ldots \epsilon_k$. Then the likelihood of $P(B_{\underline{\epsilon}'})$ is

$$\prod_1^k P(B_{\epsilon_1,\epsilon_2,\ldots,\epsilon_j}|B_{\epsilon_1,\epsilon_2,\ldots,\epsilon_{j-1}})$$

so that the posterior density of $\{P(B_{\underline{\epsilon}1}\ |B_{\underline{\epsilon}})\}$ with respect to Π is

$$C \prod_{i=1,\epsilon_i=0}^{n} P(B_{\epsilon_1\epsilon_2\ldots\epsilon_i}|B_{\epsilon_1\epsilon_2\ldots\epsilon_{i-1}}) \prod_{i=1,\epsilon_i=1}^{n} (1 - P(B_{\epsilon_1\epsilon_2\ldots\epsilon_i}|B_{\epsilon_1\epsilon_2\ldots\epsilon_{i-1}})$$

From Lemma 2.3.1

$$\{P(B_{\underline{\epsilon}1}\ |B_{\underline{\epsilon}}) : \underline{\epsilon} \in E_1\}, \{P(B_{\underline{\epsilon}1}\ |B_{\underline{\epsilon}}) : \underline{\epsilon} \in E_2\}, \ldots, \{P(B_{\underline{\epsilon}1}\ |B_{\underline{\epsilon}}) : \underline{\epsilon} \in E_{k-1}\}$$

are independent under the posterior.

In particular if $m < k$, independence holds for

$$\{P(B_{\underline{\epsilon}1}\ |B_{\underline{\epsilon}}) : \underline{\epsilon} \in E_1\}, \{P(B_{\underline{\epsilon}1}\ |B_{\underline{\epsilon}}) : \underline{\epsilon} \in E_2\}, \ldots, \{P(B_{\underline{\epsilon}1}\ |B_{\underline{\epsilon}}) : \underline{\epsilon} \in E_{m-1}\}.$$

Letting $k \to \infty$, an application of the martingale convergence theorem gives the conclusion for the posterior given X_1. □

In this section we have discussed two general methods of constructing priors on $M(\mathbb{R})$. There are several other techniques for obtaining nonparametric priors. There are priors that arise from stochastic processes. If f is the sample path of a stochastic process then $\hat{f} = k^{-1}(f)e^f$ yields a random density when $k(f) = Ee^f$ is finite. We will study a method of this kind in the context of density estimation. Or one can look at expansions of a density using some orthogonal basis and put a prior on the coefficients. A class of priors called *neutral to the right priors*, somewhat like tail free priors, will be studied in Chapter 10 on survival analysis.

2.4 Tail Free Priors and 0-1 Laws

Suppose Π is a prior on $M(\mathbb{R})$ and $\{B_{\underline{\epsilon}} : \underline{\epsilon} \in E^*\}$ is a set of partitions as described in the last section. To repeat, for each n, $\underline{\mathcal{T}}_n = \{B_{\underline{\epsilon}} : \underline{\epsilon} \in E_n\}$ is a partition of $\mathbb{R}$ and $B_{\underline{\epsilon}0}$, $B_{\underline{\epsilon}1}$ is a partition of $B_{\underline{\epsilon}}$. Further $\mathcal{B} = \sigma\{B_{\underline{\epsilon}} : \underline{\epsilon} \in E^*\}$. Unlike the last section it is not required that $B_{\underline{\epsilon}}$ be intervals. The choice of intervals as sets in the partition played a crucial role in the construction of a probability measure on $M(\mathbb{R})$. Given a probability measure on $M(\mathbb{R})$, the following notions are meaningful, even if the $B_{\underline{\epsilon}}$ are not intervals.

For notational convenience, as before, denote by $Y_{\underline{\epsilon}} = P(B_{\underline{\epsilon}0}|B_{\underline{\epsilon}})$. Formally, $Y_{\underline{\epsilon}}$ is a random variable defined on $M(\mathbb{R})$ with $Y_{\underline{\epsilon}}(P) = P(B_{\underline{\epsilon}0}|B_{\underline{\epsilon}})$. Recall that Π is said to be tail free with respect to the partition $\underline{\mathcal{T}} = \{\mathcal{T}_n\}_{n\geq 1}$ if

$$Y \perp \{Y_0, Y_1\} \perp \{Y_{00}, Y_{01}, Y_{10}, Y_{11}\} \perp \ldots$$

Theorem 2.4.1. *Let λ be any finite measure on $\mathbb{R}$, with $\lambda(B_{\underline{\epsilon}}) > 0$ for all $\underline{\epsilon}$. If $0 < Y_{\underline{\epsilon}} < 1$ for all $\underline{\epsilon}$ then*

$$\Pi\{P : P << \lambda\} = 0 \text{ or } 1$$

Proof. Assume without loss of generality that λ is a probability measure.

Let $Z_0 = Y, Z_1 = \{Y_0, Y_1\}, Z_2 = \{Y_{00}, Y_{01}, Y_{10}, Y_{11}\}, \ldots$. By assumption, $Z_1, Z_2, \ldots$ are independent random vectors. The basic idea of the proof is to show that $L(\lambda) = \{P : P << \lambda\}$ is a tail set with respect to the Z_is. The Kolmogorov $0 - 1$ law then yields the conclusion. In the next two lemmas it is shown that for each n, $L(\lambda)$ depends only on $Z_n, Z_{n+1}, \ldots$ and is hence a tail set. □

Lemma 2.4.1. *When $P(B_{\underline{\epsilon}}) > 0$, define $P(\cdot|B_{\underline{\epsilon}})$ to be the probability $P(A|B_{\underline{\epsilon}}) = P(A \cap B_{\underline{\epsilon}})/P(B_{\underline{\epsilon}})$. Define $\lambda(\cdot|B_{\underline{\epsilon}})$ similarly. Fix n; then*

$$L(\lambda) = \{P : P(\cdot|B_{\underline{\epsilon}}) << \lambda(\cdot|B_{\underline{\epsilon}}) \text{ for all } \underline{\epsilon} \in E_n \text{ such that } P(B_{\underline{\epsilon}}) > 0\}$$

Proof. Because

$$P(A) = \sum_{\underline{\epsilon}\in E_n} P(A|B_{\underline{\epsilon}})P(B_{\underline{\epsilon}}) \text{ and } \lambda(A) = \sum_{\underline{\epsilon}\in E_n} \lambda(A|B_{\underline{\epsilon}})\lambda(B_{\underline{\epsilon}})$$

the result follows immediately.

□

Lemma 2.4.2. *Let $\underline{\mathbf{Y}} = \{Y_{\underline{\epsilon}}(P) : \underline{\epsilon} \in E^*, P \in M(\mathbb{R})\}$. The elements $\underline{\mathbf{y}}$ of $\underline{\mathbf{Y}}$ are thus a collection of conditional probabilities arising from a probability. Conversely any element $\underline{\mathbf{y}}$ of $\underline{\mathbf{Y}}$ gives rise to a probability which we denote by $P_{\underline{\mathbf{y}}}$. Then for each $\underline{\epsilon} \in E_n$, for all $A \in \mathcal{B}$, and for every $\underline{\mathbf{y}}$ in $\underline{\mathbf{Y}}$*

$$P_{\underline{\mathbf{y}}}(A|B_{\underline{\epsilon}}) \textit{ depends only on } Z_n, Z_{n+1}, \ldots$$

Proof. Let

$$\mathcal{B}_0 = \left\{A : \text{ for all } \underline{\mathbf{y}},\ P_{\underline{\mathbf{y}}}(A|B_{\underline{\epsilon}}) \text{ depends only on } Z_n, Z_{n+1}, \ldots\right\}$$

Because $0 < Y_{\underline{\epsilon}} < 1$ for all $\underline{\epsilon} \in E^*$, $P_{\underline{y}}\ (B_{\underline{\epsilon}}) > 0$ for all $\underline{\epsilon} \in E^*$. Hence $\mathcal{B}_0$ contains the algebra of finite disjoint unions of elements in $\{B_{\underline{\epsilon}'} : \underline{\epsilon}' \in \cup_{m>n} E_m\}$ and is a monotone class. Hence $\mathcal{B}_0 = \mathcal{B}$.

□

Remark 2.4.1. Let Π be tail free with respect to $\underline{\mathcal{T}} = \{\mathcal{T}_n\}_{n\geq 1}$ such that $0 < Y_{\underline{\epsilon}} <$ 1; for all $\underline{\epsilon} \in E^*$. Argue that P is discrete iff $P(.|B_{\underline{\epsilon}})$ is discrete for all $\underline{\epsilon} \in E_n$. Now use the Kolmogorov 0-1 law to conclude that $\Pi\{P : P \text{ is discrete }\} = 0$ or 1.

The next theorem, due to Kraft, is useful in constructing priors concentrated on sets like $L(\lambda)$.

Let $\Pi, \{B_{\underline{\epsilon}} : \underline{\epsilon} \in E^*\}, \{Y_{\underline{\epsilon}} : \underline{\epsilon} \in E^*\}$ be as in the Theorem 2.4.1, and, as before given any realization $\underline{y}\ = \{y_{\underline{\epsilon}} : \underline{\epsilon} \in E^*\}$, let $P_{\underline{y}}$ denote the corresponding probability measure on $\mathbb{R}$.

Theorem 2.4.2. *Let λ be a probability measure on $\mathbb{R}$ such that $\lambda(B_{\underline{\epsilon}}) > 0$ for all $\underline{\epsilon} \in E^*$. Suppose*

$$f_{\underline{y}}^n\ (x) = \sum_{\underline{\epsilon}\in E_n} \frac{P_{\underline{y}}\ (B_{\underline{\epsilon}})}{\lambda(B_{\underline{\epsilon}})} I_{B_{\underline{\epsilon}}}(x) = \sum_{\underline{\epsilon}\in E_n} \frac{\prod_{i=1,\epsilon_i=0}^{k} y_{\epsilon_1\epsilon_2\ldots\epsilon_{i-1}} \prod_{i=1,\epsilon_i=1}^{k}(1 - y_{\epsilon_1\epsilon_2\ldots\epsilon_{i-1}})}{\lambda(B_{\underline{\epsilon}})}$$

If $\sup_n E_\Pi \left[f_{\underline{y}}^n\ (x)\right]^2 \leq K$ *for all x then* $\Pi\{P : P << \lambda\} = 1$

Proof. For each $\underline{y}\ \in \underline{\mathbf{Y}}$, by the martingale convergence theorem $f_{\underline{y}}^n$ converges almost surely λ to a function $f_{\underline{y}}$. Consider the measure $\Pi \times \lambda$, which is the joint distribution of $\underline{y}$ and x, on $\prod_{\underline{\epsilon}\in E^*} Y_{\underline{\epsilon}} \times \mathbb{R}$.

Because for each $\underline{y}$, $f_{\underline{y}}^n \to f_{\underline{y}}$ a.s λ, we have $f_{\underline{y}}^n \to f_{\underline{y}}$ a.s $\Pi \times \lambda$. Further, under our assumption $\left\{ f_{\underline{y}}^n (x) : n \geq 1 \right\}$ is uniformly integrable with respect to $\Pi \times \lambda$ and hence $E_{\Pi\times\lambda} \left| f_{\underline{y}}^n (x) - f_{\underline{y}} (x) \right| \to 0$. Now for each $\underline{y}$, by Fatou's lemma, $E_\lambda f_{\underline{y}} \leq 1$. On the other hand, $E_{\Pi\times\lambda} f_{\underline{y}}^n (x) = 1$ for all n, and by the L_1-convergence mentioned earlier, $E_{\Pi\times\lambda} f_{\underline{y}} (x) = 1$. Thus $E_\lambda f_{\underline{y}} = 1$ a.e. π and this shows $\pi\{L(\lambda)\} = 1$. □

The next theorem is an application of the last theorem. It shows how, given a probability measure λ, by suitably choosing both the partitions and the parameter of the $Y_{\underline{\epsilon}}$s , we can obtain a prior that concentrates on $L(\lambda)$.

Theorem 2.4.3. *Let λ be a continuous probability distribution on $\mathbb{R}$. Denote by F the distribution function of λ and construct a partition as follows:*

$$B_0 = F^{-1}(0, 1/2] \qquad B_1 = F^{-1}(1/2, 1]$$

$$B_{00} = F^{-1}(0, 1/4], B_{01} = F^{-1}(1/4, 1/2] \quad B_{10} = F^{-1}(1/2, 3/4], B_{11} = F^{-1}(3/4, 1]$$

and in general

$$B_{\epsilon_1,\epsilon_2,\ldots,\epsilon_n} = F^{-1}\left(\sum_1^n \frac{\epsilon_i}{2^n}, \sum_1^n \frac{\epsilon_i}{2^n} + \frac{1}{2^n}\right]$$

Suppose $E(Y_{\underline{\epsilon}}) = 1/2$ for all $\underline{\epsilon} \in E^*$ and $\sup_{\underline{\epsilon}\in E_n} V(Y_{\underline{\epsilon}}) \leq b_n$, with $\sum b_n < \infty$. Then the resulting prior satisfies $\Pi(L(\lambda)) = 1$.

Proof. $\lambda(B_{\underline{\epsilon}}) > 0$, because $\lambda(B_{\underline{\epsilon}0}|B_{\underline{\epsilon}}) = 1/2$, for all $B_{\underline{\epsilon}}$. Fix x. If $x \in B_{\epsilon_1\epsilon_2,\ldots\epsilon_n}$, then

$$f_Y^n(x) = \prod_{i=0}^n \frac{Y_{\epsilon_1\epsilon_2,\ldots\epsilon_{i-1}}^{1-\epsilon_i}(1 - Y_{\epsilon_1\epsilon_2,\ldots\epsilon_{i-1}})^{\epsilon_i}}{1/2}$$

and

$$\begin{aligned} E[f_Y^n(x)]^2 &= \prod_o^n 4E\left[[Y_{\epsilon_1\epsilon_2,\ldots\epsilon_{i-1}}^2]^{1-\epsilon_i}[(1 - Y_{\epsilon_1\epsilon_2,\ldots\epsilon_{i-1}})^2]^{\epsilon_i}\right] \\ &\leq \prod_0^n 4a_i \end{aligned}$$

where $a_i = \max\left(EY_{\epsilon_1\epsilon_2,\ldots\epsilon_{i-1}}^2, E(1 - Y_{\epsilon_1\epsilon_2,\ldots\epsilon_{i-1}})^2\right)$. Now

$$EY^2_{\epsilon_1\epsilon_2,\dots\epsilon_{i-1}} = V(Y_{\epsilon_1\epsilon_2,\dots\epsilon_{i-1}}) + (1/2)^2 \le b_i + 1/4$$

and

$$E\left(1 - Y_{\epsilon_1\epsilon_2,\dots,\epsilon_{i-1}})^2\right) \le b_i + 1/4$$

Thus $\prod_1^n 4a_i \le \prod_1^n(1+4b_i)$ converges, because $\sum b_n < \infty$. □

2.5 Space of Probability Measures on $M(\mathbb{R})$

We next turn to a discussion of probability measures on $M(\mathbb{R})$. To get a feeling for what goes on we begin by asking when are two probability measures Π_1 and Π_2 equal?

Clearly $\Pi_1 = \Pi_2$ if for any finite collection $B_1, B_2, \dots, B_k$ of Borel sets,

$$(P(B_1), P(B_2), \dots, P(B_k))$$

has the same distribution under both Π_1 and Π_2. This is an immediate consequence of the definition of $\mathcal{B}_M$.

Next suppose that $(C_1, C_2, \dots, C_k)$ are Borel sets. Consider all intersections of the form

$$C_1^{\epsilon_1} \cap C_2^{\epsilon_2} \cap \dots \cap C_k^{\epsilon_k}$$

where $\epsilon_i = 0, 1$, $C_i^1 = C_i$ and $C_i^0 = C_i^c$. These intersections would give rise to a partition of $\mathcal{X}$, and since every C_i can be written as a union of elements of this partition, the distribution of $(P(C_1), P(C_2), \dots, P(C_k))$ is determined by the joint distribution of the probability of elements of this partition. In other words, if the distribution of $(P(B_1), P(B_2), \dots, P(B_k))$ under Π_1 and Π_2 are the same for every finite disjoint collection of Borel sets then $\Pi_1 = \Pi_2$. Following is another useful proposition.

Proposition 2.5.1. *Let $\mathcal{B}_0 = \{B_i : i \in I\}$ be a family of sets closed under finite intersection that generates the Borel σ-algebra $\mathcal{B}$ on $\mathcal{X}$. If for every $B_1, B_2, \dots, B_k$ in $\mathcal{B}_0$, $(P(B_1), P(B_2), \dots, P(B_k))$ has the same distribution under Π_1 and Π_2, then $\Pi_1 = \Pi_2$.*

Proof. Let $\mathcal{B}_M^0 = \{E \in \mathcal{B}_M : \Pi_1(E) = \Pi_2(E)\}$. Then $\mathcal{B}_M^0$ is a λ-system. For any J finite subset of I, by our assumption Π_1 and Π_2 coincide on the σ-algebra $\mathcal{B}_M^J$—the σ-algebra generated by $\{P(B_j) : j \in J\}$ and hence $\mathcal{B}_M^J \subset \mathcal{B}_M^0$. Further the union of $\mathcal{B}_M^J$ over all finite subsets of I forms a π-system. Because these also generate $\mathcal{B}_M$, $\mathcal{B}_M^0 = \mathcal{B}_M$. □

Remark 2.5.1. A convenient choice of $\mathcal{B}_0$ is the collection of all open balls, all closed balls, etc. When $\mathcal{X} = \mathbb{R}$ a very useful choice is the collection $\{(-\infty, a] : a \in Q\}$, where Q is a dense set in $\mathbb{R}$.

As noted earlier $M(\mathbb{R})$ when equipped with weak convergence becomes a complete separable metric space with $\mathcal{B}_M$ as the Borel σ-algebra. Thus a natural topology on $M(\mathbb{R})$ is the weak topology arising from this metric space structure of $M(\mathbb{R})$. Formally, we have the following definitions.

Definition 2.5.1. A sequence of probability measure $\{\Pi\}_n$ on $M(\mathbb{R})$ is said to *converge weakly* to a probability measure Π if

$$\int \phi(P)\, d\Pi_n \to \int \phi(P)\, d\Pi$$

for all bounded continuous functions ϕ on $M(\mathbb{R})$.

Note that continuity of ϕ is with respect to the weak topology on $M(\mathbb{R})$. If f is a bounded continuous function on $\mathbb{R}$ then $\phi(P) = \int f dP$ is bounded and continuous on $M(\mathbb{R})$. However in general there is no clear description of all the bounded continuous functions on $M(\mathbb{R})$. If $\mathcal{X}$ is compact metric, then the following description is available.

If $\mathcal{X}$ is compact metric then, by Prohorov's theorem, so is $M(\mathcal{X})$ under weak convergence. It follows from the Stone-Weirstrass theorem that the set of all functions of the form

$$\sum \prod_{j=1}^{k_i} \phi_{f_{i,j}}^{r_i}$$

where $\phi_{f_{i,j}}^{r_i}(P) = \int f_{i,j}(x) dP(x)$ with $f_{i,j}(x)$ continuous on $\mathcal{X}$, is dense in the space of all continuous functions on $M(\mathcal{X})$.

The following result is an extension of a similar result in Sethuraman and Tiwari [149].

Theorem 2.5.1. *A family of probability measures $\{\Pi_t : t \in T\}$ on $M(\mathbb{R})$ is tight with respect to weak convergence on $M(\mathbb{R})$ iff the family of expectations $\{E_{\Pi_t} : t \in T\}$, where $E_{\Pi_t}(B) = \int P(B)\, d\Pi_t(P)$, is tight in $\mathbb{R}$.*

Proof. Let $\mu_t = E_{\Pi_t}$. Fix $\delta > 0$. By the tightness of $\{\mu_t : t \in T\}$, for every positive integer d there exists a sequence of compact sets K_d in $\mathbb{R}$, such that $\sup_t \mu_t(K_d^c) \leq 6\delta/(d^3\pi^2)$.

For $d = 1, 2, \ldots$, let, $M_d = \{P \in M(\mathbb{R}) : P(K_d^c) \le 1/d\}$, and let $M = \cap_d M_d$. Then, by the pormanteau and Prohorov theorems, M is a compact subset of $M(\mathbb{R})$, in the weak topology. Further, by Markov's inequality,

$$\begin{aligned} \Pi_n(M_d^c) &\le dE_{\Pi_t}(P(K_d^c)) \\ &= d\mu_t(K_d^c) \\ &\le \frac{6\delta}{d^2\pi^2} \end{aligned}$$

Hence, for any $t \in T$, $\Pi_t(M) \le \sum_d 6\delta/(d^3\pi^2) = \delta$. This proves that $\{\mu_t\}_{t\in T}$ is tight. The converse is easy. □

Theorem 2.5.2. *Suppose $\Pi, \Pi_n, n \ge 1$ are probability measures on M. If any of the following holds then Π_n converges weakly to Π.*

(i) For any $(B_1, B_2, \ldots, B_k)$ of Borel sets

$$\mathcal{L}_{\Pi_n}(P(B_1), P(B_2), \ldots, P(B_k)) \to \mathcal{L}_{\Pi}(P(B_1), P(B_2), \ldots, P(B_k))$$

(ii) For any disjoint collection $(B_1, B_2, \ldots, B_k)$ of Borel sets

$$\mathcal{L}_{\Pi_n}(P(B_1), P(B_2), \ldots, P(B_k)) \to \mathcal{L}_{\Pi}(P(B_1), P(B_2), \ldots, P(B_k))$$

(iii) For any $(B_1, B_2, \ldots, B_k)$ where for $= i = 1, 2, \ldots, k$, $B_i = (a_i, b_i]$,

$$\mathcal{L}_{\Pi_n}(P(B_1), P(B_2), \ldots, P(B_k)) \to \mathcal{L}_{\Pi}(P(B_1), P(B_2), \ldots, P(B_k))$$

(iv) For any $(B_1, B_2, \ldots, B_k)$ where for $= i = 1, 2, \ldots, k$, $B_i = (a_i, b_i]$,a_i, b_i rationals,

$$\mathcal{L}_{\Pi_n}(P(B_1), P(B_2), \ldots, P(B_k)) \to \mathcal{L}_{\Pi}(P(B_1), P(B_2), \ldots, P(B_k))$$

(v) For any $(B_1, B_2, \ldots, B_k)$ where for $= i = 1, 2, \ldots, k$, $B_i = (-\infty, t_i]$,

$$\mathcal{L}_{\Pi_n}(P(B_1), P(B_2), \ldots, P(B_k)) \to \mathcal{L}_{\Pi}(P(B_1), P(B_2), \ldots, P(B_k))$$

(vi) For any $(B_1, B_2, \ldots, B_k)$ where for $= i = 1, 2, \ldots, k$,$B_i = (-\infty, t_i]$, t_i rationals

$$\mathcal{L}_{\Pi_n}(P(B_1), P(B_2), \ldots, P(B_k)) \to \mathcal{L}_{\Pi}(P(B_1), P(B_2), \ldots, P(B_k))$$

Proof. Because (vi) is the weakest, we will show that (vi) implies $\Pi_n \stackrel{weakly}{\rightarrow} \Pi$. Note that for all rationals t, $E_{\Pi_n}(P(-\infty,t)) \to E_\Pi(P(-\infty,t))$ and hence E_{Π_n} converges weakly to E_Π. By the Theorem 2.5.1 this shows that $\{\Pi_n\}$ is tight. If Π^* is the limit of any subsequence of $\{\Pi_n\}$, then it follows, using Proposition 2.5.1, that $\Pi^* = \Pi$. □

Remark 2.5.2. Note that $\Pi_n \stackrel{weakly}{\rightarrow} \Pi$ does not imply any of the preceding. The modifications are easy, however. For example (i) would be changed to "For any $(B_1, B_2, \ldots, B_k)$ of Borel sets such that $(P(B_1), P(B_2), \ldots, P(B_k))$ is continuous a.e Π."

We have considered other topologies on $M(\mathbb{R})$ namely, total variation, setwise convergence and the supremum metric. It is tempting to consider the weak topologies on probabilities on $M(\mathbb{R})$ induced by these topologies. But as we have observed, these topologies possess properties that make the notion of weak convergence awkward to define and work with. Besides, the σ-algebras generated by these topologies, via either open sets or open balls do not coincide with $\mathcal{B}_M$ [57]. Our interests do not demand such a general theory. Our chief interest is when the limit measure Π is degenerate at P_0, and in this case we can formalize convergence via weak neighborhoods of P_0.

When $\Pi = \delta_{P_0}$, $\Pi_n \stackrel{weakly}{\rightarrow} \delta_{P_0}$ iff $\Pi_n(U) \to \Pi(U)$ for every open neighborhood U. Because weak neighborhoods of P_0 are of the form $U = \{P : \left|\int f_i\, dP - \int f_i\, dP_0\right|\}$, weak convergence to a degenerate measure δ_{P_0} can be described in terms of continuous functions of $\mathbb{R}$ rather than those on $M(\mathbb{R})$ and can be verified more easily. The next proposition is often useful when we work with weak neighborhoods of a probability P_0 on $\mathbb{R}$.

Proposition 2.5.2. *Let Q be a countable dense subset of $\mathbb{R}$. Given any weak neighborhood U of P_0 there exist $a_1 < a_2 \ldots < a_n$ in Q and $\delta > 0$ such that*

$$\{P : |P[a_i, a_{i+1}) - P_0[a_i, a_{i+1})| < \delta \text{ for } 1 \le i \le n\} \subset U$$

Proof. Suppose $U = \{P : |\int f dP - \int f dP_0| < \epsilon\}$, where f is continuous with compact support. Because Q is dense in $\mathbb{R}$ given δ there exist $a_1 < a_2 \ldots < a_n$ in Q such that $f(x) = 0$ for $x \le a_1$, $x \ge a_n$, and $|f(x) - f(y)| < \delta$ for x $\in [a_i, a_{i+1}], 1 \le i \le n-1$. Then the function f^* defined by

$$f^*(x) = f(a_i) \text{ for } x \in [a_i, a_{i+1}), i = 1, 2, \ldots n-1$$

satisfies $\sup_x |f^*(x) - f(x)| < \delta$.

For any P, $\int f^* dP = \sum f(a_i)P[a_i, a_{i+1})$,

$$|\int f^* dP - \int f^* dP_0| < ck\delta \text{ where } c = \sup_x |f(x)|$$

In addition, if P is in U then we have

$$|\int f dP - \int f dP_0| < 2\delta + ck\delta$$

Thus with $B_i = [a_i, a_{i+1}]$ for small enough δ,$\{P : |P(B_i) - P_0(B_i)| < \delta\}$ is contained in U. The preceding argument is easily extended if U is of the form

$$\{P : |\int f_i dP - \int f_i dP_0| \leq \epsilon_i, 1 \leq i \leq k, f_i \text{ continuous with compact support}\}$$

□

Following is another useful proposition.

Proposition 2.5.3. *Let* $U = \{F : \sup_{-\infty<x<\infty} |F_0(s) - F(x)| < \epsilon\}$ *be a supremum neighborhood of a continuous distribution function* F_0. *Then* U *contains a weak neighborhood of* F_0.

Proof. Choose $-\infty = x_0 < x_1 < x_2 < \ldots < x_k = \infty$ such that $F(x_{i+1}) - F(x_i) < \epsilon/4$ for $i = 1, \ldots, k-1$. Consider

$$W = \{F : |F(x_i) - F_0(x_i)| < \epsilon/4\}, i = 1, 2, \ldots, k$$

If $x \in (x_{i-1}, x_i)$,

$$\begin{aligned} |F(x) - F_0(x)| \leq & |F(x_{i-1}) - F_0(x_i)| \vee |F(x_i) - F_0(x_{i-1})| \\ \leq & |F(x_{i-1}) - F_0(x_{i-1})| + |F_0(x_{i-1}) - F_0(x_i)| \\ & + |F(x_i) - F_0(x_i)| + |F_0(x_{i-1}) - F_0(x_i)| \end{aligned}$$

which is less than ϵ if $F \in W$. □

2.6 De Finetti's Theorem

Much of classical statistics has centered around the conceptually simplest setting of independent and identically distributed observations. In this case, $X_1, X_2, \ldots$ are a sequence of i.i.d. random variables with an unknown common distribution P. In the parametric case, P would be constrained to lie in a parametric family, and in the general nonparametric situation P could be any element of $M(\mathbb{R})$. The Bayesian framework in this case consists of a prior Π on the parameter set $M(\mathbb{R})$; given P the $X_1, X_2, \ldots$ is modeled as i.i.d. P. In a remarkable theorem, De Finetti showed that a minimal judgment of exchangeability of the observation sequence leads to the Bayesian formulation discussed earlier.

In this section we briefly discuss De Finetti's theorem. A detailed exposition of the theorem and related topics can be found in Schervish [144] in the section on De Finetti's theorem and the section on Extreme models.

As before, let $X_1, X_2, \ldots$ be a sequence of $\mathcal{X}$ -valued random variables defined on $\Omega = \mathbf{R}^\infty$.

Definition 2.6.1. Let μ be a probability measure on $\mathbb{R}^\infty$. The sequence $X_1, X_2, \ldots$ is said to be *exchangeable* if, for each n and for every permutation g of $\{1, \ldots, n\}$, the distribution of $X_1, X_2, \ldots, X_n$is the same as that of $X_{g(1)}, X_{g(2)}, \ldots, X_{g(n)}$.

Theorem 2.6.1 (De Finetti). *Let μ be a probability measure on $\mathbb{R}^\infty$. Then $X_1, X_2, \ldots$ is exchangeable iff there is a unique probability measure Π on $M(\mathbb{R})$ such that for all n and for any Borel sets $B_1, B_2, \ldots, B_n$,*

$$\mu\{X_1 \in B_1, X_2 \in B_2, \ldots, X_n \in B_n\} = \int_{M(\mathbb{R})} \prod_1^n P(B_i)\, d\Pi(P) \tag{2.4}$$

Proof. We begin by proving the theorem when all the X_is take values in a finite set $\mathcal{X} = \{1, 2, \ldots, k\}$. This proof follows Heath and Sudderth [95].

So let $\mathcal{X} = \{1, 2, \ldots, k\}$ and μ be a probability measure on $\mathcal{X}^\infty$ such that $X_1, X_2, \ldots$ is exchangeable. For each n, let $T_n(X_1, X_2, \ldots, X_n) = (r_1, r_2, \ldots, r_k)$, where $r_j = \sum_{i=1}^n I_{\{j\}}(X_i)$ is the number of occurrences of js in $X_1, X_2, \ldots, X_n$. Let μ_n^* denote the distribution of $T_n/n = (r_1/n, r_2/n, \ldots, r_k/n)$ under μ. μ_n^* is then a discrete probability measure on $M(\mathcal{X})$ supported by points of the form $(r_1/n, r_2/n, \ldots, r_k/n)$, where for $j = 1, 2, \ldots, k$, $r_j \geq 0$ is an integer and $\sum r_j = n$. Because $M(\mathcal{X})$ is compact, there is a subsequence $\{n_i\}$ that converges to a probability measure Π on $M(\mathcal{X})$. We will argue that Π satisfies (2.4).

Because $X_1, X_2, \ldots, X_n$ is exchangeable, it is easy to see that the conditional distribution of $X_1, X_2, \ldots, X_n$ given T_n is also exchangeable. In particular, the conditional probability given $T_n(X_1, X_2, \ldots, X_n) = (r_1, r_2, \ldots, r_k)$ is just the uniform distribution on $T_n^{-1}(r_1, r_2, \ldots, r_k)$. In other words, the conditional distribution of $X_1, X_2, \ldots, X_n$ given $T_n = (r_1, r_2, \ldots, r_k)$ is the same as the distribution of n successive draws from an urn containing n balls with r_i of color i, for $i = 1, 2, \ldots, k$.

Fix m and $n > m$. Then, given $T_n(X_1, X_2, \ldots, X_n) = (r_1, r_2, \ldots, r_k)$, the conditional probability that

$$\left(X_1 = 1, \ldots, X_{s1} = 1, X_{s_1+1} = 2, \ldots, X_{s_1+s_2} = 2, \ldots, X_{m-s_{k-1}+1} = k, \ldots, X_m = k\right)$$

is $(r_1)_{s_1}(r_2)_{s_2} \ldots (r_k)_{s_k}/(n)_m$, where for any real a and integer b, $(a)_b = \prod_0^{b-1}(a - i)$.

Because

$$\sum_{(r_1,r_2,\ldots,r_k),\sum r_j = n} \frac{(r_1)_{s_1}(r_2)_{s_2} \ldots (r_k)_{s_k}}{(n)_m} \mu\left(\frac{T_n}{n} = (\frac{r_1}{n}, \frac{r_2}{n}, \ldots, \frac{r_k}{n})\right)$$

$$= \int_{M(\mathcal{X})} \frac{(p_1 n)_{s_1}(p_2 n)_{s_2} \ldots (p_k n)_{s_k}}{(n)_m} \, d\mu_n^*(p_1, p_2, \ldots, p_k)$$

As $n \to \infty$ the sequence of functions

$$\frac{(p_1 n)_{s_1}(p_2 n)_{s_2} \ldots (p_k n)_{s_k}}{(n)_m}$$

converges uniformly on $M(\mathcal{X})$to $\prod p_j^{s_j}$ so that by taking the limit through the subsequence $\{n_i\}$, the probability of

$$(X_i = 1, 1 \le i \le s_1; X_i = 2, s_1 + 1 \le i \le s_1 + s_2, \ldots, X_i = k, m - s_{k-1} + 1 \le i \le m)$$

is

$$\int_{M(\mathcal{X})} \prod p_j^{s_j} \, d\Pi(p_1, p_2, \ldots, p_k) \tag{2.5}$$

Uniqueness is immediate because if Π_1, Π_2 are two probability measures on $M(\mathcal{X})$ satisfying (2.5) then it follows immediately that they have the same moments.

To move on to the general case $\mathcal{X} = \mathbb{R}$, let $B_1, B_2, \ldots, B_k$ be any collection of disjoint Borel sets in $\mathbb{R}$. Set $B_0 = \left(\cup_1^k B_i\right)^c$.

Define $Y_1, Y_2, \dots$ by $Y_i = j$ if $X_i \in B_j$. Because $X_1, X_2, \dots$ is exchangeable, so are $Y_1, Y_2, \dots$. Since each Y_i takes only finitely many values, we use what we have just proved and writing $X_i \in B_j$ for $Y_i = j$, there is probability measure $\Pi_{B_1, B_2, \dots, B_k}$ on $\{p_1, p_2, \dots, p_k : p_j \geq 0, \sum p_j \leq 1\}$ such that for any m,

$$\mu\left(X_1 \in B_{i1}, X_2 \in B_{i2}, \dots, X_m \in B_{im}\right) = \int \prod_1^m P(B_{ij})\, d\Pi_{B_1, B_2, \dots, B_k}(P) \tag{2.6}$$

where $i1, i2, \dots, im$ are all elements of $\{0, 1, 2, \dots, k\}$ and $P(B_0) = 1 - \sum_1^k P(B_i)$.

We will argue that these $\Pi_{B_1, B_2, \dots, B_k}$s satisfy the conditions of Theorem 2.3.4.

If $A_1, A_2, \dots, A_l$ is a collection of disjoint Borel sets such that B_i are union of sets from $A_1, A_2, \dots, A_l$ then the distribution of $P(B_1), P(B_2), \dots, P(B_k)$ obtained from $P(A_1), P(A_2), \dots, P(A_l)$ and $\Pi_{B_1, B_2, \dots, B_k}$ both would satisfy (2.5). Uniqueness then shows that both distributions are same.

If $(B_{1n}, B_{2n}, \dots, B_{kn}) \to (B_1, B_2, \dots, B_k)$ then (2.6) again shows that moments of $\Pi_{B_{1n}, B_{2n}, \dots, B_{kn}}$ converges to the corresponding moment of $\Pi_{B_1, B_2, \dots, B_k}$.

It is easy to verify the other conditions of Theorem 2.3.4. Hence there exists a Π with $\Pi_{B_1, B_2, \dots, B_k}$s as marginals. It is easy to verify that Π satisfies (2.4). □

De Finetti's theorem can be viewed from a somewhat general perspective. Let $\mathcal{G}_n$ be the group of permutations on $\{1, 2, \dots, n\}$ and let $\mathcal{G} = \cup \mathcal{G}_n$. Every $g \in \mathcal{G}$ induces in a natural way a transformation on $\Omega = \mathcal{X}^\infty$ through the map, if, say g in $\mathcal{G}_n$, then $(x_1, \dots, x_n, \dots) \mapsto (x_{g(1)}, \dots, x_{g(n)}, \dots)$. It is easy to see that the set of exchangeable probability measures is the same as the set of probability measures on Ω that are invariant under $\mathcal{G}$. This set is a convex set, and De Finetti's theorem asserts that the set of extreme points of this convex set is $\{P^\infty : P \in M(\mathcal{X})\}$ and that every invariant measure is representable as an average over the set of extreme points. This view of exchangeable measures suggests that by suitably enlarging $\mathcal{G}$ it would be possible to obtain priors that are supported by interesting subsets of $M(\mathcal{X})$. Following is a simple, trivial example.

Example 2.6.1. Let $H = \{h, e\}$, where $h(x) = -x$ and $e(x) = x$. Set $\mathcal{H} = \cup H^n$. If $(h_1, h_2, \dots, h_n)) \in H^n$, then the action on Ω is defined by $(x_1, x_2, \dots, x_n) \mapsto (h(x_1), h(x_2), \dots, h(x_n)$. Then an exchangeable probability measure μ is $\mathcal{H}$ invariant iff it is a mixture of symmetric i.i.d. probability measures. To see this by De Finetti's theorem

$$\mu(A) = \int P^\infty(A) d\Pi(P)$$

Because by $\mathcal{H}$ invariance $\mu(X_1 \in A, X_2 \in -A) = \mu(X_1 \in A, X_2 \in A)$, it is not hard to see that $E_\Pi(P(A) - P(-A))^2 = 0$. Letting A run through a countable algebra generating the σ-algebra on $\mathcal{X}$, we have the result.

More non trivial examples are in Freedman [68]

Sufficiency provides another frame through which De Finetti's theorem can be usefully viewed. The ideas leading to such a view and the proofs involve many measure-theoretic details. Most of the interesting examples involve invariance and sufficiency in some form. We do not discuss these aspects here but refer the reader to the excellent survey in Schervish [144], the paper by Diaconis and Freedman [[44]] and Fortini, Ladelli, and Regazzini [67].

To use DeFinetti's theorem to construct a specific prior on $M(\mathbb{R})$, we need to know what to expect from the prior in terms of the observables $X_1, X_2, \ldots, X_n$. Although this method of assigning a prior is attractive from a philosophical point of view, it is not easy to either describe explicitly an exchangeable sequence or identify a prior, given such a sequence. We will not pursue this aspect here.

3

Dirichlet and Polya tree process

3.1 Dirichlet and Polya tree process

In this chapter we develop and study a very useful family of prior distributions on $M(\mathbb{R})$ introduced by Ferguson [61]. Ferguson introduced the Dirichlet processes, uncovered many of their basic properties, and applied them to a variety of nonparametric estimation problems, thus providing for the first time a Bayesian interpretation for some of the commonly used nonparametric procedures. These priors are relatively easy to elicit. They can be chosen to have large support and thus capture the nonparametric aspect. In addition they have tractable posterior and nice consistency properties. These processes are not an answer to all Bayesian nonparametric or semiparametric problems but they are important as both a large class of interpretable priors and a point of departure for more complex prior distributions.

The Dirichlet process arises naturally as an infinite-dimensional analogue of the finite-dimensional Dirichlet prior, which in turn has its roots in the one-dimensional beta distribution . We will begin with a review of the finite-dimensional case.

3.1.1 Finite Dimensional Dirichlet Distribution

In this section we summarize some basic properties of the Dirichlet distribution, especially those that arise when the Dirichlet is viewed as a prior on $M(\mathcal{X})$ -the set of probability measures on $\mathcal{X}$. Details are available in many standard texts, for example Berger [13].

First consider the simple case when $\mathcal{X} = \{1, 2\}$. Then

$$M(\mathcal{X}) = \{\underline{p} = (p_1, p_2) : p_1 \geq 0, p_2 \geq 0, p_1 + p_2 = 1\}$$

Because $p_2 = 1 - p_1$ and $0 \leq p_1 \leq 1$, any probability measure on $[0, 1]$ defines a prior distribution on $M(\mathcal{X})$. In particular say that $\underline{p}$ has a beta(α_1, α_2) prior if $\alpha_1 > 0, \alpha_2 > 0$ and if the prior has the density

$$\Pi(p_1) = \frac{\Gamma(\alpha_1 + \alpha_2)}{\Gamma(\alpha_1)\Gamma(\alpha_2)} p_1^{\alpha_1 - 1}(1 - p_1)^{\alpha_2 - 1} \qquad 0 \leq p_1 \leq 1$$

It is easy to see that

$$E(p_1) = \frac{\alpha_1}{\alpha_1 + \alpha_2}$$

$$V(p_1) = \frac{\alpha_1(\alpha_1 + 1)}{(\alpha_1 + \alpha_2)(\alpha_1 + \alpha_2 + 1)} - \left(\frac{\alpha_1}{(\alpha_1 + \alpha_2)}\right)^2 = \frac{\alpha_1\alpha_2}{(\alpha_1 + \alpha_2)^2(\alpha_1 + \alpha_2 + 1)}$$

We adopt the convention of setting the beta prior to be degenerate at $p_1 = 0$ if $\alpha_1 = 0$ and degenerate at $p_2 = 0$ if $\alpha_2 = 0$. Note that the convention goes well with the expression for $E(p_1)$. In fact the following proposition provides more justification for this convention.

Proposition 3.1.1. *If $\alpha_{1n} \to 0$ and $\alpha_{2n} \to c$, $0 < c < \infty$, then beta$(\alpha_{1n}, \alpha_{2n})$ converges weakly to δ_0.*

Proof. If p_n is distributed as beta$(\alpha_{1n}, \alpha_{2n})$, then $Ep_n \to 0, V(p_n) \to 0$ and hence $p_n \to 0$ in probability. □

The following representation of the beta is useful and well known. Let Z_1, Z_2 be independent gamma random variables with parameters $\alpha_1, \alpha_2 > 0$, i.e., the density is given by

$$f(z_i) = \frac{1}{\Gamma(\alpha_i)} e^{-z_i} z_i^{\alpha_i - 1} \qquad z_i > 0$$

then $Z_1/(Z_1 + Z_2)$ is independent of $Z_1 + Z_2$ and is distributed as beta(α_1, α_2).

If we define a gamma distribution with $\alpha = 0$ to be the measure degenerate at 0, then the representation of beta random variables remains valid for all $\alpha_1 \geq 0, \alpha_2 \geq 0$ as long as one of them is strictly positive.

Suppose $X_1, X_2, \ldots, X_n$ are $\mathcal{X}$ -valued i.i.d. random variables distributed as $\underline{p}$, then beta priors are conjugate in the sense that if $\underline{p}$ has a beta(α_1, α_2) prior distribution then the posterior distribution is also a beta, with parameters $\alpha_1 + \sum \delta_{X_i}(1)$ and $\alpha_2 + \sum \delta_{X_i}(2)$, where δ_x stands for the degenerate measure $\delta_x(x) = 1$. Moreover, the marginal distribution of $X_1, X_2, \ldots, X_n$ is exchangeable with marginal probability $\lambda(X_1 = i) = \alpha_i/(\alpha_1 + \alpha_2)$.

Next we move on to the case where $\mathcal{X} = \{1, 2, \ldots, k, \}$. The set $M(\mathcal{X})$ of probability measures on $\mathcal{X}$, is now in 1-1 correspondence with the simplex

$$S_k = \left\{\underline{p} = (p_1, p_2, \ldots, p_{k-1}) : p_i \geq 0 \text{ for } i = 1, 2, \ldots, k-1, \sum p_i \leq 1\right\}$$

and as before we set $p_k = 1 - \sum_1^{k-1} p_i$. A prior is specified by specifying a probability distribution for $(p_1, p_2, \ldots, p_{k-1})$. This distribution determines the joint distribution of the 2^k vectors $\{P(A) : A \subset \mathcal{X}\}$ through $P(A) = \sum_{i \in A} p_i$. The k- dimensional Dirichlet distribution is a natural extension of the beta distribution.

Definition 3.1.1. Let $\underline{\alpha} = (\alpha_1, \alpha_2, \ldots, \alpha_k)$ with $\alpha_i > 0$ for $i = 1, 2, \ldots, k$. $\underline{p} = (p_1, p_2, \ldots, p_k)$ is said to have *Dirichlet distribution* with parameter $(\alpha_1, \alpha_2, \ldots, \alpha_k)$, if the density is

$$\Pi(p_1, p_2, \ldots, p_{k-1}) = \frac{\Gamma(\sum_1^k \alpha_i)}{\Gamma(\alpha_1)\Gamma(\alpha_2), \ldots, \Gamma(\alpha_k)} p_1^{\alpha_1 - 1} p_2^{\alpha_2 - 1} p_{k-1}^{\alpha_{k-1} - 1} (1 - \sum_1^{k-1} p_i)^{\alpha_k - 1} \tag{3.1}$$

for $(p_1, p_2, \ldots, p_{k-1})$ in S_k.

Convention If any $\alpha_i = 0$, we still a define a Dirichlet by setting the corresponding $p_i = 0$ and interpreting the density (3.1.1) as a density on a lower-dimensional set.

The Dirichlet distribution with the vector $(\alpha_1, \alpha_2, \ldots, \alpha_k)$ as parameter will be denoted by $D(\alpha_1, \alpha_2, \ldots, \alpha_k)$. So we have a Dirichlet distribution defined for all $(\alpha_1, \alpha_2, \ldots, \alpha_k)$, as long as $\sum \alpha_i > 0$. Following are some properties of the Dirichlet distribution.

Properties.

1. Like the beta distribution, Dirichlet distributions admit a useful representation in terms of gamma variables. If $Z_1, Z_2, \ldots, Z_k$ are independent gamma random variables with parameter $\alpha_i \geq 0$, then

(a)

$$\left(\frac{Z_1}{\sum_1^k Z_i}, \frac{Z_2}{\sum_1^k Z_i}, \ldots, \frac{Z_k}{\sum_1^k Z_i} \right) \tag{3.2}$$

is distributed as $D(\alpha_1, \alpha_2, \ldots, \alpha_k)$;

(b)

$$\left(\frac{Z_1}{\sum_1^k Z_i}, \frac{Z_2}{\sum_1^k Z_i}, \ldots, \frac{Z_k}{\sum_1^k Z_i} \right) \tag{3.3}$$

is independent of $\sum_1^k Z_i$ and

(c) If $\underset{\sim}{p} = (p_1, p_2, \ldots, p_k)$ is distributed as $D(\alpha_1, \alpha_2, \ldots, \alpha_k)$, then for any partition $A_1, A_2 \ldots, A_m$ of $\mathcal{X}$, the vector $(P(A_1), P(A_2), \ldots, P(A_m)) = \left(\sum_{i \in A_1} p_i, \sum_{i \in A_2} p_i, \ldots, \sum_{i \in A_m} p_i \right)$ is a $D(\alpha'_1, \alpha'_2, \ldots, \alpha'_k)$

where $\alpha'_i = \sum_{j \in A_i} \alpha_j$. In particular, the marginal distribution of p_i is beta with parameters $(\alpha_i, \sum_{i \neq j} \alpha_j)$.

This property suggests that it would be convenient to view the parameter $(\alpha_1, \alpha_2, \ldots, \alpha_k)$ as a measure $\alpha(A) = \sum_{i \in A} \alpha_i$. Thus every non-zero measure α on $\mathcal{X}$ defines a Dirichlet distribution and the last property takes the form

$$(P(A_1), P(A_2), \ldots, P(A_m)) \text{ is } D(\alpha(A_1), \alpha(A_2), \ldots, \alpha(A_m))$$

2. (Tail Free Property) Let $M_1, M_2, \ldots, M_k$ be a partition of $\mathcal{X}$. For $i = 1, 2, \ldots, k$ with $\alpha(M_i) > 0$, let $P(.|M_i)$ be the conditional probability given M_i defined by

$$P(j|M_i) = \frac{P(j)}{P(M_i)} : \text{ for } j \in M_i$$

If $\alpha(M_i) = 0$ then take $P(.|M_i)$ to be an arbitrary fixed probability for all P.

If P the probability on $\mathcal{X}$ is $D(\alpha)$ then

(i) $(P(M_1), P(M_2), \ldots, P(M_k)), P(.|M_1), P(.|M_2), \ldots, P(.|M_k)$ are independent;

(ii) if $\alpha(M_i) > 0$ then $P(.|M_i)$ is $D(\alpha_{M_i})$, where α_{M_i} is the restriction of α to M_i, and

(iii) $(P(M_1), P(M_2), \ldots, P(M_k))$ is Dirichlet with parameter $(\alpha(M_1), \alpha(M_2), \ldots, \alpha(M_k))$

To see this, let $\mathcal{X} = \{1, 2, \ldots, n\}$ and let $\{Y_i : 1 \le i \le n\}$ be independent gamma random variables with parameter $\alpha(x_i)$. The gamma representation of the Dirichlet immediately shows that

$$P(.|M_1), P(.|M_2), \ldots, P(.|M_k) \tag{3.4}$$

are independent. Further if $Z_j = \sum_{i \in M_j} Y_i$, then

$$Z_1, Z_2, \ldots, Z_k$$

are independent, and using (3.4) it is easy to see that $(Z_1, Z_2, \ldots, Z_k)$ and hence $\sum_j Z_j$ is independent of

$$P(.|M_1), P(.|M_2), \ldots, P(.|M_k)$$

Because $P(M_j) = Z_j / \sum_j Z_j$ the result follows.

3. (Neutral to the right property) Let $B_1 \supset B_2 \supset \ldots B_k$. Then we have the independence relations given by

$$P(B_1) \perp P(B_2|B_1) \perp \ldots \perp P(B_k|B_{k-1})$$

This follows from the tail free property by successively considering partitions
B_1, B_1^c;
$B_1^c, B_2, B_1 \cap B_2^c; \ldots$

4. Let α_1, α_2 be two measures on $\mathcal{X}$ and P_1, P_2 be two independent k-dimensional Dirichlet random vectors with parameters α_1, α_2. If Y independent of P_1, P_2 is distributed as beta$(\alpha_1(\mathcal{X}), \alpha_2(\mathcal{X}))$, then $YP_1 + (1 - Y)P_2$ is $D(\alpha_1 + \alpha_2)$.

To see this, let $Z_1, Z_2, \ldots, Z_k$ be independent random variables with $Z_i \sim$ gamma$(\alpha_1\{i\})$. Similarly for $i = 1, 2, \ldots k$ let $Z_{k+i} \sim$ Gamma$(\alpha_2\{i\})$ be independent gamma random variables. Then

$$\frac{\sum_1^k Z_i}{\sum_1^{2k} Z_i}\left(\frac{Z_1}{\sum_1^k Z_i}, \ldots, \frac{Z_k}{\sum_1^k Z_i}\right) + \frac{\sum_{k+1}^k Z_i}{\sum_1^{2k} Z_i}\left(\frac{Z_{k+1}}{\sum_1^k Z_i}, \ldots, \frac{Z_{2k}}{\sum_1^k Z_i}\right)$$

has the same distribution as $YP_1 + (1-Y)P_2$. But then the last expression is equal to

$$\left(\frac{Z_1 + Z_{k+1}}{\sum_1^k Z_i}, \ldots, \frac{Z_k + Z_{2k}}{\sum_1^k Z_i}\right)$$

which is distributed as $D(\alpha_1 + \alpha_2)$. Note that the assertion remains valid even if some of the $\alpha_1\{i\}, \alpha_2\{j\}$ are zero. An interesting consequence is: If P is $D(\alpha)$ and Y is independent of P and distributed as Beta$(c, \alpha(\mathcal{X}))$, then

$$Y\delta_{(1,0,\ldots,0)} + (1-Y)P \sim D(\alpha\{1\} + c, \alpha\{2\}, \ldots, \alpha\{k\})$$

This follows if we think of $\delta_{1,0,\ldots,0}$ as Dirichlet with parameter $(c, 0, \ldots, 0)$. A corresponding statement holds if $(1, 0, \ldots, 0)$ is replaced by any vector with a 1 at one coordinate and 0 at the other coordinates.

5. For each p in $M(\mathcal{X})$, let $X_1, X_2, \ldots, X_n$ be i.i.d. P and let P itself be $D(\alpha)$. Then the likelihood is proportional to

$$\prod_1^k p_i^{\alpha_i - 1 + n_i}$$

where $n_i = \#\{j : X_j = i\}$. Hence the posterior distribution of P given $X_1, X_2, \ldots, X_n$ can be conveniently written as $D(\alpha + \sum \delta_{X_i})$.

6. The marginal distribution of each X_i is $\bar{\alpha}$ where $\bar{\alpha}(i) = \alpha(i)/\alpha(\mathcal{X})$ and also $E(P) = \bar{\alpha}$. To see this, note that for each $A \subset \mathcal{X}$, $P(A)$ is beta$(\alpha(A), \alpha(A^c))$ and hence $E(P(A) = \alpha(A)/(\alpha(A) + \alpha(A^c))$.

   Property 5 immediately leads to

7. $$D(\alpha)\,(P \in C) = \sum_1^k \frac{\alpha(i)}{\alpha(\mathcal{X})}\, D(\alpha + \delta_i)(C)$$

This follows from $D(\alpha)\,(P \in C) = E\,(E(P \in C|X_1))$; $E(P \in C|X_1)$ is by property 5, $D(\alpha + \delta_{X_1})(C)$, and the marginal of X_1 is $\bar{\alpha}$.

8. Let P be distributed as $D(\alpha)$ and X independent of P be distributed as $\bar{\alpha}$. Let Y be independent of X and P be a beta$(1, \alpha(\mathcal{X}))$ random variable. Then $Y\delta_X + (1 - Y)P$ is again a $D(\alpha)$ random probability.

   This follows from properties 4 and 7 by conditioning on $x = i$, interpreting δ_i as a $D(\delta_i)$ distribution, and then using properties 4 and 7.

9. The predictive distribution of X_{n+1} given $X_1, X_2, \ldots, X_n$ is

$$\frac{\alpha + \sum_1^n \delta_{X_i}}{\alpha(\mathcal{X}) + n}$$

10. $\alpha_1 \neq \alpha_2$ implies $D(\alpha_1) \neq D(\alpha_2)$, except when α_1, α_2 are degenerate and put all their masses at the same point.

    This can be verified by choosing an i such that $\alpha_1(i) \neq \alpha_2(i)$. Then $P(i)$ has a nondegenerate beta distribution under at least one of α_1, α_2. Next use the fact that a beta distribution is determined by its first two moments.

11. It is often convenient to write a finite measure α on $\mathcal{X}$ as $\alpha = c\bar{\alpha}$, where $\bar{\alpha}$ is a probability measure. Let $\alpha_n = c_n\bar{\alpha}_n$ be a sequence of measures on $\mathcal{X}$. Then $D(c_n\bar{\alpha}_n)$ is a sequence of probability measures on the compact set S_k and hence has limit points. The following convergence results are useful.

    (a) If $\bar{\alpha}_n \to \bar{\alpha}$ and $c_n \to c, 0 < c < \infty$, then $D(c_n\bar{\alpha}_n) \to D(c\bar{\alpha})$ weakly.
    If $\bar{\alpha}\{i\} > 0$ for all i, then the density of $D(c_n\bar{\alpha}_n)$ converges to that of $D(c\bar{\alpha})$. If $\bar{\alpha}\{i\} = 0$ for some of the is, then the result can be verified by showing that the moments of $D(c_n\bar{\alpha}_n)$ converge to the moments of $D(c\bar{\alpha})$.

    (b) Suppose that $\bar{\alpha}_n \to \bar{\alpha}$ and $c_n \to 0$. Then $D(c_n\bar{\alpha}_n)$ converges weakly to the discrete measure μ which gives mass $\bar{\alpha}_i$ to the probability degenerate at i. To see this note that $E_{D(c_n\bar{\alpha}_n)}p_i = \bar{\alpha}_n\{i\} \to \bar{\alpha}\{i\}$, and it follows from simple calculations that $E_{D(c_n\bar{\alpha}_n)}p_i^2$ also converges to $\bar{\alpha}\{i\}$. Thus each p_i is 0 or 1 almost surely with respect to any limit point of $D(c_n\bar{\alpha}_n)$. In other words, any limit point of $D(c_n\bar{\alpha}_n)$ is a measure concentrated on the set of degenerate probabilities on $\mathcal{X}$. It is easy to see that any two limit points have the same expected value and this together with the fact that they are both concentrated on degenerate measures shows that $D(c_n\bar{\alpha}_n)$ converges.

(c) $\bar{\alpha}_n \to \bar{\alpha}$ and $c_n \to \infty$. In this case also, $E_{D(c_n\bar{\alpha}_n)}p_i$ converges to $\bar{\alpha}\{i\}$. However $Var_{D(c_n\bar{\alpha}_n)}p_i \to 0$, and hence $D(c_n\bar{\alpha}_n)$ converges to the measure degenerate at $\bar{\alpha}$.

3.1.2 *Dirichlet Distribution via Polya Urn Scheme*

The following alternative view of the Dirichlet process is both interesting and a powerful tool. For a recent use of this approach, see Mauldin et al.[133].

Consider a Polya urn with $\alpha(\mathcal{X})$ balls of which $\alpha(i)$ are of color $i; i = 1, 2, \ldots, k$.[For the moment assume that $\alpha(i)$ are whole numbers or 0]. Draw balls at random from the urn, replacing each ball drawn by two balls of the same color. Let $X_i = j$ if the i th ball is of color j. Then

$$P(X_1 = j) = \frac{\alpha(j)}{\alpha(\mathcal{X})} \tag{3.5}$$

$$P(X_2 = j|X_1) = \frac{\alpha(j) + \delta_{X_1}(j)}{\alpha(\mathcal{X}) + 1} \tag{3.6}$$

and in general

(3.7)

$$P(X_{n+1} = j|X_1, X_2, \ldots, X_n) = \frac{\alpha(j) + \sum_1^n \delta_{X_i}(j)}{\alpha(\mathcal{X}) + n} \tag{3.8}$$

Thus we are reproducing the joint distribution of $X_1, X_2, \ldots$ that would be obtained from property 9 in the last section. The joint distribution of $X_1, X_2, \ldots$ is exchangeable. In fact, if λ_α denotes the joint distribution

$$\lambda_\alpha (X_1 = x_1, X_2 = x_2, \ldots, X_n = x_n)$$

$$= \frac{\alpha(x_1)}{\alpha(\mathcal{X})} \prod_{i=1}^{n-1} \frac{\alpha + \delta_{\sum_1^{i-1} x_j}}{\prod_1^{i-1}(\alpha(\mathcal{X}) + j)}(x_{i+1})$$

setting $n_i = \#\{X_j = i\}$

$$= \frac{\{\alpha(1)(\alpha(1)+1)\ldots(\alpha(1)+n_1-1)\}\{\alpha(2)(\alpha(2)+1)\ldots(\alpha(2)+n_2-1)\}\ldots}{\alpha(\mathcal{X})(\alpha(\mathcal{X})+1)\ldots(\alpha(\mathcal{X})+n-1)}$$

$$= \frac{[\alpha(1)]^{[n_1]}\ldots[\alpha(k)]^{[n_k]}}{[\alpha(\mathcal{X})]^{[n]}} \tag{3.9}$$

where $m^{[n]}$ is the ascending factorial given by $m^{[n]} = m(m+1)\dots(m+n-1)$.

It is clear that (3.5) defines successive conditional distributions even when $\alpha\{i\}$ is not an integer but only ≥ 0. The scheme (3.5) thus leads to a sequence of exchangeable random variables and the corresponding mixing measure Π coming out of De Finetti's theorem is precisely D_α. What we need to show is that if D_α is the prior on $M(\mathcal{X})$ and if given P, $X_1, X_2, \dots$ are i.i.d P, then the sequence $X_1, X_2, \dots$ has the distribution given in (3.9). In fact, (3.9) is equal to

$$\int_{M(\mathcal{X})} [P(1)]^{n_1} \dots [P(k)]^{n_k}\, D_\alpha(dP)$$

which is equal to

$$\int_{M(\mathcal{X})} [P(1)]^{n_1} \dots [P(k)]^{n_k}\, \Pi(dP)$$

Since the finite-dimensional Dirichlet is determined by its moments, this shows $\Pi = D_\alpha$.

The posterior given $X_1, X_2, \dots, X_n$ can also be recovered from this approach. For a given X_1, (3.5) defines a scheme of conditional distributions with α replaced by $\alpha + \delta_{X_1}$. Once again DeFinetti's theorem leads to the prior $D(\alpha + \delta_{X_1})$, this is also the posterior given X_1.

We end this section with the question of interpretation and elicitation of α. From property 6, $\bar{\alpha} = \alpha(\cdot)/\alpha(\mathcal{X}) = E(P)$. So $\bar{\alpha}$ is the prior guess about the expected P.

If we rewrite property 10 in terms of the Bayes estimate $E(p_i|X_1, X_2, \dots, X_n)$ of p_i given $X_1, X_2, \dots, X_n$

$$E(p_i|X_1, X_2, \dots, X_n) = \frac{\alpha(\mathcal{X})}{\alpha(\mathcal{X}) + n}\bar{\alpha}(i) + \frac{n}{\alpha(\mathcal{X}) + n}(\frac{n_i}{n})$$

which shows the Bayes estimate can be viewed as a convex combination of the "prior guess" and the empirical proportion. Because the weight of the "prior guess" is determined by $\alpha(\mathcal{X})$, this suggests interpreting $\alpha(\mathcal{X})$ as a measure of strength of the prior belief. This ease in interpretation and elicitation is a consequence of the fact that Dirichlet is a conjugate prior for i.i.d. sampling from $\mathcal{X}$. We will show that all these properties hold when $\mathcal{X} = \mathbb{R}$. The fact that variability of P is determined by a single parameter $\alpha(\mathcal{X})$ can be a problem when $k > 2$.

3.2 Dirichlet Process on $M(\mathbb{R})$

3.2.1 Construction and Properties

Dirichlet process priors are a natural generalization to $M(\mathbb{R})$ of the finite-dimensional distributions considered in the last section. Let $(\mathbb{R}, \mathcal{B})$ be the real line with the Borel σ-algebra $\mathcal{B}$ and let $M(\mathbb{R})$ be the set of probability measures on $\mathbb{R}$, equipped with the σ-algebra $\mathcal{B}_M$.

The next theorem asserts the existence of a Dirichlet process and also serves as a definition of the process D_α.

Theorem 3.2.1. *Let α be a finite measure on $(\mathbb{R}, \mathcal{B})$. Then there exists a unique probability measure D_α on $M(\mathbb{R})$ called the Dirichlet process with parameter α satisfying*

$$\text{For every partition } B_1, B_2, \ldots, B_k \text{ of } \mathbb{R} \text{ by Borel sets}$$
$$(P(B_1), P(B_2) \ldots, P(B_k)) \text{ is } D\left(\alpha(B_1), \alpha(B_2) \ldots, \alpha(B_k)\right)$$

Proof. The consistency requirement in Theorem 2.3.4 follows from property 2 in the last section. Continuity requirement 3 follows from the fact that if $B_n \downarrow B$ then $\alpha(B_n) \downarrow \alpha(B)$ and from property 11 of the last section. □

Note that finite additivity of α is enough to ensure the consistency requirements. The countable additivity is required for the continuity condition.

Assured of the existence of the Dirichlet process, we next turn to its properties. These properties motivate other constructions of D_α via De Finetti's theorem and an elegant construction due to Sethuraman. These constructions are not natural unless one knows what to expect from a Dirichlet process prior.

If $P \sim D(\alpha)$, then it follows easily that $E(P(A)) = \bar{\alpha}(A) = \alpha(A)/\alpha(\mathbb{R})$. Thus one might write $E(P) = \bar{\alpha}$ as the prior expectation of P.

Theorem 3.2.2. *For each P in $M(\mathbb{R})$, let $X_1, X_2, \ldots, X_n$ be i.i.d. P and let P itself be distributed as D_α, where α is finite measure. (A version of) the posterior distribution of P given $X_1, X_2, \ldots, X_n$ is $D_{\alpha + \sum_1^n \delta_{X_i}}$.*

Proof. We prove the assertion when $n = 1$; the general case follows by repeated application. A similar proof appears in Schervish[144].

To show that $D_{\alpha+\delta_X}$ is a version of the posterior given X, we need to verify that for each $B \in \mathcal{B}$ and C a measurable subset of $M(\mathbb{R})$,

$$\int_B D_{\alpha+\delta_x}(C)\ \bar{\alpha}(dx) = \int_C P(B)\ D_\alpha(dP)$$

As C varies each side of this expression defines a measure on $M(\mathbb{R})$, and we shall argue that these two measures are the same. It is enough to verify the equality on σ-algebras generated by functions $P \mapsto (P(B_1), P(B_2) \ldots, P(B_k))$, where $B_1, B_2, \ldots, B_k$ is a measurable partition of $\mathbb{R}$. We do this by showing that the moments of the vector $(P(B_1), P(B_2) \ldots, P(B_k))$ are same under both measures.

First suppose that $\alpha(B_i) > 0$ for $i = 1, 2, \ldots, k$. For any nonnegative $r_1, r_2, \ldots, r_n$, look at

$$\int_B \left[\int \prod_1^k [P(B_i)]^{r_i} D_{\alpha+\delta_x}(dP) \right] \bar{\alpha}(dx) \tag{3.10}$$

If we denote by $D_{\alpha'+\delta_i}$ and $D_{\alpha'}$ the k-variate Dirichlet distributions with parameters $(\alpha(B_1), \ldots, \alpha(B_i) + 1, \ldots, \alpha(B_k))$ and $(\alpha(B_1), \ldots, \alpha(B_i), \ldots, \alpha(B_k))$, then (3.10) is equal to

$$\sum_1^k \frac{\alpha(B \cap B_i)}{\alpha(B)} \int y_1^{r_1} \ldots y_i^{r_i} \ldots y_k^{r_n} D_{\alpha'+\delta_i}(dy_1 \ldots dy_{k-1}).$$

which in turn is equal to

$$= \sum_1^k \frac{\alpha(B \cap B_i)}{\alpha(B)} \int y_1^{r_1} \ldots y_i^{r_i+1} \ldots y_k^{r_n} D_{\alpha'}(dy_1 \ldots dy_{k-1}).$$

On the other hand because $P(B) = \sum P(B \cap B_i)$,

$$\begin{aligned} &\int \prod_1^k [P(B_i)]^{r_i} P(B) D_\alpha(dP) \\ &= \sum_1^k \int \prod_1^k [P(B_i)]^{r_i} P(B \cap B_i) D_\alpha(dP) \\ &= \sum_1^k \int P(B_1)^{r_1} \ldots P(B_i)^{r_i+1} \ldots P(B_k)^{r_k} \ldots \frac{P(B \cap B_i)}{P(B_i)} D_\alpha(dP) \end{aligned}$$

Since $\frac{P(B \cap B_i)}{P(B_i)}$ is a Beta random variable and independent of $(P(B_1), P(B_2) \ldots, P(B_k))$, the preceding equals

$$\sum_1^k \frac{\alpha(B_i) \cap B}{\alpha(B)} \int P(B_1)^{r_1} \ldots P(B_i)^{r_i+1} \ldots P(B_k)^{r_k} \ldots D_\alpha(dP)$$

which is equal to the expression obtained earlier. To take care of the case when some of the $\alpha(B_i)$ may be 0, consider the simple case when, say $\alpha(B_1) = 0, r_1 > 0$ and the rest of the $\alpha(B_i)$ are positive. In this case

$$\int_B \left[\int \prod_1^k [P(B_i)]^{r_i} D_{\alpha+\delta_x}(dP) \right] \bar{\alpha}(dx) = 0$$

Because in $\sum_1^k (\alpha(B \cap B_i)/\alpha(B)) \int y_1^{r_1} \dots y_i^{r_i} \dots y_k^{r_n} D_{\alpha'+\delta_i}(dy_1 \dots dy_{k-1}), \alpha(B \cap B_1) =$ 0 and for $i \neq 1$, $y_1 = 0$ a.e.$D_{\alpha'+\delta_i}$,

$$\int y_1^{r_1} \dots y_i^{r_i} \dots y_k^{r_n} D_{\alpha'+\delta_i}(dy_1 \dots dy_{k-1}) = 0$$

A Similar argument applies when $\alpha(B_i)$ is 0 for more than one i. □

Remark 3.2.1 *(Tail Free Property).* Fix a partition $B_1, B_2, \dots, B_k$ of $\mathcal{X}$. Consider a sequence $\{\underline{\underline{T}}\}_{n:n\geq 1}$ of nested partitions with $\underline{\underline{T}}_1 = \{B_1, B_2, \dots, B_k\}$ and $\sigma\{\{\underline{\underline{T}}\}_{n:n\geq 1}\} = \mathcal{B}$. Then D_α is tail free with respect to this partition. And we leave it to the reader to verify that with Dirichlet as the prior and with given P, $X \sim P$,

$$(P(B_1), P(B_2) \dots, P(B_k)) \text{ and } X$$

are conditionally independent given $\{I_{B_i}(X); 1 \leq i \leq k\}$. Consequently, the conditional distribution of the vector $(P(B_1), P(B_2) \dots, P(B_k))$ given $\underline{\underline{T}}_n$ is the same for all n and is equal to the marginal distribution of

$$(P(B_1), P(B_2) \dots, P(B_k))$$

under the measure $D_{\alpha+\delta_X}$.

The last remark provides an alternative and more natural approach to demonstrate that $D_{\alpha+\delta_X}$ is indeed the posterior given X. For, by the martingale convergence theorem, the conditional distribution of $(P(B_1), P(B_2) \dots, P(B_k))$ given $\underline{\underline{T}}_n$ converges to the conditional distribution of $(P(B_1), P(B_2) \dots, P(B_k))$ given X, and this limit is the marginal distribution of the vector $(P(B_1), P(B_2) \dots, P(B_k))$ arising out of $D_{\alpha+\delta_X}$. This is true for any partition $B_1, B_2, \dots, B_k$ and since a measure on $M(\mathbb{R})$ is determined by the distribution of finite partitions, we can conclude that $D_{\alpha+\delta_X}$ is indeed the posterior.

Remark 3.2.2 *(Neutral to the Right property).* Another useful independence property follows immediately from Property 4 of the last section. If $t_1 < t_2, \ldots < t_k$, then

$$(1 - F(t_1)), \frac{1 - F(t_2)}{1 - F(t_1)}, \ldots, \frac{1 - F(t_k)}{1 - F(t_{k-1})}$$

are independent.

Many of the properties of the Dirichlet process on $M(\mathbb{R})$ either easily follow from, or are suggested by the corresponding property for the finite-dimensional Dirichlet distribution. One major difference is that in the case of $M(\mathbb{R})$ the measure α can be continuous. This leads to some interesting consequences, some of which are explored next.

Denote by λ_α the joint distribution of $P, X_1, X_2, \ldots$. Suppose $P \sim D(\alpha)$ and given P, $X_1, X_2, \ldots$ are i.i.d. P. From Theorem 3.2.2 it immediately follows that the predictive distribution of X_{n+1} given $X_1, X_2, \ldots, X_n$ is

$$\frac{\alpha + \sum_1^n \delta_{X_i}}{\alpha(\mathbb{R}) + n}$$

and hence that
X_1 is distributed as $\bar{\alpha}$
Conditional distribution of X_2 given X_1 is $\frac{\alpha+\delta_{X_1}}{\alpha(\mathbb{R})+1}$
Conditional distribution of X_3 given X_1, X_2 is $\frac{\alpha+\delta_{X_1}+\delta_{X_2}}{\alpha(\mathbb{R})+2}$
Conditional distribution of X_{n+1} given $X_1, X_2, \ldots, X_n$ is $\frac{\alpha+\sum_1^n \delta_{X_i}}{\alpha(\mathbb{R})+n}$, etc.

Suppose that α is a discrete measure and let $\mathcal{X}_0$ be the countable subset of $\mathbb{R}$ such that $\alpha(\mathcal{X}_0) = \alpha(\mathbb{R})$ and $\alpha\{x\} > 0$ for all $x \in \mathcal{X}_0$. D_α can then be viewed as a prior on $M(\mathcal{X}_0)$. Further the joint distribution of $X_1, X_2, \ldots, X_n$ can be written explicitly.

For each $(x_1, x_2, \ldots, x_n)$ and for each $x \in \mathcal{X}_0$, let $n(x)$ be the number of is such that $x_i = x$. Note that $n(x)$ is nonzero for at most n many xs. If α_n denotes the joint distribution of $X_1, X_2, \ldots, X_n$, then

$$\alpha_n(x_1, x_2, \ldots, x_n) = \prod_{x \in \mathcal{X}_0} \alpha(x)^{[n(x)]} \tag{3.11}$$

where $a^{[b]} = a(a+1)\ldots(a+b-1)$.

The case when α is continuous is a bit more involved. Even if α has density with respect to Lebesgue measure, for $n \geq 2$, because $P\{X_1 = X_2\} \neq 0$, α_2 is no longer absolutely continuous with respect to the two-dimensional Lebesgue measure. To see this formally, note that

$$\alpha_2 \{X_1 = X_2\} = \int \frac{(\alpha + \delta_{x_1})}{\alpha(\mathbb{R}) + 1} \{x_1\} \, d\bar{\alpha}(x_1) = \frac{1}{\alpha(\mathbb{R}) + 1}$$

On the other hand the Lebesgue measure of $\{(x, x) : x \in \mathbb{R}\}$ is 0.

While α_n is not dominated by the n-dimensional Lebesgue measure, it is dominated by a measure λ_n^* composed of Lebesgue measure in lower-dimensional spaces, and with respect to this measure, it is possible to obtain a fairly explicit form of the density of α_n. We will look at the case $n = 3$ in some detail and then extend these ideas to general n.

We will begin by calculating $\alpha_n(A \times B \times C)$ when α is a continuous measure. Let

$$\mathbf{R}_{1,2,3} = \{(x_1, x_2, x_3) : x_1, x_2, x_3 \text{ are all distinct }\}$$

Then

$$\begin{aligned} &\alpha_3 \left((A \times B \times C) \cap \mathbf{R}_{1,2,3}\right) \\ &\quad = \alpha_3 \{X_1 \in A, X_2 \in B - \{X_1\}, X_3 \in C - \{X_1, X_2\}\} \\ &\quad = \frac{\alpha(A)}{\alpha(\mathbb{R})} \frac{\alpha(B)}{(\alpha(\mathbb{R}) + 1)} \frac{\alpha(C)}{(\alpha(\mathbb{R}) + 2)} \end{aligned}$$

where the last equality follows from the fact that for each x_1, by continuity of α, $\alpha(B - \{x_1\}) = \alpha(B)$ and $\delta_{x_1}(B - \{x_1\}) = 0$. Consequently

$$\Pr\{X_2 \in B - \{x_1\} = \frac{[\alpha + \delta_{x_1}]}{\alpha(\mathbb{R}) + 1}(B - \{x_1\}) = \frac{\alpha(B)}{\alpha(\mathbb{R}) + 1}$$

Similarly for $\Pr\{X_3 \in C - \{x_1, x_2\}$.

Next, let

$$\mathbf{R}_{12,3} = \{(x, x, x_3) : x \neq x_3\}$$

Then

$$\begin{aligned} &\alpha_3 \left((A \times B \times C) \cap \mathbf{R}_{12,3}\right) \\ &\quad = \alpha_3 \{X_1 \in A, X_2 = \{X_1\}, X_3 \in C - \{X_1\}\} \end{aligned}$$

Because $Pr\{X_2 = x|X_1 = x\} = [\alpha + \delta_x]/(\alpha(\mathbb{R})+1)(\{x\}) = 1/(\alpha(\mathbb{R})+1)$, again by continuity of α, we have the preceding is equal to

$$\frac{\alpha(A \cap B)}{\alpha(\mathbb{R})} \frac{1}{(\alpha(\mathbb{R})+1)} \frac{\alpha(C)}{(\alpha(\mathbb{R})+2)}$$

Similarly, if

$$\mathbf{R}_{13,2} = \{(x, x_2, x) : x \neq x_2\}$$

then by exchangeability

$$\alpha_n (A \times B \times C \cap \mathbf{R}_{13,2}) = \alpha_n (A \times C \times B \cap \mathbf{R}_{12,3})$$
$$= \frac{\alpha(A \cap C)}{\alpha(\mathbb{R})} \frac{1}{(\alpha(\mathbb{R})+1)} \frac{\alpha(B)}{(\alpha(\mathbb{R})+2)}$$

A similar expression holds for $\mathbf{R}_{1,23}$.

Let $\mathbf{R}_{123} = \{(x,x,x)\}$. Then $A \times B \times C \cap \mathbf{R}_{123} = \{(x,x,x) x \in A \cap B \cap C\}$. We then have

$$\alpha_n (A \times B \times C \cap \mathbf{R}_{123}) = \frac{2\alpha(A \cap B \cap C)}{\alpha(\mathbb{R})} \frac{1}{(\alpha(\mathbb{R})+1)} \frac{\alpha(B)}{(\alpha(\mathbb{R})+2)}$$

where the factor 2 in the numerator arises from $P(X_3 = x|X_1 = X_2 = x) = (\delta_x + \delta_x)\alpha(B)/(\alpha(\mathbb{R})(\alpha(\mathbb{R})+1)(\alpha(\mathbb{R})+2))(x)$.

Suppose that α has a density $\tilde{\alpha}$ with respect to Lebesgue measure. Define a measure λ_3^* as follows:

λ_3^* restricted to $\mathbf{R}_{1,2,3}$ is the three-dimensional Lebesgue measure

λ_3^* restricted to $\mathbf{R}_{12,3}$ is the two-dimensional Lebesgue measure obtained from $\mathbb{R}^2$ via the map $(x,y) \mapsto (x,x,y)$.

Define the restriction on $\mathbf{R}_{1,23}$ and $\mathbf{R}_{13,2}$ similarly.

λ_3^* restricted to $\mathbf{R}_{12,3}$ is the one-dimensional Lebesgue measure obtained from $x \mapsto (x,x,x)$.

Note that the function on $\mathbf{R}_{1,2,3}$ defined by

$$\tilde{\alpha}_3(x_1, x_2, x_3) = \frac{\tilde{\alpha}(x_1)\tilde{\alpha}(x_2)\tilde{\alpha}(x_3)}{\alpha(\mathbb{R})(\alpha(\mathbb{R})+1)(\alpha(\mathbb{R})+2)}$$

when viewed as a density with respect to λ_3^* restricted to $\mathbf{R}_{1,2,3}$ gives, for any $(A \times B \times C)$, $\alpha_n (A \times B \times C \cap \mathbf{R}_{1,2,3})$. Similarly the function on $\mathbf{R}_{12,3}$ defined by

$$\tilde{\alpha}_3(x_1, x_1, x_3) = \frac{\tilde{\alpha}(x_1)\tilde{\alpha}(x_3)}{\alpha(\mathbb{R})(\alpha(\mathbb{R})+1)(\alpha(\mathbb{R})+2)}$$

corresponds to the density of α_3 with respect to λ_3^* restricted to $\mathbf{R}_{12,3}$ and

$$\tilde{\alpha}_3(x_1, x_1, x_1) = \frac{2\tilde{\alpha}(x_1)}{\alpha(\mathbb{R})(\alpha(\mathbb{R})+1)(\alpha(\mathbb{R})+2)}$$

corresponds to the density of α_3 with respect to λ_3^* restricted to $\mathbf{R}_{123}$.

The general case is similar but notationally cumbersome. For a partition $\{C_1, \dots, C_k\}$ of $\{1, 2, \dots, n\}$, let

$$\mathbf{R}_{C_1, C_2, \dots, C_k} = \{(x_1, x_2, \dots, x_n) : x_i = x_j \text{ iff } i, j \in C_m \text{ for some } m, 1 \leq m \leq k\}$$

The measure λ_n^* is defined by setting its restriction on $\mathbf{R}_{C_1, C_2, \dots, C_k}$ to be the k-dimensional Lebesgue measure. As before if we set $I_1 = 1$ and

$$\begin{cases} I_j = 1 & \text{if } , x_j \notin \{x_1, x_2, \dots, x_n\} \\ 0 & \text{otherwise.} \end{cases}$$

the density of α_n with respect to λ_n^* on $\mathbf{R}_{C_1, C_2, \dots, C_k}$ is given by

$$\tilde{\alpha}_n(x_1, x_2, \dots, x_n) = \frac{\prod_j \tilde{\alpha}(x_j)^{I_j}(e_j - 1)!}{(\alpha(\mathbb{R}))^{[n]}} \tag{3.12}$$

where $e_j = \#c_j$.

The verification follows essentially the same ideas, for example

$$\alpha_n(A_1 \times A_2 \times \dots \times A_n \cap \mathbf{R}_{C_1, C_2, \dots, C_k}) = \frac{\alpha(B_1)\alpha(B_2)\dots\alpha(B_k)}{(\alpha(\mathbb{R}))^{[n]}}$$

where $B_j = \cap_{i \in C_j} A_i$.

Theorem 3.2.3. $D_\alpha\{P : P \textit{ is discrete }\} = 1$.

Proof. Let $\tilde{E} = \{(P, x) : P\{x\} > 0\}$. Note that P is a discrete probability measure if $\sum_{\{x:(P,x) \in \tilde{E}\}} P(x) = 1$. We saw in the last chapter that $\tilde{E}$ is a measurable set. Let

$$\tilde{E}_x = \{P : P\{x\} > 0\} \qquad \tilde{E}_P = \{x : P\{x\} > 0\}$$

Then

$$\begin{aligned} \lambda_\alpha(\tilde{E}) &= E_{\lambda_\alpha}\left(\lambda_\alpha(\tilde{E}|X_1)\right) \\ &= E_{\lambda_\alpha}\left(\lambda_\alpha(\tilde{E}_{X_1}|X_1)\right) = E_{\lambda_\alpha}\left(D_{\alpha+\delta_{X_1}}(\tilde{E}_{X_1})\right) \\ &= 1 \end{aligned}$$

Because $P\{x_1\}$ is beta with positive parameter $\alpha\{x_1\}+1$, $P\{x_1\} > 0$ with probability 1. Now

$$\lambda_\alpha(\tilde{E}) = E_{\lambda_\alpha}\left(\lambda_\alpha(\tilde{E}|P\right) = E_{\lambda_\alpha}\left(P(\tilde{E}_P)\right) = 1$$

so $P(\tilde{E}_P) = 1$ almost everywhere D_α. □

The preceding proof is based on a presentation in Basu and Tiwari[10] . A variety of proof for this interesting fact is available. See Blackwell & Mcqueen [25], and Blackwell [23], Berk and Savage [17]. Another nice proof is due to Hjort [99]

3.2.2 The Sethuraman Construction

Sethuraman [148] introduced and elaborated on a useful and clever construction of D_α, which provides insight into these processes and helps in simulation of the process.

As before let α be a finite measure and $\bar{\alpha} = \alpha/\alpha(\mathbb{R})$. Let Ω be a probability space with a probability μ such that

$$\theta_1, \theta_2, \ldots \text{ defined on } \Omega \text{ are i.i.d. beta}(1, \alpha(\mathbb{R}))$$

$Y_1, Y_2, \ldots$ are also defined on Ω such that they are i.i.d. $\bar{\alpha}$ and independent of the θ_is

Set $p_1 = \theta_1$ and for $n \geq 2$, let $p_n = \theta_n \prod_1^{n-1}(1-\theta_i)$. Easy computation shows that $\sum_1^\infty p_n = 1$ almost surely. Now define an $M(\mathbb{R})$ valued random variable on Ω by

$$P(\omega, A) = \sum_1^\infty p_n(\omega)\delta_{Y_n(\omega)}(A) \tag{3.13}$$

Because $\sum_1^\infty p_n = 1$, the function $\omega \mapsto P(\omega, \cdot)$ takes values in $M(\mathcal{X})$. It is not hard to see that this map is also measurable. This random measure is a discrete measure that puts weight p_i on Y_i. Sethuraman showed that this random measure is distributed as D_α. Formally, if Π is the distribution of $\omega \mapsto P(\omega, \cdot)$ then $\Pi = D_\alpha$. We will establish this by showing that for every partition $B_1, B_2, \ldots, B_k$ of $\mathbb{R}$ by Borel sets $(P(\omega, B_1), P(\omega, B_2), \ldots, P(\omega, B_k))$ is distributed as $D(\alpha(B_1), \ldots, \alpha(B_k))$.

Denote by $\delta^k_{Y_i}$ the element of S_k given by $(I_{B_1}(Y_i), I_{B_2}(Y_i), \ldots, I_{B_k}(Y_i))$. Then for each ω, $(P(\omega, B_1), P(\omega, B_2), \ldots, P(\omega, B_k))$ can be written as $\sum_1^\infty p_i(\omega)\delta^k_{Y_i(\omega)}$.

Let P be an S_k valued random variable, independent of the Ys and θs, and distributed as $D(\alpha(B_1), \ldots, \alpha(B_k))$.

Consider the S_k valued random variable

$$P^1 = p_1\delta^k_{Y_1} + (1-p_1)P$$

where $Y_1 \in B_i$, $\delta^k_{Y_i}$ is the vector with a 1 in the ith coordinate and 0 elsewhere. Hence by property 4 from Section 3.1, given $Y_1 \in B_i$, P^1 is distributed as a Dirichlet with parameter $(\alpha(B_1),\ldots,\alpha(B_i)+1,\ldots,\alpha(B_k))$. Since $\mu(Y_1 \in B_i) = \alpha(B_i)$, by property 8 in Section 3.1, P^1 is distributed as $D(\alpha(B_1),\ldots,\alpha(B_k))$.

It follows by easy induction that for all n, $1-\sum_1^n p_i = \prod_1^n(1-\theta_i)$. Using this fact, a bit of algebra gives

$$\begin{aligned}
&\sum_1^n p_i\delta^k_{Y_i} + (1-\sum_1^n p_i)P \\
&= \sum_1^{n-1} p_i\delta^k_{Y_i} + (1-\sum_1^{n-1} p_i)(\theta_n\delta^k_{Y_n} + (1-\theta_n)P)
\end{aligned}$$

Because our earlier argument showed that $\theta_n\delta^k_{Y_n}+(1-\theta_n)P$ has the same distribution as P, a simple induction argument shows that, for all n,

$$\sum_1^n p_i\delta^k_{Y_i} + (1-\sum_1^n p_i)P$$

is distributed as $D(\alpha(B_1),\ldots,\alpha(B_k))$. Letting $n\to\infty$ and observing that $(1-\sum_1^n p_i)$ goes to 0, we get the result.

Note that we have not assumed the existence of a D_α prior. Because $P(\omega,\cdot)$ is $M(\mathcal{X})$ valued, the argument also shows the existence of the Dirichlet prior.

3.2.3 *Support of D_α*

We begin by recalling that $M(\mathbb{R})$ under the weak topology is a complete separable metric space, and hence for any probability measure Π on $M(\mathbb{R})$ the support—the smallest closed set of measure 1— exists. Note that support is not meaningful if we consider the total variation metric or setwise convergence.

Theorem 3.2.4. *Let α be a finite measure on $\mathbb{R}$ and let E be the support of α. Then*

$$M_\alpha = \{P : \text{ support of } P \subset E\}$$

is the weak support of D_α

Proof. M_α is a closed set by the Portmanteau theorem, since E is closed and if $P_n \to P$ then $P(E) \geq \limsup_n P_n(E)$. Further, because $P(E)$ is beta$(\alpha(\mathbb{R}), 0)$, $D_\alpha(M_\alpha) = 1$.

Let P_0 belong to M_α and let U be a neighborhood of P_0. Our theorem will be proved if we show that $D_\alpha(U) > 0$.

Choose points $a_0 < a_1 < \ldots < a_{T-1} < a_T$ and let $W_j = (a_j, a_{j+1}] \cap E$ and $J = \{j : \alpha(W_j) > 0\}$. Then depending on whether $\alpha(\cup_{j \in J} W_j) = \alpha(\mathbb{R})$ or $\alpha(\cup_{j\in J} W_j) < \alpha(\mathbb{R})$, $(P(W_j) : j \in J)$ or $\left(P(W_j) : j \in J, 1 - \sum_{j \in J} P(W_j)\right)$ has a finite-dimensional Dirichlet distribution with all parameters positive. And in either case, for any $\eta > 0$,

$$D_\alpha\{P \in M(\mathbb{R}) : |P(W_j) - P_0(W_j)| < \delta : j \in J\} > 0$$

By Propositon 2.5.2 for small enough δ, U contains a set of the above form. Hence $D_\alpha(U) > 0$.

□

3.2.4 Convergence Properties of D_α

Many of the theorems in this section are adapted from Sethuraman and Tiwari [149].

Because under D_α, $E(P) = \bar{\alpha}$, Theorem 2.5.1 in Chapter 2 immediately yields the following.

Theorem 3.2.5. *Let $\{\alpha_t : t \in T\}$ be a family of finite measures on $\mathbb{R}$. Then the family $\{D_{\alpha_t} : t \in T\}$ is tight iff $\{\bar{\alpha}_t : t \in T\}$ is tight.*

Theorem 3.2.6. *Suppose $\{\alpha_m\}, \alpha$ are finite measures on $\mathbb{R}$ such that $\bar{\alpha}_m \to \bar{\alpha}$ weakly.*

(i) If $\alpha_m(\mathbb{R}) \to \alpha(\mathbb{R})$ where $0 < \alpha(\mathbb{R}) < \infty$, then $D_{\alpha_m} \to D_\alpha$ weakly.

(ii) If $\alpha_m(\mathbb{R}) \to 0$. Then D_{α_m} converges weakly to D^*, where

$$D^*\{P : P \textit{ is degenerate}\} = 1$$

(iii) If $\alpha(\mathbb{R}) \to \infty$ then D_{α_m} converges weakly to δ_α.

Proof. By Theorem 3.2.5, $\{D_{\alpha_m}\}$ is tight and hence any subsequence has a further subsequence that converges to, say, D^*.

(i) We will argue that the limit D^* is D_α and is the same for all subsequences. By (iii) of Theorem 2.5.2 and (a) of property 11 of the finite-dimensional Dirichlet (see Section 3.1) it follows that $D^* = D_\alpha$.

(ii) From property 11 for any $\bar{\alpha}$ continuity set A, $D^*\{P : P(A) = 0, \text{ or } 1\} = 1$. By using a countable collection of $\bar{\alpha}$ continuity sets that generate the Borel σ-algebra, the result follows.

(iii) (iii) Recall that $E(P(A)) = \bar{\alpha}(A)$. Because $\alpha_n(\mathbb{R}) \to \infty$, $\mathrm{Var}(P(A)) \to 0$ for all A. Hence $P(A)$ converges in probability to $\bar{\alpha}(A)$. This holds for any finite collection of sets. The result now follows as in the preceding case.

□

As a consequence of the theorem we have the following results.

Theorem 3.2.7. *(i) Let α be a finite measure. Then for each P_0 the posterior $D_{\alpha+\sum_1^n \delta_{X_i}} \to \delta_{P_0}$ weakly, almost surely P_0.*

(ii) As $\alpha(\mathbb{R})$ goes to 0, the posterior converges weakly to $D_{\sum_1^n \delta_{X_i}}$.

Proof. Because a.e. P_0, $\alpha + \sum_1^n \delta_{X_i} = \alpha_n$ satisfies $\bar{\alpha}_n \to P_0$ and $\alpha_n \to \infty$, (iii) of Theorem 3.2.6 yields the result. □

Remark 3.2.3. Note that posterior consistency holds for all P_0, not necessarily in the weak support of D_α. This is possible because the version of the posterior chosen behaves very nicely. This version is not unique even for P_0 in the weak support of D_α. One sufficient condition for uniqueness up to P_0 null sets is that P_0 be dominated by α.

Remark 3.2.4. Assertion (ii) has been taken as a justification of the use of $D_{\sum_1^n \delta_{X_i}}$ as a noninformative (completely nonsubjective in the terminology of Chapter 1) posterior. Note that Theorem 3.2.6 shows that the corresponding prior is far from a noninformative prior.

The posterior $D_{\sum_1^n \delta_{X_i}}$ has been considered as a sort of Bayesian bootstrap by Rubin [142]. For an interesting discussion of the Bayesian bootstrap and Efron's bootstrap, see Schervish [144].

We would like to remark that all the theorems in this section go through if $\mathbb{R}$ is replaced by any complete separable metric space. The existence aspect of the Dirichlet process can be handled via the famous Borel isomorphism theorem, which says that there is a 1-1, bimeasurable function form $\mathbb{R}$ onto $\mathcal{X}$. The proofs of other results require only trivial modifications.

3.2.5 Elicitation and Some Applications

We have seen that with a D_α prior the posterior given $X_1, X_2, \ldots, X_n$ is $D_{\alpha+\sum \delta_{X_i}}$. As $\alpha(\mathbb{R})$ goes to 0, $(\alpha + \sum \delta_{X_i})/(\alpha(\mathbb{R}) + n)$ converges to $\sum \delta_{X_i}/n$, the empirical distribution, further $\alpha(\mathbb{R}) + n$ converges to n. Hence as observed in the last section $D_{\alpha+\sum \delta_{X_i}}$ converges weakly to $D_{\sum \delta_{X_i}}$. In particular if the $X_1, X_2, \ldots, X_n$ are distinct then $D_{\delta_{X_i}}$ is just the uniform distribution on the n-dimensional probability simplex S_n^*. This phenomenon suggests an interpretation of $\alpha(\mathbb{R})$ goes to 0, as leading to a "noninformative"prior. In this section we investigate a few examples, all taken from Ferguson [61], where as $\alpha(\mathbb{R})$ goes to 0, the Bayes procedure converges to the corresponding frequentist nonparametric method.

While these examples corroborate the feeling that $\alpha(\mathbb{R})$ goes to 0 leads to a non-informative prior, (ii) of Theorem 3.2.6 points out the need to be careful with such an interpretation. As $\alpha(\mathbb{R})$ goes to 0 the posterior leads to an intuitive noninformative limit. However the corresponding prior cannot be considered noninformative. We believe these applications are justified in the completely non-parametric context of making inference about P because the Dirichlet is conjugate in that setting. Similar assessments of conjugate prior in finite-dimensional problems is well known.

However, the Dirichlet is often used in problems where it is not a conjugate prior. In such problems the interpretation of $\alpha(\mathbb{R})$ as a sort of sample size or a measure of prior variability is of doubtful validity. See Newton et al. [136] in this connection.

Estimation of F . Suppose that we want to estimate the unknown distribution function under the loss $L(F,G) = \int (F(t) - G(t))^2\, dt$. If Π is a prior on $M(\mathbb{R})$, equivalently on the space of distribution functions $\mathcal{F}$ on $\mathbb{R}$, it is well known that the no-sample Bayes estimate is given by $\hat{F}_\Pi(t) = \int F(t)\ d\Pi(F)$. If Π is D_α then because the posterior is $D_{\alpha+\sum \delta_{X_i}}$, the Bayes estimate of F given $X_1, X_2, \ldots, X_n$ is $(\alpha + \sum \delta_{X_i})\,(-\infty, t]/(\alpha(\mathbb{R}) + n)$. Setting F_n as the empirical distribution, we rewrite this as

$$\frac{\alpha(\mathbb{R})}{\alpha(\mathbb{R}) + n}\bar{\alpha}(-\infty, t] + \frac{n}{\alpha(\mathbb{R}) + n}F_n$$

which is a convex combination of the prior guess and a frequentist nonparametric estimate.

This property makes it clear how α is to be chosen. If the prior guess of the distribution of X is, say, $N(0,1)$ then that is $\bar{\alpha}$. The value of $\alpha(\mathbb{R})$ determines how certain one feels about the prior guess. This interpretation of $\alpha(\mathbb{R})$ as a measure of one's faith in a prior guess is endorsed by the fact that if $\alpha(\mathbb{R}) \to \infty$ then the prior goes to $\delta_{\bar{\alpha}}$.

If $\alpha(\mathbb{R}) \to 0$ the Bayes estimate of P converges to the empirical distribution and the posterior converges weakly to D_{nF_n}. Since the prior has no role any more, D_{nF_n} is called a noninformative posterior and F_n the corresponding noninformative Bayes estimate. These intuitive ideas are helpful in calibrating $\alpha(\mathbb{R})$ as a cost of sample size and $\alpha(\mathbb{R}) = 1$ is sometimes taken as a prior with low information.

Estimation of mean of F. The problem here is to estimate the mean μ_F of the unknown distribution function F, the loss function being the usual squared error loss, i.e., $L(F,a) = (\mu_F - a)^2$. If Π is a prior on $\mathcal{F}$ such that $\hat{F}_\Pi$ has finite mean, then the Bayes estimate $\hat{\mu}$ is $\int \mu_F \, d\Pi(F)$ and with probability 1 this is the same as the mean of $\hat{F}_\Pi$. This follows because

$$\int \left[\int x dF\right] \Pi(dF)$$

$$= \lim \int \left[\int x I_{[0,n]} dF\right] \Pi(dF)$$

$$= \int x d\hat{F}_\Pi(x) = \int x d\hat{F}_\Pi(x) < \infty$$

Thus if α has finite mean then

$$D_\alpha\{F : F \text{ has finite mean}\} = 1$$

and given $X_1, X_2, \ldots, X_n$, the Bayes estimate of μ_F is the mean of $\alpha + \sum \delta_{X_i}$. This is easily seen to be a convex combination of the mean of $\bar{\alpha}$ and $\bar{X}$ and goes to $\bar{X}$ as $\alpha(\mathbb{R}) \to 0$.

Estimation of median of F. We next turn to the estimation of the median of the unknown distribution F. For any $F \in \mathcal{F}$, t is a median if

$$F(t-) \le \frac{1}{2} \le F(t)$$

If α has support $[K_1, K_2], -\infty \le K_1 < K_2 \le \infty$ then with D_α probability 1, F has unique median. If $t_1 < t_2$ are both medians of F, then for any rational $a, b; t_1 < a < b < t_2$ we have $F(a) = F(b)$. On the other hand $D_\alpha\{F : F(a) = F(b)\} = 0$. By considering all rationals a, b in the interval (K_1, K_2) we have the result.

In the context of estimating the median the absolute deviation loss is more natural and convenient than the squared error loss. Formally, $L(F, m) = |m_F - m|$. If Π is a prior on $\mathcal{F}$ then the "no-sample" Bayes estimate is just the median of the distribution of m_F.

If the prior is $D_\alpha$then any median of m_F is also a median of $\bar{\alpha}$. This may be seen as follows: t is a median of m_F iff

$$D_\alpha\{m_F < t\} \le \frac{1}{2} \le D_\alpha\{m_F \le t\}$$

Now $m_F \le t$ iff $F(t) \ge 1/2$. Because $F(t)$ is beta $(\alpha(-\infty,t], \alpha(t,\infty)$, $D_\alpha\{F(t) \ge 1/2\} \ge 1/2$ iff $\alpha(t,\infty)/\alpha(\mathbb{R}) \ge 1/2$ (see exercise 11.0.2). On the other hand $m_F < t$ iff $F(t-) > 1/2$. This yields $\alpha(-\infty,t)/\alpha(\mathbb{R}) \le 1/2$ and such a t is a median of $\bar{\alpha}$.

Consequently, the Bayes estimate of the median given $X_1, X_2, \ldots, X_n$ is a median of $(\alpha + \sum \delta_{X_i})/(\alpha(\mathbb{R})+n))$. If $\bar{\alpha}$ is continuous then the median of $(\alpha + \sum \delta_{X_i})/(\alpha(\mathbb{R})+n))$ is unique. As $\alpha(\mathbb{R})$ goes to 0 the limit points of the Bayes estimates of m_F are medians of the empirical distribution.

Testing for median of F. Consider the problem of testing the hypotheses that the median of F is less than or equal to 0 against the alternative that the median is greater than 0. If we view this as a decision problem with 0-1 loss, for a D_α prior on $\mathcal{F}$ the Bayes rule is

$$\text{decide median is} \le 0 \qquad \text{if } D_\alpha\{F(0) > \frac{1}{2}\} > \frac{1}{2}$$

Because $D_\alpha\{F(0) > 1/2\} = 1/2$ iff the two parameters are equal this reduces to "accept the hypotheses that the median is 0 iff

$$\frac{\alpha(-\infty,0]}{\alpha(\mathbb{R})} > \frac{1}{2}''$$

Given $X_1, X_2, \ldots, X_n$ this condition becomes "accept the hypotheses that the median is 0 iff

$$W_n > \frac{1}{2}n + \alpha(\mathbb{R})\left(\frac{1}{2} - \bar{\alpha}(-\infty,0)\right)''$$

where W_n is the number $X_i \le 0$.

Estimation of $P(X \le Y)$. Suppose that $X_1, \ldots, X_n$ are i.i.d. F and $Y_1, \ldots, Y_m$ are independent of the X_is and are i.i.d G. We want to estimate $P(X_1 \le Y_1) = \int F(t)\, dG(t)$ under squared error loss. Suppose that the prior for (F,G) is of the form $\Pi_1 \times \Pi_2$. The Bayes estimate is then $\int \hat{F}_{\Pi_1}(t)\, d\hat{F}_{\Pi_2}(dt)$, where for $i = 1,2$, $\hat{F}_{\Pi_i}(t)$ is the distribution function $\int F(t)\, d\Pi_i(t)$.

If the prior is D_α then the Bayes estimate given $X_1, X_2, \ldots, X_n$ becomes

$$\int \frac{(\alpha_1 + \sum \delta_{X_i})}{\alpha_1(\mathbb{R}) + n}(-\infty,t]\, d\left(\frac{\alpha_2 + \sum \delta_{Y_i}}{\alpha_2(\mathbb{R}) + n}\right)(dt)$$

This can be written as

$$p_{1,n}p_{2,m}\int\bar{\alpha_1}(-\infty,t)d\bar{\alpha_2}(t)+p_{1,n}(1-p_{2,m})\frac{1}{n}\sum_1^m\bar{\alpha_1}(-\infty,Y_j]$$
$$+(1-p_{1,n})p_{2,m})\frac{1}{m}\sum_1^n(1-\bar{\alpha_2}(-\infty,X_i))+(1-p_{1,n})(1-p_{2,m})\frac{1}{mn}U$$

where $p_{1,n}=\alpha_1(\mathbb{R})/(\alpha_1(\mathbb{R})+n)$, $p_{2,m}=\alpha_2(\mathbb{R})/(\alpha_2(\mathbb{R})+m)$ and U, is the number of pairs for which $X_i\le Y_j$, i.e.,

$$U=\sum_1^n\sum_1^m I_{(\infty,Y_j]}(X_i).$$

As $\alpha_1(\mathbb{R})$ and $\alpha_2(\mathbb{R})$ go to 0, the nonparametric estimate converges to $(mn)^{-1}U$, which is the familiar Mann-Whitney statistic.

3.2.6 Mutual Singularity of Dirichlet Priors

As before, we have a D_α prior on $M(\mathbb{R})$, given P, $X_1,X_2,\ldots,X_n$ is i.i.d. P, and λ_α is the joint distribution of P and $X_1,X_2,\ldots$. The main result in this section is ' If α_1 and α_2 are two nonatomic measures on $\mathbb{R}$, then λ_{α_1} and λ_{α_2} are mutually singular and hence so are D_{α_1} and D_{α_2}'. Mutual singularity of all priors in a family being used is undesirable. It shows that the family is too small to be flexible enough to represent prior opinion, which is based on information and judgment and is independent of the data. To clarify, consider a simple example of this sort. Let $X_1,X_2,\ldots,X_n$ be i.i.d. $N(\theta,1)$ and suppose we are allowed only $N(\mu,1)$ priors and the only values of μ allowed are finite and widely separated as 0 and 10. Then for a large n if we get $\bar{X}$, it is clear that with high probability the data can be reconciled with only one prior in the family. The result proved next is of this kind but stronger. It follows from a curious result of Korwar and Hollander [116], who show that the prior D_α can be estimated consistently from $X_1,X_2,\ldots$. We begin with their result.

Lemma 3.2.1. *Define $\tau_1,\tau_2,\ldots$ and $Y_1,Y_2,\ldots$ by $\tau_1=1$ and $\tau_n=k$ if the number of distinct elements in $\{X_1,X_2,\ldots,X_k\}$ is n and the number of distinct elements in $\{X_1,X_2,\ldots,X_{k-1}\}$ is $n-1$. In other words, τ_n is the number of observations needed to get n distinct elements.*

Set $Y_n = X_{\tau_n}$ *and set*

$$D_n = \begin{cases} 1 & \text{if } X_n \notin \{X_1, X_2, \ldots, X_{n-1}\} \\ 0 & \text{otherwise} \end{cases}$$

Note that $\sum_1^n D_i$ *is the number of distinct units in the first* n *observations. If* α *is nonatomic then*

(i) *for any Borel set* $U, 1/n \sum_1^n \delta_{Y_n}(U) \to \bar{\alpha}(U)$ *a.e.* λ_α*;*

(ii) $1/\log n \sum_1^n (D_i - E(D_i)) \to 0$ *a.e.* λ_α*; and*

(iii) $1/\log n \sum_1^n E(D_i) \to \alpha(\mathcal{X})$*.*

Proof. Note that $\tau_i < \infty$ a.e.

To prove (i) it is enough to show that $Y_1, Y_2, \ldots$ are i.i.d. $\bar{\alpha}$.

We start with a finer conditioning than $Y_1, \ldots, Y_{n-1}$. Consider for $t_1 < t_2, \ldots < t_{n-1}, t_n,$,

$$\begin{aligned} Pr\left\{Y_n \in A | X_1, X_2, \ldots X_{t_{n-1}}, \tau_{n-1} = t_{n-1}, \tau_n = t_n\right\} \\ = \frac{Pr\left\{Y_n \in | X_1, \ldots X_{t_{n-1}}, \tau_{n-1} = t_{n-1}, \tau_n \geq t_n\right\}}{Pr\left\{\tau_n = t_n | X_1, \ldots X_{t_{n-1}}, \tau_{n-1} = t_{n-1}, \tau_n \geq t_n\right\}} \end{aligned} \tag{3.14}$$

After cancelling out $\alpha(\mathcal{X}) + t_n - 1$ from the numerator and denominator this becomes

$$\frac{\left(\alpha + \sum_1^{t_n - 1} \delta_{X_i}\right)(A - \{Y_1, \ldots, Y_n\})}{\left(\alpha + \sum_1^{t_n - 1} \delta_{X_i}\right)(\mathcal{X} - \{Y_1, \ldots, Y_n\})}$$

and by nonatomicity this reduces to $\bar{\alpha}$. Thus $Y_1, Y_2, \ldots$ are i.i.d and (i) follows.

For the second assertion, it is easy to see that the D_n are independent with $\lambda_\alpha(D_n = 1) = \alpha(\mathbb{R})/(\alpha(\mathbb{R}) + n - 1)$.

By Kolomogorov's SLLN for independent random variables

$$\frac{1}{\log n} \sum_1^n (D_i - E(D_i)) \to 0 \text{ a.s. } \lambda_\alpha \text{ if } \sum_1^\infty \frac{V(D_i)}{(\log i)^2} < \infty$$

Here $V(D_i) = \alpha(\mathbb{R})(i-1)/((\alpha(\mathbb{R}) + i - 1)^2)$ and the preceding condition holds.

Moreover

$$\frac{1}{\log n}\sum_{1}^{n} E(D_i) = \frac{1}{\log n}\sum_{1}^{n}\frac{\alpha(\mathbb{R})}{\alpha(\mathbb{R})+i-1} \to \alpha(\mathbb{R})$$

because

$$\sum_{2}^{n}\frac{\alpha(\mathbb{R})}{\alpha(\mathbb{R})+i-1} = \sum_{2}^{n}\frac{\alpha(\mathbb{R})}{i-1} - \alpha(\mathbb{R})\sum_{2}^{n}\frac{\alpha(\mathbb{R})}{\alpha(\mathbb{R})+i-1}\frac{1}{i-1}$$

and as $n \to \infty$, the second term on the right converges, so that

$$\sum_{2}^{n}\frac{\alpha(\mathbb{R})}{\alpha(\mathbb{R})+i-1} = \alpha(\mathbb{R})\,[\log n + O(1)]$$

□

Theorem 3.2.8. *If α_1 and α_2 are two nonatomic measures on $\mathbb{R}$, $\alpha_1 \neq \alpha_2$, then λ_{α_1} and λ_{α_2} are mutually singular and hence so are D_{α_1} and D_{α_2}.*

Proof. Let U be a Borel set such that $\alpha_1(U) \neq \alpha_2(U)$, and set

$$E = \left\{\omega : \frac{1}{n}\sum_{1}^{n}\delta_{Y_i}(U) \to \bar{\alpha}_1(U) \text{ and } \frac{1}{\log n}\sum_{1}^{n} D_i \to \alpha_1(\mathbb{R})\right\}$$

By Lemma 3.2.1, $\lambda_{\alpha_1}(E) = 1$ and $\lambda_{\alpha_2}(E) = 0$.
Further, because $E \subset \mathbb{R}^\infty$, we also have

$$\lambda_{\alpha_1}(E) = \int P^\infty(E)\; D_{\alpha_1}(dP) = 1$$

so that, $D_{\alpha_1}\{P : P^\infty(E) = 1\} = 1$. Similarly $D_{\alpha_2}\{P : P^\infty(E) = 1\} = 0$.

□

Remark 3.2.5. To handle the general case, consider the decomposition of α_1, α_2 into $\alpha_i = \alpha_{i1} + \alpha_{i2}$, where α_{i1} is the nonatomic part of α_i and α_{i2} is the discrete part.

Let M_1, M_2 be the support of α_{12} and α_{22}. Then if $\alpha_{11} \neq \alpha_{21}$ but $M_1 = M_2$, then also λ_{α_1} and λ_{α_2} are singular.

If $\alpha_{11} = \alpha_{21}$ and $M_1 = M_2$; λ_{α_1} and λ_{α_2} may not be orthogonal. Sethuraman gives necessary and sufficient condition for the orthogonality using Kakutani's well-known criteria based on Hellinger distance.

Remark 3.2.6. The Theorem 3.2.8 shows that Dirichlet process used as priors display a curious behavior. Suppose α is a continuous measure, then for every sample sequence $X_1, X_2, \ldots$ the continuous part of the successive posterior base measures changes from $\alpha/\alpha(\mathcal{X})+n$ to $\alpha/(\alpha(\mathcal{X})+n+1)$ and hence the sequence of the posteriors are mutually singular.

3.2.7 *Mixtures of Dirichlet Process*

Dirichlet process requires specification of the base measure α, which itself can be viewed as consisting of the prior expectation $\bar{\alpha}$ and the strength of the prior belief $\alpha(\mathbb{R})$. In order to achieve greater flexibility Antoniak [4] proposed mixtures of Dirichlet process which arise by considering a family α_θ of base measures indexed by a hyperparameter θ and a prior for the parameter θ.

Because the Dirichlet processes sit on discrete measures, so does any mixture of these and hence they are unsuitable as priors for densities. For the same reason, it is also inappropriate for the parametric part of a semiparametric problem. For example, Diaconis and Freedman [46] show that the Dirichlet prior in a location parameter problem can lead to pathologies as well as inconsistency of the posterior for even reasonable "true" densities.

Usually one will not have as a prior a completely specified $\bar{\alpha}$ but an α_θ—like $N(\eta, \sigma^2)$—with $\theta = (\eta, \sigma^2)$ unknown but having a prior μ so that the distribution of P given θ is D_{α_θ}. Suppose that $X_1, X_2, \ldots, X_n$ are—given P—i.i.d P. Because given θ,$X_1, X_2, \ldots, X_n$; P is distributed as $D_{\alpha_\theta + \sum \delta_{X_i}}$, the distribution of P given $X_1, X_2, \ldots, X_n$is obtained by integrating $D_{\alpha_\theta + \sum \delta_{X_i}}$ with the conditional distribution of θ given $X_1, X_2, \ldots, X_n$.

For simplicity let $\Theta = \mathbb{R}$ let μ be the prior on Θ with density $\tilde{\mu}$; for each θ, α_θ is a finite measure on $\mathbb{R}$ with density $\tilde{\alpha_\theta}$ with respect to Lebesgue measure. Using equation (3.12) the joint density of θ and $X_1, X_2, \ldots, X_n$ is

$$\tilde{\mu}(\theta) \frac{\prod_j \tilde{\alpha}_\theta(x_j)^{I_j}(e_j - 1)!}{(\alpha_\theta(\mathbb{R}))^{[n]}} \tag{3.15}$$

The conditional density of θ given $X_1, X_2, \ldots, X_n$ is thus

$$C(x_1, x_2, \ldots, x_n)\tilde{\mu}(\theta) \frac{\prod_j \tilde{\alpha}_\theta(x_j)^{I_j}(e_j - 1)!}{(\alpha_\theta(\mathbb{R}))^{[n]}} \tag{3.16}$$

If the "true" distribution P_0 is continuous then with probability 1, the I_js are all equal to 1 and the conditional density becomes

$$C(x_1, x_2, \ldots, x_n)\tilde{\mu}(\theta)\frac{\prod_1^n \tilde{\alpha}(\theta)(x_j)}{(\alpha(\theta)(\mathbb{R}))^{[n]}} \tag{3.17}$$

Newton et al. [137] provides an interesting heuristic approximation to Bayes estimates in this context.

3.3 Polya Tree Process

Polya tree process are a large class of priors that include Dirichlet processes and provide a flexible framework for Bayesian analysis of nonparametric problems. Like the Dirichlet, Polya tree priors form a conjugate class with a tractable expression for the posterior. However they differ from Dirichlet process in two important aspects. The Polya tree process are determined by a large collection of parameters and thus provide means to incorporate a wide range of beliefs. Further, by suitably choosing the parameters, the Polya tree priors can be made to sit on continuous, even absolutely continuous, distributions.

Polya tree priors were explicitly constructed by Ferguson [62] as a special case of tail free processes discussed in the Chapter 2. A formal mathematical development using De Finetti's theorem is given in Mauldin et al. [133], Lavine [118, 119], indicates the construction and discusses the choice of various components that go into the construction of Polya tree priors. Here we briefly explore the properties of Polya tree priors. The basic references for these are Ferguson [62], Mauldin et al. [133] and Lavine[118, 119].

3.3.1 The Finite Case

The construction in this section is a special case of the discussion in Section 2.3.1. To briefly recall, let $\mathcal{X} = \{x_1, x_2, \ldots, x_{2^k}\}$. Let $B_0 = \{x_1, x_2, \ldots, x_{2^{k-1}}\}$ and $B_1 = \{x_{2^{k-1}}, \ldots, x_{2^k}\}$ be a partition of $\mathcal{X}$. For any j let E_j stand for all sequences of 0s and 1s of length j and $E_j^* = \cup_{i\leq j} E_i$. For each $j \leq k$, we consider a partition $\{B_{\underline{\epsilon}} : \underline{\epsilon} \in E_j\}$ of $\mathcal{X}$ such that $B_{\underline{\epsilon}0}$, $B_{\underline{\epsilon}1}$ is a partition of $B_{\underline{\epsilon}}$. If $\underline{\epsilon} \in E_k$, clearly $B_{\underline{\epsilon}}$ is a singleton.

Definition 3.3.1. A prior Π on $M(\mathcal{X})$ is said to be a *Polya tree prior* with parameter $\underline{\alpha} = \{\alpha_{\underline{\epsilon}} : \underline{\epsilon} \in E_k^*\}$ if $\alpha_{\underline{\epsilon}} \geq 0$ and

(i) $\{P(B_{\underline{\epsilon}0}|B_{\underline{\epsilon}}) : \underline{\epsilon} \in E_{k-1}^*\}$ are all independent and

(ii) $P(B_{\underline{\epsilon}0}|B_{\underline{\epsilon}})$ is a beta$(\alpha_{\underline{\epsilon}0}, \alpha_{\underline{\epsilon}1})$ random variable

when $\underline{\epsilon} = \emptyset$ take, $P(B_{\underline{\epsilon}0}|B_{\underline{\epsilon}})$ to be $P(B_{\underline{\epsilon}0})$. (i) and (ii) uniquely determine a Π, for if $x = B_{\epsilon_1,\epsilon_2,\ldots,\epsilon_k}$, then

$$P(x) = P(B_{\epsilon_1\epsilon_2\ldots\epsilon_k}) = \prod_{i:\epsilon_i=0} P\left(B_{\epsilon_1\epsilon_2\ldots\epsilon_{i-1}0}|B_{\epsilon_1\epsilon_2\ldots\epsilon_{i-1}}\right) \prod_{i:\epsilon_i=1} \left(P\left(B_{\epsilon_1\epsilon_2\ldots\epsilon_{i-1}1}|B_{\epsilon_1\epsilon_2,\ldots\epsilon_{i-1}}\right)\right) \tag{3.18}$$

Because

$$P\left(B_{\epsilon_1,\epsilon_2,\ldots,\epsilon_{i-1}1}|B_{\epsilon_1,\epsilon_2,\ldots,\epsilon_{i-1}}\right) = 1 - P\left(B_{\epsilon_1,\epsilon_2,\ldots,\epsilon_{i-1}0}|B_{\epsilon_1,\epsilon_2,\ldots,\epsilon_{i-1}}\right)$$

$\left\{P(B_{\underline{\epsilon}0}|B_{\underline{\epsilon}}) : \underline{\epsilon} \in E^*_{k-1}\right\}$ determines the distribution of $P(x)$.

Suppose Π is a Polya tree prior on $M(\mathcal{X})$ and given P, X is distributed as P. For any x let $\epsilon_1(x) = 0$ if $x \in B_0$ and 1 otherwise, and let $\epsilon_k(x) = 0$ if $x \in B_{\epsilon_1(x)\ldots\epsilon_{k-1}(x)0}$ and 1 otherwise. The joint density of P and X is given, up to a constant, by

$$\prod_{\underline{\epsilon}\in E^*_k} \left[P(B_{\underline{\epsilon}0}|B_{\underline{\epsilon}})\right]^{\alpha_{\underline{\epsilon}0}-1} \left[1 - P(B_{\underline{\epsilon}0}|B_{\underline{\epsilon}})\right]^{\alpha_{\underline{\epsilon}1}-1} \prod_{i:\epsilon_i(x)=0} P(B_{\underline{\epsilon}0}|B_{\underline{\epsilon}}) \prod_{i:\epsilon_i(x)=1} \left(1 - P(B_{\underline{\epsilon}0}|B_{\underline{\epsilon}})\right)$$
$$= \prod_{\underline{\epsilon}\in E^*_k} \left[P(B_{\underline{\epsilon}0}|B_{\underline{\epsilon}})\right]^{\alpha'_{\underline{\epsilon}0}-1} \left[1 - P(B_{\underline{\epsilon}0}|B_{\underline{\epsilon}})\right]^{\alpha'_{\underline{\epsilon}1}-1}$$

where

$$\alpha'_{\underline{\epsilon}} = \begin{cases} 1 + \alpha_{\underline{\epsilon}} & \text{if } x \in B_{\underline{\epsilon}} \\ \alpha_{\underline{\epsilon}} & \text{otherwise} \end{cases}$$

We summarize this discussion as the following theorem

Theorem 3.3.1. *If the prior on $M(\mathcal{X})$ is $PT(\underline{\alpha})$ where $\underline{\alpha} = \{\alpha_{\underline{\epsilon}} : \underline{\epsilon} \in E^*_k\}$ and if given P, $X_1, X_2, \ldots, X_n$ are i.i.d. P, then*

(i) the posterior distribution on $M(\mathcal{X})$ given $X_1, X_2, \ldots, X_n$ is a Polya tree with parameters $\{\alpha_{\underline{\epsilon},X_1,X_2,\ldots,X_n} : \underline{\epsilon} \in E^*_k\}$ where

$$\alpha_{\underline{\epsilon},X_1,X_2,\ldots,X_n} = \alpha_{\underline{\epsilon}} + \sum_1^n I_{B_{\underline{\epsilon}}}(X_i)$$

(ii) the marginal distribution of X_1 is given by

$$Pr\{X = x\} = \prod_1^k \frac{\alpha_{\epsilon_1(x)\epsilon_2(x)\ldots\epsilon_i(x)}}{\alpha_{\epsilon_1(x)\epsilon_2(x)\ldots\epsilon_{i-1}(x)0} + \alpha_{\epsilon_1(x)\epsilon_2(x)\ldots\epsilon_{i-1}(x)1}}$$

and the predictive distribution of X_{n+1} *given* $X_1, X_2, \ldots, X_n$ *is of the same form with* $\alpha_{\underline{\epsilon}}$ *replaced by* $\alpha_{\underline{\epsilon}, X_1, X_2, \ldots, X_n}$.

To prove (ii), $\Pr(X = x) = \int P(x) d\Pi(P)$ and is the integral of the terms in (3.18). The components in the product are independent beta random variables and a direct computation yields the result.

The distribution of $X_1, X_2, \ldots, X_n$ defined via Theorem 3.3.1 can be thought of as a Polya urn scheme, though not as easy to describe as that for a Dirichlet. This is done in Mauldin et al. (92) and we refer the interested reader to their paper.

Remark 3.3.1. The assumption that $\mathcal{X}$ contains 2^k elements and that partitions are into two halves is not really necessary. All we need is $\underline{\underline{\mathcal{T}}}_i = \{B_{\underline{\epsilon}} : \underline{\epsilon} \in E_i\}$ for $i = 1, 2, \ldots, k$ be a nested sequence of partitions. The equal halves can be relaxed by allowing empty sets to be in the partition and setting the corresponding parameter to be 0.

Remark 3.3.2. The form of the posterior distribution shows that X and the vector $\{P(B_{\underline{\epsilon}0} | B_{\underline{\epsilon}}) : \underline{\epsilon} \in E_i^*\}$ are conditionally independent given $\{I_{B_{\underline{\epsilon}}} : \underline{\epsilon} \in E_i\}$.

3.3.2 $\mathcal{X} = \mathbb{R}$

Motivated by the $\mathcal{X}$ is finite case, we define a Polya tree prior on $M(\mathbb{R})$ as follows: Recall that E_j is the set of all sequences of 0s and 1s of length j and $E^* = \cup_j E_j$ is all sequences of 0s and 1s of finite length. Also E is the set of all infinite sequences of 0s and 1s.

Definition 3.3.2. For each n, let $\underline{\mathcal{T}}_n = \{B_{\underline{\epsilon}} : \underline{\epsilon} \in E_n\}$ be a partition of $\mathbb{R}$ such that for all $\underline{\epsilon}$ in E^*, $B_{\underline{\epsilon}0}$, $B_{\underline{\epsilon}1}$ is a partition of $B_{\underline{\epsilon}}$.

Let $\underline{\alpha} = \{\alpha_{\underline{\epsilon}} : \underline{\epsilon} \in E^*\}$ be a set of nonnegative real numbers.

A prior Π on $M(\mathbb{R})$ is said to be a Polya tree (with respect to the partition $\underline{\mathcal{T}} = \{\mathcal{T}_n\}_{n \geq 1}$) with parameter $\underline{\alpha}$, denoted by $PT(\underline{\alpha})$, if under Π

1. $\{P(B_{\underline{\epsilon}0} | B_{\underline{\epsilon}}) : \underline{\epsilon} \in E^*\}$ are a set of independent random variables
2. for all $\underline{\epsilon} \in E^*$, $P(B_{\underline{\epsilon}0} | B_{\underline{\epsilon}})$ is beta$(\alpha_{\underline{\epsilon}0}, \alpha_{\underline{\epsilon}1})$.

The first question, of course, is do such priors exist? We have already discussed this in Chapter 2.

Theorem 3.3.2. *A Polya tree with parameter* $\underline{\alpha} = \{\alpha_{\underline{\epsilon}} : \underline{\epsilon} \in E^*\}$ *exists if for all* $\underline{\epsilon} \in E^*$

$$\left(\frac{\alpha_{\underline{\epsilon}0}}{\alpha_{\underline{\epsilon}0} + \alpha_{\underline{\epsilon}1}}\right) \left(\frac{\alpha_{\underline{\epsilon}00}}{\alpha_{\underline{\epsilon}00} + \alpha_{\underline{\epsilon}01}}\right) \left(\frac{\alpha_{\underline{\epsilon}000}}{\alpha_{\underline{\epsilon}000} + \alpha_{\underline{\epsilon}001}}\right) \cdots = 0 \tag{3.19}$$

and

$$\left(\frac{\alpha_{10}}{\alpha_{10}+\alpha_{11}}\right)\left(\frac{\alpha_{110}}{\alpha_{110}+\alpha_{111}}\right)\ldots = 0$$

Proof. This is an immediate consequence of Theorem 2.3.5 . We noted there that if we set $Y = P(B_0), Y_{\underline{\epsilon}} = P(B_{\underline{\epsilon}0}|B_{\underline{\epsilon}})$ then $\{Y_{\underline{\epsilon}} : \underline{\epsilon} \in E^*\}$ induces a measure on $M(\mathbb{R})$- if it satisfies the continuity condition

$$Y_{\underline{\epsilon}} Y_{\underline{\epsilon}0} Y_{\underline{\epsilon}00} \ldots = 0 \text{ almost surely}$$

Because $\prod_1^n Y_{\underline{\epsilon}0\ldots0}$ is decreasing in n and bounded by 0 and 1, this happens iff $E(\prod_1^n Y_{\underline{\epsilon}0\ldots0}) \to 0$. The $Y_{\underline{\epsilon}}$ are independent beta random variables and the condition translates precisely to (3.19). □

Marginal distribution of X Let P$\sim$ $PT(\underline{\alpha})$ and given P, X be distributed as P and let m be the marginal distribution of X. Because the finite union of sets in $\cup_n \underline{\mathcal{T}}_n$ is an algebra it is enough to calculate $m(B_{\underline{\epsilon}})$ for all $\underline{\epsilon}$ in E^*.

If $\underline{\epsilon} = \epsilon_1\epsilon_2\ldots\epsilon_k$,

$$\begin{aligned} m(X \in B_{\epsilon_1\epsilon_2\ldots,\epsilon_k}) &= E\left[\prod_{i\le k-1} P\left(B_{\epsilon_1\epsilon_2\ldots\epsilon_{i-1}\epsilon_i}|B_{\epsilon_1\epsilon_2\ldots\epsilon_{i-1}}\right)\right] \\ &= \prod_{\{i:\epsilon_i=0, i\le k-1\}} Y_{\epsilon_1\epsilon_2\ldots\epsilon_{i-1}} \prod_{\{i:\epsilon_i=1, i\le k-1\}} (1 - Y_{\epsilon_1\epsilon_2\ldots\epsilon_{i-1}}) \end{aligned} \tag{3.20}$$

The factors inside the expectation are independent beta random variables, and hence we have

$$= \prod_1^k \frac{\alpha_{\epsilon_1\epsilon_2\ldots\epsilon_i}}{\alpha_{\epsilon_1\epsilon_2\ldots\epsilon_i 0} + \alpha_{\epsilon_1\epsilon_2\ldots\epsilon_i 1}}$$

Theorem 3.3.3. *Suppose that X is distributed as P and P itself has a $PT(\underline{\alpha})$ prior. Then the posterior distribution of P given X is $PT(\underline{\alpha}_X)$, where $\underline{\alpha}_X = \underline{\alpha} + I_{B_{\underline{\epsilon}}}(X)$.*

Based on the corresponding result for the finite case, it is reasonable to expect the posterior to be $PT(\underline{\alpha}_X)$. In fact the posterior distribution of $\{P(B_\epsilon) : \epsilon \in E_n\}$ given X is same as the posterior of $\{P(B_\epsilon) : \epsilon \in E_n\}$ given $\{I_{B_\epsilon}(X) : \epsilon \in E_n\}$. The calculation in the finite case done in the last section shows that this posterior distribution is a Polya tree with parameters $\{\alpha_{\epsilon,X} = \alpha_\epsilon + I_{B_{\underline{\epsilon}}}(X) : \epsilon \in \cup_1^n E_i\}$. The proof is completed by recognizing that posterior distributions of $\{P(B_\epsilon) : \epsilon \in E_n : n \geq 1\}$ determine the posterior distribution $\Pi(\cdot|X)$.

Repeatedly applying the last theorem we get the following.

Theorem 3.3.4. *If $PT(\underline{\alpha})$ is the prior on $M(\mathbb{R})$ and given P; if $X_1, X_2, \ldots, X_n$ are i.i.d. P, then the posterior distribution of P given $X_1, X_2, \ldots, X_n$ is a polya tree with parameter $\underline{\alpha}_{X_1,X_2,\ldots,X_n}$ where*

$$\alpha_{\underline{\epsilon},X_1,X_2,\ldots,X_n} = \alpha_{\underline{\epsilon}} + \sum_1^n I_{B_{\underline{\epsilon}}}(X_i)$$

Predictive distribution and Bayes estimate

It is immediate from the last two properties that if $X_1, X_2, \ldots, X_n$ are i.i.d. Pgiven P, and $P$is has $PT(\underline{\alpha})$ prior, then the predictive distribution of X_{n+1} given $X_1, X_2, \ldots, X_n$ is

$$P\{X_{n+1} \in B_{\epsilon_1\epsilon_2\ldots\epsilon_k}\}$$
$$\frac{\alpha_{\epsilon_1} + \sum_1^n I_{B_{\epsilon_1}}(X_i)}{\alpha_0 + \alpha_1 + n} \frac{\alpha_{\epsilon_1\epsilon_2} + \sum_1^n I_{B_{\epsilon_1\epsilon_2}}(X_i)}{\alpha_{\epsilon_1 0} + \alpha_{\epsilon_1 1} + n_{\epsilon_1}} \cdots \frac{\alpha_{\epsilon_1\ldots\epsilon_k} + \sum_1^n I_{B_{\epsilon_1\ldots\epsilon_k}}(X_i)}{\alpha_{\epsilon_1\ldots\epsilon_{k-1}0} + \alpha_{\epsilon_1\ldots\epsilon_{k-1}1} + n_{\epsilon_1\ldots\epsilon_{k-1}}}$$

where n_ϵ is the number of X_is falling in B_ϵ.

In view of the calculations done so far, $\hat{P} = E(P|X_1, X_2, \ldots, X_n)$ is the measure satisfying

$$\hat{P}(B_{\epsilon_1\epsilon_2\ldots\epsilon_k}) = \prod_1^k \frac{\alpha_{\epsilon_1\ldots\epsilon_j} + \sum_1^n I_{B_{\epsilon_1\ldots\epsilon_j}}(X_i)}{\alpha_{\epsilon_1\ldots\epsilon_j 0} + \alpha_{\epsilon_1\ldots\epsilon_j 1} + n_{\epsilon_1\ldots\epsilon_{j-1}}}$$

Like the Dirichlet, here also the posterior is consistent. Formally, we have the following theorem.

Theorem 3.3.5. *Let P be distributed as $PT(\underline{\alpha})$ and given P, let $X_1, X_2, \ldots, X_n$ be i.i.d. P. Then for any P_0, as $n \to \infty$, the posterior*

$$PT(\underline{\alpha}_{X_1,X_2,\ldots,X_n}) \to \delta_{P_0} \quad \textit{weakly a.s } P_0$$

.

The result would follow as a particular case of a more general theorem proved later for tail free priors. However one can give proof along the same lines as that for the Dirichlet process and follows from the following lemmas.

Lemma 3.3.1. *Let $\bar{\alpha}_m = E_{\underline{\alpha}_m(P)}$, where $E_{\bar{\alpha}_m}$ is the expectation taken under $PT(\underline{\alpha}_m)$. If $\{\bar{\alpha}_m : m \in M\}$ is tight, then so is $\{PT(\underline{\alpha}_m) : m \in M\}$.*

Proof. Easily follows from corollary to Theorem 2.5.1 □

Lemma 3.3.2. *If $\bar{\alpha}_m \to P_0$ and if for all $\underline{\epsilon} \in E^*$,*

$$\mathcal{L}\left(P(B_{\underline{\epsilon}})|PT(\underline{\alpha}_m)\right) \to \mathcal{L}\left(P(B_{\underline{\epsilon}})|\delta_{P_0}\ \right)$$

then $PT(\underline{\alpha}_m)$ converges weakly to δ_{P_0} .

Proof. The tightness of $\underline{\alpha}_m$ ensures that $PT(\underline{\alpha}_m)$ has a limit point. This limit point can be identified as δ_{P_0} using calculations similar to Theorem 3.2.6. □

To prove the theorem, let $\Omega = \{\omega : 1/n\sum_1^n I_{B_{\underline{\epsilon}}}(X_i) \to P_0(B_{\underline{\epsilon}}) \text{ for all } \underline{\epsilon} \in E^*\}$. $P_0(\Omega) = 1$, and further for each $\omega \in \Omega$ it is easily verified that $\bar{\alpha}_m = \alpha_{\underline{\epsilon},X_1,X_2,\ldots,X_n(\omega)}$ satisfies the assumptions of the Lemma 3.3.2.

Support of PT($\underline{\alpha}$)

Our next theorem is on the topological support of $PT(\underline{\alpha})$. Recall that the support is the smallest closed set of $PT(\underline{\alpha})$ measure 1. Here we assume that $\{a_\epsilon : \epsilon \in E^*\}$ is a dense set of numbers and induce a nested sequence of partitions.

Theorem 3.3.6. *$PT(\underline{\alpha})$ has all of $M(\mathbb{R})$ as support iff $\alpha_{\underline{\epsilon}} > 0$ for all $\underline{\epsilon} \in E^*$.*

Proof. The proof follows along the same lines as for the Dirichlet (see Theorem 3.2.4). □

Mauldin et al. [133] show that, unlike the Dirichlet, we can find $\underline{\alpha}$ which will ensure that $PT(\underline{\alpha})$ sits on the space of continuous measures. Because Polya tree priors are tail free, we can use Theorem 2.4.3 to show that by suitably choosing the partitions and parameters the Polya tree can be made to sit on, not just continuous distributions but even absolutely continuous distributions. The theorem is an application of Theorem 2.4.3 to Polya tree processes. The proof is just a verification of the conditions of Theorem 2.4.3.

Theorem 3.3.7. *Let λ be a continuous probability measure on $\mathbb{R}$ with distribution function λ. Define $B_{\epsilon_1\epsilon_2\ldots\epsilon_i} = \lambda^{-1}(\sum \frac{\epsilon_i}{2^i}, \sum \frac{\epsilon_i}{2^i} + \frac{1}{2^i})$. If $\alpha_{\epsilon_1\epsilon_2\ldots\epsilon_i} = a_i$, and $\sum a_i^{-1} < \infty$ then $PT(\underline{\alpha})(L(\lambda)) = 1$.*

In particular when $\alpha_{\epsilon_1\epsilon_2\ldots\epsilon_i} = i^2$, the polya tree gives mass 1 to probabilities that are absolutely continuous with respect to λ.

A few concluding remarks about Polya tree priors: The Polya tree prior depends on the underlying partition $\underline{\mathcal{T}} = \{\mathcal{T}_n\}_{n\geq 1}$ and is tail free with respect to this partition. In fact a prior which is tail-free with respect to every sequence of partitions is, except for trivial cases, a Dirichlet process [Doksum [48]].

We have seen that Polya tree priors, unlike the Dirichlet, can be made to sit on densities. One unpleasant feature of this construction is that absolute continuity of P is ensured by controlling the variability of P around the chosen absolutely continuous λ. We have seen that for the Dirichlet the prior and posterior are mutually singular. Dragichi and Ramamoorthi [56] have shown that if the parameters are as in the Theorem 3.3.7, then the posterior given distinct observations is absolutely continuous with respect to the prior.

Lavine suggests that, if the prior expectation is F, then the partitions of the form $F^{-1}(\sum \epsilon_i/2^i, \sum \epsilon_i/2^i + 1/2^i)$ would be appropriate. For then the ratios

$$\alpha_{\epsilon_1\epsilon_2\ldots\epsilon_{i-1}0}/(\alpha_{\epsilon_1\epsilon_2\ldots\epsilon_{i-1}0} + \alpha_{\epsilon_1\epsilon_2\ldots\epsilon_{i-1}1}) = 1/2$$

and this would ensure that the marginal of X is F, which may then be treated as a "prior guess" of the "mean" of the random P. As to the magnitude of the α_ϵs (as distinct from their ratios), their role is somewhat similar to that of $\alpha(\mathbb{R})$ for the Dirichlet, except that the availability of more parameters introduces more flexibility. It is expected that for moderate k a choice of the magnitude would be on the basis of prior belief and for higher k, a conventional choice would be made. A conventional choice might be to ensure that the prior sits on densities. For example, one may take $\alpha_{\epsilon_1,\ldots,\epsilon_k} = 1/k^2$. Lavine [118] has expressed well what the main considerations are; we refer the reader to his paper.

4 Consistency Theorems

4.1 Introduction

We briefly discussed consistency of the posterior in Chapters 1, 2 and 3. To recall, our setup consists of:
a (unknown) parameter θ that lies in a parameter space Θ;
a prior distribution Π for θ, equivalently, a probability measure on Θ; and
$X_1, X_2, \ldots, X_n$, which are given θ, i.i.d. with common distribution P_θ.

Our interest centers on the consistency of the posterior distribution, and as discussed in Chapter 1, this is a requirement that if indeed θ_0 is the "true " distribution of $X_1, X_2, \ldots, X_n$ then the posterior should converge to δ_{θ_0} almost surely. In other words, as $n \to \infty$, the posterior probability of every neighborhood of θ_0 should go to 1 with P_{θ_0} probability 1.

We noted that posterior consistency can be viewed as

- a sort of frequentist validation of the Bayesian method;
- merging of posteriors arising from two different priors; and
- as an expression of "data eventually swamps the prior".

In Chapter 1 we saw that when Θ is a subset of a finite-dimensional Euclidean space and if $\theta \mapsto P_\theta$ is smooth, then for smooth priors the posterior is consistent in the support of the prior. In Chapter 1 we also saw an example showing that inconsistency cannot be ruled even when $\Theta = \mathbb{R}$.

The example in Chapter 1 may be dismissed as a technical pathology, but in the nonparametric case inconsistency can scarcely be called pathological. This has led some to question the role of consistency in Bayesian inference. The argument is that it is well known that the prior and the posterior given by Bayes theorem are imperatives arising out of axioms of rational behavior–and since we are already rational why worry about one more criteria? In other words inconsistency does not warrant the abandonment of a prior. We would argue that in the nonparametric context typically one would have many priors that would be consistent with one's prior beliefs, and it does make sense to choose among these priors that are consistent at a large number of parameter values, among which we expect the true parameter to lie.

In the nonparametric context Θ is $M(\mathbb{R})$ or large subset of it. $M(\mathbb{R})$ has various kinds of convergence, namely, total variation, setwise , weak, etc. Each of these leads to a corresponding notion of consistency. The issue of consistency has been approached from different point of view by [143]. We begin with a formal definition of these.

4.2 Preliminaries

Definition 4.2.1. $\{\Pi(\cdot|X_1, X_2, \ldots, X_n)\}$ is said to be *strongly or L_1-consistent* at P_0 if there is a $\Omega_0 \subset \Omega$ such that $P_0^\infty(\Omega_0) = 1$ and for $\omega \in \Omega_0$

$$\Pi(U|X_1, X_2, \ldots, X_n) \to 1$$

for all total variation neighborhoods of P_0.

Definition 4.2.2. $\{\Pi(\cdot|X_1, X_2, \ldots, X_n)\}$ is said to be *weakly consistent* at P_0 if there is a $\Omega_0 \subset \Omega$ such that $P_0^\infty(\Omega_0) = 1$ and for $\omega \in \Omega_0$

$$\Pi(U|X_1, X_2, \ldots, X_n) \to 1$$

for all weak neighborhoods of P_0.

Before we proceed to the study of consistency, we note that Bayes estimates inherit the convergence property of the posterior. Recall that we denote $X_1, X_2, \ldots, X_n$ by $\mathbf{X_n}$.

Proposition 4.2.1. *Define the Bayes estimate $\hat{P}_n(\cdot|\mathbf{X_n})$ to be the probability measure $\hat{P}_n(A|\mathbf{X_n}) = \int P(A)\ \Pi(dP|X_1, X_2, \ldots, X_n) = E(P(A)|\mathbf{X_n})$. Then*

1. *if* $\{\Pi(\cdot|X_1, X_2, \ldots, X_n)\}$ *is strongly consistent at* P_0*, then* $||\hat{P}_n - P_0|| \to 0$*, almost surely* P_0*.*

2. *If* $\{\Pi(\cdot|X_1, X_2, \ldots, X_n)\}$ *is weakly consistent at* P_0*, then* $\hat{P}_n \to P_0$ *weakly, almost surely* P_0*.*

Proof. By Jensen's inequality

$$||\hat{P}_n - P_0|| \leq \int ||P - P_0|| \; \Pi(dP|\mathbf{X_n})$$
$$= \int_U ||P - P_0|| \; \Pi(dP|\mathbf{X_n}) + \int_{U^c} ||P - P_0|| \; \Pi(dP|\mathbf{X_n})$$

and if $U = \{P : ||P - P_0|| < \epsilon\}$ then

$$\leq \epsilon \Pi(U|\mathbf{X_n}) + \Pi(U^c|\mathbf{X_n}) \leq \epsilon + o(1)$$

as $n \to \infty$.

A similar argument works for assertion (ii) by considering $|\int f \; d\hat{P}_n - \int f \; dP_0|$, f bounded continuous.

□

It is worth pointing out that the conventional Bayes estimate considered earlier is a Bayes estimate only for the squared error loss for $P(A)$. The Bayes estimate $\tilde{P}(A)$ for, say, the absolute deviation loss, will be the posterior median. Unfortunately, the $\tilde{P}(\cdot)$ so obtained will not be a probability measure.

As far as the prior is on $M(\mathbb{R})$, weak consistency is intimately related to the consistency of the Bayes estimates of $F(t)$.

Theorem 4.2.1. *Suppose* Π *is a prior on* $\mathcal{F}$*, the space of distribution functions on* $\mathbb{R}$*, and* $X_1, X_2, \ldots, X_n$ *be given* F*; i.i.d.* F*. Then the posterior is weakly consistent at* F_0*, iff there is a dense subset* Q *of* $\mathbb{R}$ *such that for* t *in* Q

(i) $\lim_{n\to\infty} E(F(t)|\mathbf{X_n}) = F_0(t)$*; and*

(ii) $\lim_{n\to\infty} V(F(t)|\mathbf{X_n}) = 0$*.*

Proof. If (i) and (ii) hold, it follows from a simple use of Chebychev's inequality that for every t in Q,

$$\Pi((F(t) - F_0(t)| < \delta)|\mathbf{X_n}) \to 1 \text{ a.s } F_0$$

and hence it follows that

$$\Pi((F(t_i) - F_0(t_i)| < \delta|\mathbf{X_n}) \text{ for } 1 \le i \le k) \to 1 \text{ a.s } F_0$$

By Proposition 2.5.2, any weak neighborhood U of F_0 contains a set of the above form for a suitable δ. Hence $\Pi(U|\mathbf{X_n}) \to 1$ a.e. F_0.

On the other hand, if the posterior is weakly consistent, then it is easy to see that (i) and (ii) hold for any t that is a continuity point of F_0. □

Since strong consistency is desirable, it is natural to seek a prior Π for which the posterior would be strongly consistent at all P in $M(\mathbb{R})$. Such a prior can be thought of as more diffuse than priors that do not have this property. However such a prior does not exist. If it did, the corresponding Bayes estimates, by the last Proposition 4.2.1, would give a sequence of estimates of P that is consistent in the total variation metric and such estimates do not exist [41]. On the other hand the Dirichlet priors considered earlier provide an example of a prior that is weakly consistent at all P. We note that Doob's theorem is applicable also to strong consistency.

If U is a neighborhood of P_0 with prior probability 0, then any reasonable version of the posterior will assign mass 0 to U and consequently it is unreasonable to expect consistency at such a P_0. Thus it is appropriate to confine the search for points of consistency to the (topological) support of the prior.

4.3 Finite and Tail free case

When $\mathcal{X}$ is a finite set, $M(\mathcal{X})$ is a subset of the Euclidean space, and all the topologies coincide on $M(\mathcal{X})$, and we have the following pleasing theorem. This theorem can also be proved from Theorem 1.3.4. Here is a direct proof that in a way is related to the Schwartz theorem discussed later in this chapter.

Theorem 4.3.1. *Let Π be a prior on $M(\mathcal{X})$, where $\mathcal{X} = \{1, 2, \ldots, k\}$. Then the posterior is consistent at all points in the support of* Π.

Proof. Let

$$V = \{P : \|P - P_0\| < \delta\}$$

be a neighborhood of P_0.

$$\Pi(V^c|\mathbf{X_n}) = \frac{\int_{V^c} e^{-n\sum_1^k (n_i/n)\log(P_0(i)/P(i))} d\Pi(p)}{\int_{\mathcal{X}} e^{-n\sum_1^k (n_i/n)\log(P_0(i)/P(i))} d\Pi(p)}$$

where n_i is the number of $\mathbf{X_n}$ equal to i. Writing it as

$$\frac{I_1(\mathbf{X_n})}{I_2(\mathbf{X_n})}$$

we will show that

(i) for all $\beta > 0$, $\liminf_{n\to\infty} e^{n\beta} I_2(\mathbf{X_n}) = \infty$ a.s P_0; and

(ii) there exists a $\beta_0 > 0$ such that $e^{n\beta_0} I_1(\mathbf{X_n}) \to 0$ a.s P_0.

condition (i) follows from the strong law of large numbers.

As for (ii)

$$\sum_1^k \frac{n_i}{n} \log \frac{P_0(i)}{P(i)} = \sum_1^k \frac{n_i}{n} \log \frac{n_i/n}{P(i)} + \sum_1^k \frac{n_i}{n} \log \frac{P_0(i)}{n_i/n}$$

which gives

$$\lim_{n\to\infty} \sum_1^k \frac{n_i}{n} \log \frac{P_0(i)}{P(i)} = \lim_{n\to\infty} \sum_1^k \frac{n_i}{n} \log \frac{n_i/n}{P(i)}$$

If F_n stands for the empirical distribution

$$\sum_i^k \frac{n_i}{n} \log \frac{n_i/n}{P(i)} = K(F_n, P)$$

and by Proposition 1.2.2

$$K(F_n, P) \geq \frac{||F_n - P||^2}{4} = \frac{(||P - P_0|| - ||F_n - P_0||)^2}{4}$$

If $P \in V^c$ and n is large so that $||F_n - P_0|| < \delta/2$, we have

$$K(F_n, P) \geq \frac{(\delta - \delta/2)^2}{4} = \delta_0$$

In other words,

$$\inf_{P\in V^c} K(F_n, P) > \delta_0 \text{ a.s } P_0$$

Consequently

$$\lim_{n\to\infty} e^{n\beta} I_1(\mathbf{X_n}) \leq \lim_{n\to\infty} e^{n\beta} \int_{V^c} e^{-nK(F_n,P)} d\Pi(p) \leq e^{n(\beta-\delta_0)}$$

which goes to 0 if $\beta < \delta_0$. The proof of the theorem is easily completed by taking $\beta_0 < \delta_0$. $\square$

When $\mathcal{X}$ is infinite, even weak consistency can fail to occur in the weak support of Π. Freedman [69] provided dramatic examples when $\mathcal{X} = \{1, 2, 3, \ldots, \}$. Another elegant example, due to Ferguson, is described in [65].

Theorem 4.3.2. *For $k = 1, 2, \ldots,$ let $\mathcal{T}_k = \{B_{\underline{\epsilon}} : \underline{\epsilon} \in E_k\}$ be a partition of $\mathbb{R}$ into intervals. Further assume that $\{\mathcal{T}_k : k \geq 1\}$ are nested. If Π is a prior on $M(\mathbb{R})$, tail free with respect to $\{\mathcal{T}_k : k \geq 1\}$ and with support all of $M(\mathbb{R})$ then (there exits a version of) the posterior which is weakly consistent at every P_0.*

Proof. By Theorem 2.5.2, enough to show that for each n the posterior distribution of $\{P(B_\epsilon) : \epsilon \in E_n\}$ given $\mathbf{X_n}$ converges a.e. P_0 to $\{P_0(B_\epsilon) : \epsilon \in E_n\}$. Proposition 2.3.6 ensures that the posterior distribution of $\{P(B_\epsilon) : \epsilon \in E_n\}$ given $X_1, X_2, \ldots, X_n$ is the same as that given $\{n_\epsilon : \epsilon \in E_n\}$, where n_ϵ is the number of $X_1, X_2, \ldots, X_n$ in B_ϵ. A little reflection will show that we are now in the same situation as Theorem 4.3.1. □

4.4 Posterior Consistency on Densities

4.4.1 Schwartz Theorem

In the last section we looked at priors on $M(\mathbb{R})$. An important special case is when the prior is concentrated on L_μ, the space of densities with respect to a σ-finite measure μ on $\mathbb{R}$. This case is important because of its practical relevance. In addition this is a situation when one has a natural posterior given by the Bayes theorem. Our (conventional) Bayes estimate is the expectation of f with respect to the posterior.

We begin the discussion with a theorem of Schwartz [145]. Our later applications will show that Schwartz's theorem is a powerful tool in establishing posterior consistency. Barron [8] provides insight into the role of Schwartz's theorem in consistency.

Our setup, then, is $L_\mu = \left\{f : f \text{ is measurable}, f \geq 0, \int f \, d\mu = 1\right\}$. We tacitly identify the μ equivalence classes in L_μ and equip L_μ with the total variation or L_1-metric $||f - g|| = \int |f - g| \, d\mu$. Every f in L_μ corresponds to a probability measure P_f, and it is easy to see that the Borel σ-algebra generated by the L_1-metric and the σ-algebra $\mathcal{B}_M \cap L_\mu$ are the same.

Let Π be a prior on L_μ. Recall that $K(f, g)$ stands for the Kullback-Leibler divergence $\int f \log(f/g) \, d\mu$. $K_\epsilon(f)$ will stand for the neighborhood $\{g : K(f, g) < \epsilon\}$.

Definition 4.4.1. Let f_0 be in L_μ. f_0 is said to be in the *K-L support* of the prior Π, if for all $\epsilon > 0$, $\Pi(K_\epsilon(f_0)) > 0$.

As before, $X_1, X_2, \ldots$ are given f, i.i.d. P_f. P_f^n will stand for the joint distribution of $X_1, X_2, \ldots, X_n$ and P_f^∞ for the joint distribution of the entire sequence $X_1, X_2, \ldots$. We will, when needed, view P_f^∞ as a measure on $\Omega = \mathbb{R}^\infty$.

Let U be a set containing f_0. In order for the posterior probability of U given $\mathbf{X_n}$ to go to 1, it is necessary that f_0 and U^c can be separated. This idea of separation is conveniently formalized through the existence of appropriate tests for testing $H_0 : f = f_0$ versus $H_1 : f \in U^c$. Recall that a test function is a nonnegative measurable function bounded by 1.

Let $\{\phi_n(\mathbf{X_n}) : n \geq 1\}$ be a sequence of test functions.

Definition 4.4.2. $\{\phi_n(\mathbf{X_n}) : n \geq 1\}$ is *uniformly consistent* for testing $H_0 : f = f_0$ versus $H_1 : f \in U^c$, if as $n \to \infty$,

$$E_{f_0}(\phi_n(\mathbf{X_n})) \to 0$$

$$\inf_{f \in U^c} E_f(\phi_n(\mathbf{X_n})) \to 1$$

Definition 4.4.3. A test $\phi(\mathbf{X_n})$ is *strictly unbiased* for $H_0 : f = f_0$ versus $H_1 : f \in U^c$, if

$$E_{f_0}(\phi_n(\mathbf{X_n})) < \inf_{f \in U^c} E_f(\phi_n(\mathbf{X_n}))$$

Definition 4.4.4. $\{\phi_n(\mathbf{X_n}) : n \geq 1\}$ is *uniformly exponentially consistent* for testing $H_0 : f = f_0$ versus $H_1 : f \in U^c$, if there exist C, β positive such that for all n,

$$E_{f_0}(\phi_n(\mathbf{X_n})) \leq Ce^{-n\beta}$$

and

$$\inf_{f \in U^c} E_f(\phi_n(\mathbf{X_n})) \geq 1 - Ce^{-n\beta}$$

The next proposition relates these three definitions. The proposition is itself interesting, and the ideas involved in the proof surface again in later arguments.

Proposition 4.4.1. *The following are equivalent*

(i) There exists a uniformly consistent sequence of tests for testing $H_0 : f = f_0$ versus $H_1 : f \in U^c$.

(ii) for some $n \geq 1$, there exists a strictly unbiased test $\phi(\mathbf{X_n})$ for $H_0 : f = f_0$ versus $H_1 : f \in U^c$.

(iii) There exists a uniformly exponentially consistent sequence of test functions for testing $H_0 : f = f_0$ versus $H_1 : f \in U^c$.

Proof. Clearly, (i) implies (ii) and (iii) implies(i). So all that needs to be established is that (ii) implies (iii).

Consider first the simple case when $m = 1$, i.e., there exists $\phi(X)$ such that $E_{f_0}\phi = \alpha < \inf_{f\in U^c} E_f\phi = \gamma$.

Let

$$A_k = \left\{(x_1, x_2, \ldots, x_k) : \frac{1}{k}\sum \phi(X_i) > \frac{(\alpha+\gamma)}{2}\right\}$$

Then $P_{f_0}^k(A_k) = P_{f_0}^k\left(\sum \phi(X_i) - kE_{f_0}\phi > k(\gamma-\alpha)/2\right)$, and by Hoeffeding's inequality,

$$P_{f_0}^k\left(\sum \phi(X_i) - kE_{f_0}\phi > \frac{k(\gamma-\alpha)}{2}\right) \leq e^{\frac{-k^2(\gamma-\alpha)^2}{4k}} = e^{\frac{-k(\gamma-\alpha)^2}{4}}$$

On the other hand, for $f \in U^c$

$$P_f^k(A_k) \geq P_f^k\left(\sum \phi(X_i) - kE_f\phi > \frac{k(\alpha-\gamma)}{2}\right)$$

Because $\alpha - \gamma < 0$, by applying Hoeffeding's inequality to $-\phi$, we get

$$P_f(A_k) \geq 1 - e^{\frac{-k(\gamma-\alpha)^2}{4}}$$

and thus $\phi_k = I_{A_k}$ provides the required sequence of tests.

To move on to the general case, suppose

$$E_{f_0}\phi_m(X_1, X_2, \ldots, X_m) = \alpha < \inf_{f\in U^c} E_f\phi_m(X_1, X_2, \ldots, X_m) = \gamma$$

From what we have just seen, if $n = km$, then there is a set A_k with $P_{f_0}^n(A_k) \leq e^{-n(\gamma-\alpha)^2/4m}$. If $km < n \leq (k+1)m$, then

$$\begin{aligned} P_{f_0}^n(A_k) &\leq e^{\frac{-nkm(\gamma-\alpha)^2}{n4m}} \\ &\leq e^{\frac{-nk(\gamma-\alpha)^2}{(k+1)4m}} \leq e^{\frac{-n(\gamma-\alpha)^2}{8m}} \end{aligned}$$

Thus, setting $\beta = (\gamma-\alpha)^2/8m$, we have the exponential bound for $\phi_n = I_{A_k}$ with respect to P_{f_0}. A similar argument yields the corresponding inequality for $\inf_{f\in U^c} P_f(A_k)$. $\square$

Corollary 4.4.1. *Let ν be any probability measure on U^c. When there is a $\phi_n(\mathbf{X_n})$ such that $E_{f_0^n}\phi_n(\mathbf{X_n}) \leq Ce^{-n\beta}$ and $\inf_{f\in U^c} E_f\phi_n(\mathbf{X_n}) \geq 1 - Ce^{-n\beta}$, we have $||f_0 - \int f^n\, \nu(df)|| \geq 2(1-2Ce^{-n\beta})$, where f^n is the n-fold product density $\prod_1^n f(x_i)$.*

Theorem 4.4.1 (Schwartz). *Let Π be a prior on L_μ. If $f_0 \in L_\mu$, and U satisfy*

(i) f_0 is in the K-L support of $\Pi$and

(ii) there exists a uniformly consistent sequence of tests for testing $H_0 : f = f_0$ versus $H_1 : f \in U^c$,

then $\Pi(U|X_1, X_2, \dots, X_n) \to 1$ a.s $P_{f_0}^\infty$

Proof. Because

$$\Pi(U^c|X_1, X_2, \dots, X_n) = \frac{\int_{U^c} \prod_1^n f(X_i)\, \Pi(df)}{\int_{L_\mu} \prod_1^n f(X_i)\, \Pi(df)} = \frac{\int_{U^c} \prod_1^n \frac{f(X_i)}{f_0(X_i}\, \Pi(df)}{\int_{L_\mu} \prod_1^n \frac{f(X_i)}{f_0(X_i)}\, \Pi(df)}$$

it is enough to show that the last term in this expression goes to 0 a.s. $P_{f_0}^\infty$.
We will show in Lemma 4.4.1 that condition (i) implies

$$\text{for every } \beta > 0, \liminf_{n\to\infty} e^{n\beta} \int_{L_\mu} \prod_1^n \frac{f(X_i)}{f_0(X_i)}\, \Pi(df) = \infty \text{ a.e.} P_{f_o}^\infty \tag{4.1}$$

By Proposition 4.4.1, there exist exponentially consistent tests for testing f_0 against U^c. Using these we invoke Lemma 4.4.2, by taking $V_n = U^c$ for all n to show that

$$\text{for some } \beta_0 > 0, \lim_{n\to\infty} e^{n\beta_0} \int_{U^c} \prod_1^n \frac{f(X_i)}{f_0(X_i)}\, \Pi(df) = 0 \text{ a.e.} P_{f_o}^\infty \tag{4.2}$$

By taking $\beta = \beta_0$ in (4.1) it easily follows that the ratio in (4.4.1) goes to 0 a.e. □

Lemma 4.4.1. *If f_0 is in the Kullback-Leibler support of Π then*

$$\textit{for every } \beta > 0, \liminf_{n\to\infty} e^{n\beta} \int_{L_\mu} \prod_1^n \frac{f(X_i)}{f_0(X_i)}\, \Pi(df) = \infty \textit{ a.e.} P_{f_o}^\infty$$

Proof.

$$\int_{L_\mu} \prod_1^n \frac{f(X_i)}{f_0(X_i)}\, \Pi(df) \geq \int_{K_\epsilon(f_0)} e^{-\sum_1^n \log \frac{f_0}{f}(X_i)}$$

For each f in $K_\epsilon(f_0)$, by the law of large numbers

$$\frac{1}{n} \log \frac{f_0}{f}(X_i) \to -K(f_0, f) > -\epsilon \text{ a.s } P_{f_0}^\infty$$

Equivalently, for each f in $K_\epsilon(f_0)$,

$$e^{n\left(2\epsilon-\frac{1}{n}\log\frac{f_0}{f}(X_i)\right)} \to \infty \text{ a.s } P_{f_0}^\infty \tag{4.3}$$

Hence by Fubini, there is a $\Omega_0 \subset \Omega$ of $P_{f_0}^\infty$ measure 1 such that, for each $\omega \in \Omega_0$, for all f in $K_\epsilon(f_0)$, outside a set of Π measure 0, (4.3) holds. Using Fatou's lemma,

$$\begin{aligned}\liminf e^{n2\epsilon}\int_{L_\mu}\prod_1^n\frac{f(X_i)}{f_0(X_i)}\Pi(df) &\geq \liminf e^{n2\epsilon}\int_{K_\epsilon(f_0)}\prod_1^n\frac{f(X_i)}{f_0(X_i)}\Pi(df)\\ &\geq \int_{K_\epsilon(f_0)} e^{n\left(2\epsilon-\frac{1}{n}\sum\log\frac{f_0}{f}(X_i)(\omega)\right)}\,\Pi(df) \to \infty\end{aligned}$$

□

We will state the next lemma in a form slightly stronger than what we need.

Lemma 4.4.2. *If there exist tests $\phi_n(\mathbf{X_n})$ and sets V_n with $\liminf_n \Pi(V_n) > 0$, such that for some $\beta > 0$,*

$$E_{f_0}\phi_n(\mathbf{X_n}) \leq Ce^{-n\beta}$$

and

$$\inf_{f\in V_n} E_f\phi_n(\mathbf{X_n}) \geq 1 - Ce^{-n\beta}$$

then

$$\textit{for some } \beta_0 > 0,\ \lim_{n\to\infty} e^{n\beta_0}\int_{V_n}\prod_1^n\frac{f(X_i)}{f_0(X_i)}\,\Pi(df) = 0 \ \textit{a.e. } P_{f_o}^\infty$$

Proof. Set $q_n(x_1, x_2, \ldots, x_n) = (1/\Pi(V_n))\int_{V_n}\prod_1^n f(X_i)\,\Pi(df)$. Denoting by $A(f_0^n, q_n) = \int\sqrt{\prod f_0(x_i)}\sqrt{q_n(x_i)}\,d\mu$, by Corollaries 4.4.1 and 1.2.1 , there is $0 < r < 1$ such that

$$A(f_0^n, q_n) \leq \sqrt{(1 - \frac{||P-Q||^2}{4})} \leq 2Ce^{-nr}$$

Thus

$$P_{f_0}^n\left\{\frac{q_n(\mathbf{X_n})}{\prod f_0(X_i)} \geq e^{-nr}\right\} = P_{f_0}^n\left\{\sqrt{\frac{q_n(\mathbf{X_n})}{\prod f_0(X_i)}} \geq e^{-n\frac{r}{2}}\right\} \leq 2Ce^{n\frac{r}{2}}e^{-nr}$$

An application of Borel-Cantelli yields

$$\frac{q_n(\mathbf{X_n})}{\prod f_0(X_i)} \leq e^{-nr} \text{ a.s } P_{f_0}^\infty$$

and we have

$$\frac{1}{\Pi(V_n)} e^{n\frac{r}{2}} \int_{V_n} \frac{\prod_1^n f(X_i)}{\prod_1^n f_0(X_i)} \Pi(df) \to 0 \text{ a.s } P_{f_0}^{\infty}$$

Since $\liminf \Pi(V_n) > 0$, we have the conclusion.

□

Remark 4.4.1. The role of the assumption that f_0 is in the Kullback-Leibler support is to ensure that (4.1) holds. Sometimes it might be possible to verify it by direct calculation without invoking the K-L support assumption. We will see an example of this kind in the next chapter.

Let f_0 be in the K-L support of Π. In order to apply the Schwartz theorem, we need to identify neighborhoods of f_0 for which there exists a uniformly consistent test for $H_0 : f = f_0$ vs $H_1 : f \in U^c$.

Let U be a weak neighborhood of the form

$$U = \int f dP - \int f dP_0 < \epsilon, f \text{ bounded continuous} \tag{4.4}$$

Because f is bounded, by adding a constant we make it nonnegative and multiplying by a positive constant we can make $0 \leq f \leq 1$. Then U has the same expression in terms of this transformed f, with perhaps a different ϵ. Now f is a test function and which separates P_0 and U^c. Thus for neighborhoods of the form displayed we have an unbiased test and consequently a uniformly consistent sequence of tests for

$$H_0 : P = P_0 \qquad\qquad H_1 : P \in U^c$$

For any test function f , $|\int f dP - \int f dP_0| < \epsilon$ iff $\int f dP - \int f dP_0 < \epsilon$ and $\int(1-f)dP - \int(1-f)dP_0 < \epsilon$. In other words $U = \{P : |\int f dP - \int f dP_0| < \epsilon\}$ can be expressed as intersections of sets of the type in (4.4).

Theorem 4.4.2. *Let Π be a prior on L_μ. If f_0 is in the K-L support of Π, then the posterior is weakly consistent at f_0 .*

Proof. If $U = \{P : |\int f_i dP - \int f_i dP_0| < \epsilon_i : 1 \leq i \leq k\}$ then

$$U = \cap_1^k \{P : |\int f_i dP - \int f_i dP_0| < \epsilon_i\}$$

Hence it is enough to show that the posterior probability of each of the sets in the intersection goes to 1 a.s f_0. By the discussion preceding the theorem, $\{P : |\int f_i dP -$

$\int f_i dP_0| < \epsilon_i\}$ is an intersection of two sets of the type displayed in (4.4). Since the Schwartz condition is satisfied for these sets

$$\Pi(U|X_1, X_2, \ldots, X_n) \to 1 \text{ a.s } P_{f_0}^{\infty}.$$

Further, using a countable base for weak neighborhoods, we can ensure that almost surely $P_{f_0}^{\infty}$, for all U, $\Pi(U|X_1, X_2, \ldots, X_n) \to 1$. □

If we have a tail free prior on densities, like a suitable Polya tree prior, then we do not need a condition like "f_0 is in the K-L support of Π" to prove weak consistency of the posterior. On the other hand, consistency is proved for a tail free prior by using a Schwartz like argument for finite-dimensional multinomials, which tacitly uses the condition of f_0 being in the K-L support. See also the result in the next section that establishes posterior consistency without invoking Schwartz's condition.

Applications of Schwartz's theorem appear in Chapters 5, 6 and 7.

4.4.2 L_1-Consistency

What if U is a total variation neighborhood of f_0? LeCam [122] and Barron [7] show that in this case, if f_0 is nonatomic, then a uniformly consistent test for $H_0 : f = f_0$ versus $H_1 : f \in U^c$ will not exist.

Barron investigated the connection between posterior consistency and existence of uniformly consistent tests. The next two results are adapted from an unpublished technical report of Barron. Some of these appear in [8].

Proposition 4.4.2. *Suppose for some $\beta_0 > 0$, $\Pi(W_n) < Ce^{-n\beta_0}$. If f_0 is in the K-L support of Π then*

$$\Pi(W_n|\mathbf{X_n}) \to 0 \ a.s. P_{f_0}^{\infty}$$

Proof. By the Markov inequality

$$P_{f_0}\left\{\int_{W_n} \prod_1^n \frac{f}{f_0}(X_i)\ \Pi(df) > e^{-n\beta}\right\}$$

$$\leq e^{n\beta} \int_{\mathbb{R}^n} \int_{W_n} \prod_1^n \frac{f}{f_0}(X_i)\ \Pi(df) \prod_1^n f_0(X_i)\ \mu^n(dx_1, dx_2, \ldots, dx_n)$$

$$= e^{n\beta} \int_{W_n} \Pi(df)$$

$$\leq e^{n\beta} C e^{-n\beta_0}$$

and if $\beta < \beta_0$

$$P_{f_0}^{\infty}\left\{\int_{W_n}\prod_1^n \frac{f}{f_0}(X_i)\ \Pi(df) > e^{-n\beta} \text{ i.o }\right\} = 0$$

By Lemma 4.4.1, for all $\beta > 0$,

$$e^{n\beta}\int_{L_\mu}\prod_1^n \frac{f(X_i)}{f_0(X_i)}\ \Pi(df) \to \infty \text{ a.s } P_{f_0}^{\infty}.$$

The argument is now easily completed. □

Theorem 4.4.3 (Barron). *Let Π be a prior on L_μ, f_0 in L_μ and U be a neighborhood of f_0. Assume that $\Pi(K_\epsilon(f_0)) > 0$ for all $\epsilon > 0$. Then the following are equivalent.*

(i) There exists a β_0 such that

$$P_{f_0}\{\Pi(U^c|X_1, X_2, \ldots, X_n) > e^{-n\beta_0} \textit{ infinitely often}\} = 0$$

(ii) There exist subsets V_n, W_n of L_μ, positive numbers $c_1, c_2, \beta_1, \beta_2$ and a sequence of tests $\{\phi_n(\mathbf{X_n})\}$ such that

(a) $U^c \subset V_n \cup W_n$,

(b) $\Pi(W_n) \le C_1 e^{-n\beta_1}$, and

(c) $P_{f_0}\{\phi_n(\mathbf{X_n}) > 0$ infinitely often$\} = 0$ and $\inf_{f\in V_n} E_f\phi_n \ge 1 - c_2 e^{-n\beta_2}$.

Proof. (i) $\implies$ (ii): Set $S_n = \{(x_1, x_2, \ldots, x_n) : \Pi(U^c|x_1, x_2, \ldots, x_n) > e^{-n\beta_0}\}$ and $\phi_n = I_{S_n}$. Let $\beta < \beta_0$

$$V_n = \{f : P_f(S_n) > 1 - e^{-n\beta}\}$$
$$W_n = \{f : P_f(S_n^c) \ge e^{-n\beta}\} \cap U^c$$

By assumption $P_{f_0}^{\infty}\{\phi_n = 1 \text{ infinitely often }\} = 0$ and by construction

$$\inf_{f\in V_n} E_f\phi_n > 1 - e^{-n\beta}$$

Now,

$$\Pi(W_n) = \Pi\left(\{f : P_f(S_n^c) > e^{-n\beta}\} \cap U^c\right)$$
$$\leq e^{n\beta} \int_{U^c} P_f(S_n^c)\ \Pi(df)$$

and by Fubini

$$= e^{n\beta} \int_{S_n^c} \pi\left(U^c|\mathbf{x_n}\right)\ d\lambda_n(\mathbf{x_n})$$
$$\leq e^{n\beta} e^{-n\beta_0} = e^{-n(\beta_0-\beta)}$$

where λ_n is the marginal distribution of $\mathbf{X_n}$.

$(ii) \Longrightarrow (i)$:

$$\Pi(U^c|\mathbf{X_n}) = \Pi(U^c \cap V_n|\mathbf{X_n}) + \Pi(U^c \cap W_n|\mathbf{X_n})$$

Since W_n has exponentially small prior probability, by Proposition 4.4.2

$$\Pi(W_n|\mathbf{X_n}) \to 0 \text{ a.s } P_{f_0}^{\infty}$$

The proof actually shows that for some $\beta_0 > 0$, writing i.o. for "infinitely often"

$$P_{f_0}^{\infty}\left\{\Pi\left(W_n|\mathbf{X_n}\right) > e^{-n\beta_0} \text{ i.o }\right\} = 0$$

Because $\Pi(U^c \cap V_n|\mathbf{X_n}) \leq \Pi(V_n|\mathbf{X_n})$, it is enough to show that, for some $\beta > 0$,

$$P_{f_0}^{\infty}\left\{\Pi(V_n|\mathbf{X_n}) > e^{-n\beta} \text{ i.o }\right\} = 0$$

Now,

$$\Pi(V_n|\mathbf{X_n})$$
$$= \phi_n(\mathbf{X_n})\Pi(V_n|\mathbf{X_n}) + (1 - \phi_n(\mathbf{X_n}))\Pi(V_n|\mathbf{X_n})$$

Since $P_{f_0}^{\infty}\{\phi_n > 0 \text{ i.o. }\} = 0$, for any $\beta > 0$, $P_{f_0}^{\infty}\{\phi_n\Pi(V_n|\mathbf{X_n}) > 0 \text{ i.o. }\} = 0$.

For any β an application of Markov's inequality and Borel-Cantelli lemma shows that

$$\begin{aligned}
&P_{f_0}\left\{\int_{V_n}\prod_1^n \frac{f}{f_0}(x_i)\ \Pi(df)(1-\phi_n(\mathbf{x_n})) > e^{-n\beta}\right\}\\
&\leq e^{n\beta}\int_{\mathbb{R}^n}\int_{V_n}\prod_1^n \frac{f}{f_0}(x_i)(1-\phi_n(\mathbf{x_n}))\ \Pi(df)\prod_1^n f_0(x_i)\mu^n(d\mathbf{x_n})\\
&= e^{n\beta}\int_{V_n} E_f(1-\phi_n)\ \Pi(df)\\
&\leq e^{n\beta}C_2e^{-n\beta_2}
\end{aligned}$$

and if $\beta < \beta_2$

$$P_{f_0}\left\{\int_{V_n}\prod_1^n \frac{f}{f_0}(x_i)\ \Pi(df)(1-\phi_n(\mathbf{x_n})) > e^{-n\beta} \text{ i.o}\right\} = 0.$$

As before by Lemma 4.4.1 for any β,

$$e^{n\beta}\int_{L_\mu}\prod_1^n \frac{f(X_i)}{f_0(X_i)}\ \Pi(df) \to \infty \text{ a.s } P_{f_0}^{\infty}.$$

The argument is now easily completed. □

This last theorem can be used to develop sufficient conditions for posterior consistency on L_1-neighborhoods. Barron, Schervish and Wasserman [5] provide such a condition using bracketing metric entropy. Motivated by their result, we prove the following.

Definition 4.4.5. Let $\mathcal{G} \subset L_\mu$. For $\delta > 0$, the L_1-*metric entropy* $J(\delta, \mathcal{G})$ is defined as the logarithm of the minimum of all n such that there exist $f_1, f_2, \ldots, f_n$ in L_μ with the property $\mathcal{G} \subset \cup_1^n\{f : \|f - f_i\| < \delta\}$.

Theorem 4.4.4. *Let Π be a prior on L_μ. Suppose $f_0 \in L_\mu$ and $\Pi(K_\epsilon(f_0)) > 0$ for all $\epsilon > 0$. If for each $\epsilon > 0$, there is a $\delta < \epsilon$, $c_1, c_2 > 0$, $\beta < \epsilon^2/2$, and $\mathcal{F}_n \subset L_\mu$ such that, for all n large,*

1. $\Pi(\mathcal{F}_n^c) < C_1 e^{-n\beta_1}$,

2. $J(\delta, \mathcal{F}_n) < n\beta$,

then the posterior is strongly consistent at f_0.

Proof. Let $U = \{f : \|f - f_0\| < \epsilon\}$, $V_n = \mathcal{F}_n \cap U^c$, and $W_n = \mathcal{F}_n^c$. We will argue that the pair (V_n, W_n) satisfy (ii) of Theorem 4.4.3. Here $U^c \subset V_n \cup W_n$ and $\Pi(W_n) < c_1 e^{-n\beta_1}$.

Let $g_1, g_2, \ldots, g_k$ in L_μ be such that $V_n \subset \cup_1^k G_i$ where $G_i = \{f : \|f - g_i\| < \delta\}$. Let $f_i \in V_n \cap G_i$. Then for each $i = 1, 2, \ldots, k$, $\|f_0 - f_i\| > \epsilon$ and if $f \in G_i$, then $\|f_0 - f\| > \epsilon - \delta$. Consequently for each $i = 1, 2, \ldots, k$, there exists a set A_i such that

$$P_{f_0}(A_i) = \alpha \text{ and } P_{f_i}(A_i) = \gamma > \alpha + \epsilon$$

Hence if $f \in G_i$, then $P_f(A_i) > \gamma - \delta > \alpha + \epsilon - \delta$.

Let

$$B_i = \left\{(x_1, x_2, \ldots, x_n) : \frac{1}{n}\sum_{j=1}^n I_{A_i}(x_j) \geq (\gamma + \alpha)/2\right\}$$

A straightforward application of Hoeffeding's inequality shows that

$$P_{f_0}(B_i) \leq \exp[-n\epsilon^2/2]$$

On the other hand, if $f \in G_i$,

$$\begin{aligned} P_f(B_i) &\geq P_f\left\{\frac{1}{n}\sum_{j=1}^n I_{A_i}(x_j) - P_f(A_i) \geq \frac{(\alpha - \gamma)}{2} + \delta\right\} \\ &\geq P_f\left\{n^{-1}\sum_{j=1}^n I_{A_i}(x_j) - P_f(A_i) \geq \frac{-\epsilon}{2} + \delta\right\} \end{aligned} \tag{4.5}$$

Applying Hoeffeding's inequality to $-n^{-1}\sum_{j=1}^n I_{A_i}(x_j)$, the preceding probability is greater than or equal to

$$1 - \exp[-(n/2)(\epsilon/2 - \delta)^2]$$

If we set

$$\phi_n(X_1, X_2, \ldots, X_n) = \max_{1 \leq i \leq k} I_{B_i}(X_1, X_2, \ldots, X_n)$$

then

$$E_{f_0}\phi_n \leq k\exp[-n\epsilon^2/2]$$

and

$$\inf_{f \in V_n} E_f\phi_n \geq 1 - \exp[-(n/2)(\epsilon/2 - \delta)^2]$$

By choosing $\log k \leq J(\delta, \mathcal{F}_n) < n\beta$, we have $E_{f_0}\phi_n \leq \exp[-n(\epsilon^2/2 - \beta)]$. Since $\beta < \epsilon^2/2$, all that is left to show is

$$P_{f_0}\{\phi_n > 0 \text{ infinitely often}\} = 0$$

This follows easily from an application of the Borel Cantelli lemma and from the fact that ϕ_n takes only values 0 or 1. □

This last theorem is very much in the spirit of Barron et al. [5]. Their theorem is in terms of bracketing entropy. If $\mathcal{G} \subset L_\mu$, for $\delta > 0$, the L_1-bracketing entropy $J_1(\delta, \mathcal{G})$ is defined as (here we use a weaker notion that suffices for our purpose) the logarithm of the minimum of all n such that there exist $g_1, g_2, \ldots, g_n$ satisfying

1. $\int g_i \leq 1 + \delta$,

2. for every $g \in \mathcal{G}$ there exists an i such that $g \leq g_i$.

We feel that in many examples the L_1 entropy is easier to apply than bracketing entropy.

4.5 Consistency via LeCam's inequality

It is of technical interest that one can prove posterior consistency without assuming that the prior is tail free or satisfies the condition of f_0 being in the K-L support. An inequality of LeCam [121] is useful to do this.

Let Π be a prior on $M(\mathcal{X})$. For any measurable subset U of $M(\mathcal{X})$, let λ_U be the probability measure on $\mathcal{X}$ given by

$$\lambda_U(B) = \frac{1}{\Pi(U)} \int_U P(B) d\Pi(P)$$

We will let λ stand for the marginal on $\mathcal{X}$.

If given P, $X \sim P$, and $\Pi(U|\mathbf{X_n})$ is the posterior probability of U, then

$$\begin{aligned}\Pi(U|\cdot) =& \Pi(U)\frac{d\lambda_U}{d\lambda}(\cdot) = \frac{\Pi(U)d\lambda_U}{\Pi(U)d\lambda_U + \Pi(U^c)d\lambda_{U^c}}(\cdot) \\ &\leq \frac{\Pi(U)}{\Pi(V)}\frac{d\lambda_U}{d\lambda_V}(\cdot) \text{ if } V \subset U^c\end{aligned}$$

Also recall that the L_1-distance satisfies

$$\|P - Q\| = 2 \sup_B |P(B) - Q(B)| = 2 \sup_{0 \le f \le 1} \left| \int f dP - \int f dQ \right|$$

where of course Bs and fs are measurable.

Lemma 4.5.1 (LeCam). *Let U, V be disjoint subsets of $\mathcal{X}$. For any P_0 and any test function ϕ*

$$\int \Pi(V|x) dP_0(x) \le \|P_0 - \lambda_U\| + \int \phi dP_0 + \frac{\Pi(V)}{\Pi(U)} \int (1 - \phi) d\lambda_V \tag{4.6}$$

Proof.

$$\begin{aligned}
\int \Pi(V|x) dP_0(x) &= \int \phi(x) \Pi(V|x) dP_0(x) + \int (1 - \phi(x)) \Pi(V|x) dP_0(x) \\
&\qquad \text{adding and subtracting } \int (1 - \phi(x)) \Pi(V|x) d\lambda_U(x) \\
&\le \int \phi(x) dP_0(x) + \left[\int (1 - \phi(x)) \Pi(V|x) dP_0(x) - \int (1 - \phi(x)) \Pi(V|x) d\lambda_U(x) \right] \\
&\qquad + \int (1 - \phi(x)) \Pi(V|x) d\lambda_U(x) \\
&\le \int \phi(x) \Pi(V|x) dP_0(x) + \|P_0 - \lambda_U\| + \frac{\Pi(V)}{\Pi(U)} \int (1 - \phi) d\lambda_V
\end{aligned}$$

where the first term comes from observing

$$0 \le \Pi(V|x) \le 1$$

and the second from

$$0 \le (1 - \phi)(x) \Pi(V|x) \le 1$$

The third term follows by noting that

$$\Pi(V|x) \le (\Pi(V)/\Pi(U))(d\lambda_V / d\lambda_U)$$

□

Our interest is when V is the complement of a neighborhood of P_0 and we have $X_1, X_2, \ldots, X_n$ which are given P, i.i.d. P. If $U_n \cap V = \emptyset$ and ϕ_n are test functions, then we can write LeCam's inequality as

$$\Pi(V|\mathbf{X_n}) \leq \|P_0^n - \lambda_{U_n}^n\| + \int \phi_n dP_0^n + \frac{\Pi(V)}{\Pi(U_n)} \int (1-\phi_n) d\lambda_V$$

where of course P^n is the n-fold product of P and $\lambda_U^n = (\int_U P^n d\Pi(P))/\Pi(U)$.

Theorem 4.5.1. *Let $U_n^\delta = \{P : \|P_0 - P\| < \delta/n\}$. If for every δ, $\{\Pi(U_n^\delta) : n \geq 1\}$ is not exponentially small, i.e.,*

$$\textit{for all } \beta > 0, e^{n\beta}\Pi(U_n^\delta) \to \infty \tag{4.7}$$

then the posterior is weakly consistent at P_0

Proof. It is not hard to see that

$$\|P_0 - P\| < \delta/n \Rightarrow \|P_0^n - P^n\| < \delta$$

Consequently the first term goes to δ. Since for any weak neighborhood we can choose an exponentially consistent test ϕ_n for testing $H_0 : f = f_0$ against $H_1 : f \in V_n^c$, and by assumption for all $\beta > 0, e^{n\beta}\Pi(U_n^\delta) \to \infty$, it is not hard to see that the third term goes to 0. Because δ is arbitrary, the result follows. □

Remark 4.5.1. By Proposition 1.2.1, $\|P - Q\| \leq 2H(P,Q)$. Hence Theorem 4.5.1 holds if we take $U_n^\delta = \{P : H(P_0, P) < \delta/n\}$

Suppose (4.7) holds and V_n are sets such that for some $\beta_0 > 0, \Pi(V_n)e^{n\beta_0} \to 0$; then choosing $\phi_n \equiv 0$ it follows easily that $\Pi(V_n|X_1, X_2, \ldots, X_n) \to 0$. In other words, we have an analog of Proposition 4.4.2. Consequently, we also have an analog of Theorem 4.4.4.

Theorem 4.5.2. *Let Π be a prior on L_μ. If for each $\epsilon > 0$, there is a $\delta < \epsilon$, $c_1, c_2 > 0$, $\beta < \epsilon^2/2$, and $\mathcal{F}_n \subset L_\mu$ such that for all n large,*

1. *$\Pi(\mathcal{F}_n^c) < C_1 e^{-n\beta_1}$ and*
2. *$J(\delta, \mathcal{F}_n) < n\beta$*

Further if with $U_n^\delta = \{P : \|P_0 - P\| < \delta/n\}$,

$$\textit{for every } \delta, \textit{ for all } \beta > 0, e^{n\beta}\Pi(U_n^\delta) \to \infty$$

then the posterior is strongly consistent at f_0.

5
Density Estimation

5.1 Introduction

As the name suggests, density estimation is the problem of estimating the density of a random variable X using observations of X. In this chapter we discuss some Bayesian approaches to density estimation.

Density estimation has been extensively studied from the non-Bayesian point of view. These include many methods of estimation starting from simple histogram estimates to more sophisticated kernel estimates, estimates through Fourier series expansions, and more recently wavelet-based methods. In addition, the asymptotics of many of these methods, including minimax rates of convergence are available. There are many good references; Silverman [151] and Van der Vaart [160] provide a good starting point.

Consider the simple case when the density is to be estimated through a histogram. Important features of the histogram are number of bins, their location and their width. In order to reflect the true density, these features of the histogram estimate need to be dependent not just on the number of observations but on the observations themselves. The need for such a dynamic choice has been recognized and there have been many reasonable, ad hoc, prescriptions. This issue persists in one form or another with the other methods of estimation such as kernel estimates. The Bayesian approach, via the posterior provides a rational method for choosing these features.

In this chapter we discuss histogram priors of Gasperini and mixtures of normal densities which were introduced by Lo [130] and further developed by Escobar, Mueller and West [[168],[59] and [170]]. Gaussian process priors developed by Leonard [[126],[127]] and studied by Lenk [125] are some what different in sprit and are also discussed. See also Hjort [98] and Hartigan [94].

Consistency is dealt with at some length for the histogram and the mixture of normal kernel priors. These partly demonstrate different techniques to show consistency. For the priors on histograms direct calculation is easier than invoking the Schwartz theorem whereas for the mixture of normal kernels Schwartz's theorem is a convenient tool. This chapter is beset with long computations. To an extent they are both natural and necessary.

5.2 Polya Tree Priors

A prerequisite for Bayesian density estimation is, of course, a prior on densities. Since the Dirichlet process and their mixtures sit on discrete measures, these are clearly unsuitable. On the other hand we have saw in Chapter 3 that by choosing the parameters appropriately we can get Polya tree priors that are supported by densities. Since the posterior for these priors involves simple updating rules, it is natural to consider Polya trees as a candidate in density estimation.

Recall that if we have a Polya tree with partitions $\{B_{\underline{\epsilon}} : \underline{\epsilon} \in E_j : j \geq 1\}$ and parameters $\{\alpha_{\underline{\epsilon}} : \underline{\epsilon} \in E_k^*\} : k \geq 1\}$, the predictive density at x is given by

$$\alpha(x) = \lim_{k\to\infty} \prod_1^k \frac{1}{\lambda(B_{\epsilon_1(x)\epsilon_2(x)\ldots\epsilon_i(x)})} \frac{\alpha_{\epsilon_1(x)\epsilon_2(x)\ldots\epsilon_i(x)}}{\alpha_{\epsilon_1(x)\epsilon_2(x)\ldots\epsilon_i(x)0} + \alpha_{\epsilon_1(x)\epsilon_2(x)\ldots\epsilon_i(x)1}}$$

where $\epsilon_i(x) = 1$ if $x \in B_{\epsilon_1(x)\epsilon_2(x)\ldots\epsilon_i(x)}$ and 0 otherwise.

If $X_1 = x_1$ is observed and $x_1 \in B_{\epsilon'_1,\epsilon'_2,\ldots\epsilon'_k}$ for a sequence $(\epsilon'_1, \epsilon'_2, \ldots)$ of 0s and 1s, and if $\underline{\epsilon}$ and $\underline{\epsilon}'$ differ for the first time at the $(j+1)$th coordinate, then the predictive density $\alpha(x|X_1 = x_1)$ is

$$\left[\alpha(x|X_1 = x_1) = \prod_1^j \frac{1}{\lambda(B_{\epsilon_1(x)\epsilon_2(x)\ldots\epsilon_i(x)})} \frac{\alpha_{\epsilon_1(x)\epsilon_2(x)\ldots\epsilon_i(x)} + 1}{\alpha_{\epsilon_1(x)\epsilon_2(x)\ldots\epsilon_i(x)0} + \alpha_{\epsilon_1(x)\epsilon_2(x)\ldots\epsilon_i(x)1}}\right]$$
$$\left[\prod_{j+1}^\infty \frac{1}{\lambda(B_{\epsilon_1(x)\epsilon_2(x)\ldots\epsilon_i(x)})} \frac{\alpha_{\epsilon_1(x)\epsilon_2(x)\ldots\epsilon_i(x)}}{\alpha_{\epsilon_1(x)\epsilon_2(x)\ldots\epsilon_i(x)0} + \alpha_{\epsilon_1(x)\epsilon_2(x)\ldots\epsilon_i(x)1}}\right]$$

As is to be expected the predictive density depends on the partition. While a general expression for the predictive density given $X_1, X_2, \ldots, X_n$ is cumbersome to write down, it is clear that sequential updating is possible.

The density estimates from Polya tree priors have no obvious relation with classical density estimates. Further, the priors lead to estimates that lack smoothness at the endpoints of the defining partition. Lavine [118] observes that this disadvantage can be overcome by considering a mixture of $\{PT(\underline{\Pi}(\theta), \underline{\alpha}(\theta))\}$ processes, where the partitions themselves depend on the hyperparameter θ. One advantage of the Polya tree priors is the relative ease with which one can conduct robustness studies; see Lavine [119].

If we have a prior on densities, as discussed in Chapter 4 the consistency of interest is L_1-consistency. It is shown in Barron et al. [5] that if $\alpha_n = 8^n$, the posterior is L_1-consistent. Such a high value of α_n implies that the random Ps are highly concentrated around the prior guess $E(P)$, so that posterior consistency will be an extremely slow process. Hjort and Walker [165] have used a some what curious argument and show that with $\alpha_n = n^{2+\delta}$ the Bayes estimate is L_1-consistent.

5.3 Mixtures of Kernels

While Polya tree priors can be made to sit on densities, it is not possible to constrain the support to have smoothness properties. Much before Polya tree priors became popular, Lo [131] had developed a useful construction of priors on densities. Much of this section is based on Lo [131] and Ferguson [63].

Let Θ be a parameter set, typically $\mathbb{R}$ or $\mathbb{R}^2$. Let $K(x, \tau)$ be a kernel, i.e.,for each $\tau, K(\cdot, \tau)$ is a probability density on $\mathcal{X}$ with respect to some σ-finite measure. For any probability P on Θ, let

$$K(x, P) = \int K(x, \tau) dP(\tau)$$

For each P, $K(\cdot, P)$ is a density on $\mathcal{X}$ and Lo's method consists of choosing a mixture $K(\cdot, P)$ at random by choosing P according to a Dirichlet process. These would be referred to as Dirichlet mixtures of $K(\cdot, P)$.

Formally the model consists of $P \sim D_\alpha$, given P; $X_1, X_2, \ldots, X_n$ are i.i.d. $K(\cdot, P)$.
If $\alpha = M\bar{\alpha}$, where $\bar{\alpha}$ is a probability measure, then the prior expected density is

$$f_0 = \int K(\cdot, P) D_\alpha(dP) = \int K(\cdot, \tau) \bar{\alpha}(d\tau)$$

It is convenient to view the $X_1, X_2, \ldots, X_n$ as arising in the following way: $P \sim D_\alpha$ given P; $\tau_1, \tau_2, \ldots, \tau_n$ are i.i.d P and given P, $\tau_1, \tau_2, \ldots, \tau_n$; $X_1, X_2, \ldots, X_n$ are independent with $X_i \sim K(\cdot, \tau_i)$.

The latent variables $\tau_1, \tau_2, \ldots, \tau_n$ although unobservable, provide insight into the structure of the posterior and are useful in describing and simulating the posterior.

A simple kernel would be to take $\tau = (i, h) : h > 0$

$$K(x, (i, h)) = \frac{I_{(ih,(i+1)h]}}{h}(x)$$

With this kernel one gets random histograms.

Another very useful kernel is the normal kernel. Here $\tau = (\theta, \sigma)$ and $K(x, \theta, \sigma) = (1/\sigma)\phi((x-\theta)/\sigma)$ where ϕ is the standard normal density. In this case the prior picks a random density that is a mixture of normal densities. The weak closure of such mixtures is all of $M(\mathbb{R})$.

The prior is a probability measure on the space of densities $\{K(\cdot, P) : P \in M(\mathbb{R})\}$ and so is the posterior given $X_1, X_2, \ldots, X_n$. For the normal kernel P is in general not identifiable. It is known from [156] that if P_1 and P_2 are discrete measures with finite support, then $K(\cdot, P_1) = K(\cdot, P_2)$ iff $P_1 = P_2$. It is easy to see that if $P_1 = N(0,1) \times \delta_{(0,\sigma_0)}$ and $P_2 = \delta_{(0,\sqrt{1+\sigma_0^2})}$, then $K(\cdot, P_1) = K(\cdot, P_2) = N(0, \sqrt{1+\sigma_0^2})$. Thus in general, P is not identifiable. Identifiability of P when restricted to discrete measures is still unresolved [63].

If we denote by $\Pi(\cdot|X_1, X_2, \ldots, X_n)$ the posterior distribution of P given $X_1, \ldots, X_n$ and by $H(\cdot|X_1, X_2, \ldots, X_n)$ the posterior distribution of $\tau_1, \ldots, \tau_n$ given $X_1, \ldots, X_n$ then

$$\Pi(\cdot|X_1, X_2, \ldots, X_n) = \int \Pi(\cdot|(\tau_1, X_1), \ldots, (\tau_n, X_n))H(d\underline{\tau}|X_1, X_2, \ldots, X_n)$$

Since P and $X_1, X_2, \ldots, X_n$ are conditionally independent given $\tau_1, \tau_2, \ldots, \tau_n$,

$$\Pi(\cdot|(\tau_1, X_1), \ldots, (\tau_n, X_n)) = \Pi(\cdot|(\tau_1, \tau_2, \ldots, \tau_n)) = D_\alpha + \sum \delta_{\tau_i}$$

and

$$\Pi(\cdot|X_1, X_2, \ldots, X_n) = \int D_{\alpha+\sum \delta_{\tau_i}} H(d\underline{\tau}|X_1, X_2, \ldots, X_n)$$

The evaluation of these quantities depend on $H(\cdot|X_1, X_2, \ldots, X_n)$. If α has a density, the joint density $\tilde{\alpha}(\tau_1, \tau_2, \ldots, \tau_n)$ is discussed in Chapter 3 (see equation 3.15).

Recall that if $C_1, C_2, \ldots, C_{N(P)}$ is a partition of $\{1, 2, \ldots, n\}$ then the density (with respect to the Lebesgue measure on $\mathbb{R}^k$) at

$$\underline{\tau} = (\tau_1, \tau_2, \ldots, \tau_n) : \tau_i = \tau_{i'}, i, i' \in C_j, j = 1, 2, \ldots N(P)$$

is

$$\prod_1^{N(P)} \frac{\alpha(\tau_j)(e_j - 1)!}{\prod_1^n (M + i)} \tag{5.1}$$

where $e_j = \#C_j$ and hence the joint density of the xs and τs at

$$\underline{\tau} = (\tau_1, \tau_2, \ldots, \tau_n) : \tau_i = \tau_{i'}, i, i' \in C_j, j = 1, 2, \ldots N(P)$$

is

$$\prod_1^{N(P)} \frac{\alpha(\tau_j)(e_j - 1)! \prod_{l \in C_j} K(x_l, \tau_j)}{\prod_1^n (M + i)}$$

Consequently, the posterior density of $\underline{\tau}$ is

$$\frac{\prod_1^{N(P)} \alpha(\tau_j)(e_j - 1)! \prod_{l \in C_j} K(x_l, \tau_j)}{\sum_P \int \prod_1^{N(P)} \alpha(\tau_j)(e_j - 1)! \prod_{l \in C_j} K(x_l, \tau_j) d(\tau_j)}$$

Thus

$$\int \left(\frac{1}{n} \sum K(x, \tau_i) \right) H(d\underline{\tau} | X_1, X_2, \ldots, X_n) \tag{5.2}$$

$$= \frac{1}{n} \sum_{\underline{P}} \frac{\sum_1^{N(\underline{P})} (e_j - 1)! \int K(x, \tau_j) \prod_{l \in C_i} K(x_l, \tau_j) \alpha(\tau_j) d\tau_j}{\sum_{\underline{P}} \int \sum_1^{N(\underline{P})} (e_j - 1)! \int \prod_{l \in C_i} K(x_l, \tau_j) \alpha(\tau_j) d\tau_j} \tag{5.3}$$

Since the Bayes estimate$\hat{f}$ of f is, by 5.2, this reduces to

$$\frac{M}{M+n} f_0(x) + \frac{n}{M+n} \int \left(\sum K(x, \tau_i) \right) H(d\underline{\tau} | X_1, X_2, \ldots, X_n)$$

Hence, we have that the Bayes estimate of f is

$$\frac{M}{M+n} \int K(x, \tau) \bar{\alpha}(d\tau)$$
$$+ \frac{n}{M+n} \sum_{\underline{P}} W(\underline{P}) \frac{e_i}{n} \frac{\int K(x, \tau) \prod_{l \in C_i} K(x_l, \tau) \alpha(\tau) d\tau}{\int \prod_{l \in C_i} K(x_l, \tau) \alpha(\tau) d\tau} \tag{5.4}$$

where $\underline{P} = \{C_1, C_2, \ldots, C_{N(\underline{P})}\}$ is a partition of $\{1, 2, \ldots, n\}$, e_i is the number of elements in C_i, and

$$W(\underline{P}) = \frac{\Phi(\underline{P})}{\sum \Phi(\underline{P})}, \Phi(\underline{P}) = \prod_{1}^{N(\underline{P})} \{(e_i - 1)! \prod_{l \in C_i} K(x_l, \tau)\alpha(\tau)d\tau\}$$

The Bayes estimate is thus composed of a part attributable to the prior and a part attributable to the observations. Since for the Dirichlet, $M \to 0$ corresponds to removing the influence of the prior, it is tempting to consider the estimate

$$\int \left(\frac{1}{n} \sum K(x, \tau_i) \right) H(d\underline{\tau} | X_1, X_2, \ldots, X_n)$$

as a partially Bayesian estimate with the influence of the prior removed. Unfortunately, this interpretation is quite misleading. As $M \to 0$ the Bayes estimate (5.4) goes to

$$\frac{\int K(x, \tau_1)\tilde{\alpha}(\tau_1) \prod_1^n K(x_i, \tau_1) d\tau_1}{\int \tilde{\alpha}(\tau_1) \prod_1^n K(x_i, \tau_1) d\tau_1} \tag{5.5}$$

corresponding to a partition in which all τ_i are equal to τ_1. All other terms have a power of M and tend to 0. The term (5.5) corresponds to assuming that all the X_is came from a single parametrized population with density $K(x, \tau)$ and so is highly parametrized.

The apparent paradox is resolved by the fact that role of the hyperparameters depends on the context. Here M decides the likelihood of different clusters and in fact relatively large values of M help bring the Bayes estimate close to a data-dependent kernel density estimate. For a penetrating discussion of the role of M, see discussion by Escobar [66] and West et al. [170].

Clearly to calculate quantities like $\int K(x, \tau)\alpha(d\tau)$ it would be convenient if α is conjugate to $K(.,.)$. Thus if K is the normal kernel a convenient choice for $\bar{\alpha}$ is a prior conjugate to $N(\tau, \sigma)$. Hence an appropriate choice for $\bar{\alpha}$ is the inverse normal-gamma prior, i.e., the precision $\rho = 1/\sigma^2$ has a gamma distribution and given ρ, τ is $N(\mu, 1/\rho)$. Ferguson [63] has interesting guidelines for choosing the parameters of $\bar{\alpha}$ and M.

The expression for the Bayes estimate, even though it has an explicit expression, involves enormous computation. The posterior for Dirichlet mixtures of normal densities is amenable to MCMC methods. Gibbs methods are based on successive simulations from one-dimensional conditional distributions of τ_i given $\tau_j, j \neq i, X_1, X_2, \ldots, X_n$.

For a good exposition see Schervish [144] and Chen et al. [32]. The MCMC methods were developed in the present context by Escobar, Mueller and West ([59], [169],[170]). A good survey of the issues underlying MCMC issues is given by Escobar and West in [60].

To implement MCMC one essentially works with the conditional distributions of τ_i given $\tau_j, j \neq i, X_1, X_2, \dots, X_n$, which may be written explicitly from the posterior distribution of the τs given earlier or directly [32]. In practice, α has a location and scale parameter (μ, σ), which leads to some complications. In the joint distribution of τs one replaces $\tilde{\alpha}$ by $\alpha_{\mu,\sigma}$ and multiplies by the prior $\Pi(\mu, \sigma)$. Starting from this, one can calculate all the relevant posterior distributions needed in MCMC. See also Neal [135].

Since no explicit expressions are available for the Bayes estimate of $f(x)$, it would be worth exploring whether approximations like Newton [137] can be developed.

The next issue would be to do the asymptotics. In Section 5.4 we do this for a slightly modified version of the mixture model. While formal asymptotics is yet to be done for the priors discussed in this section, we expect that the results and techniques of the next section will go through with minor modifications.

5.4 Hierarchical Mixtures

This method is a slight variation of the method discussed in the last section.

Let $K(x)$ be a density on $\mathbb{R}$. For each $h > 0$ consider the kernel $K_h(x, \theta) = (1/h)K((x-\theta)/h)$. For any $P \in M(\mathbb{R})$, let

$$K_{h,P} = \int K_h(x, \theta) dP(\theta)$$

Note that $K_{h,P}$ is just the convolution $K_h * P$. If $P \sim D_\alpha$, then we get a prior on

$$\mathcal{F}_h = \{K_{h,P} : P \in M(\mathbb{R})\}$$

We now view h as the smoothing "window" and think of h as a hyperparameter and put a prior μ for h. The calculations are very similar to those of the last section except that we need to incorporate the hyperparameter h.

As before, the observations can be thought of as arising from: $h \sim \mu$, given h; $P \sim D_\alpha$; given h, P; $\theta_1, \theta_2, \dots, \theta_n$ are i.i.d. P and given h, P,and $\theta_1, \theta_2, \dots, \theta_n$; $X_1, X_2, \dots, X_n$ are independent with $X_i \sim K_h(\cdot, \theta_i)$.

The posterior distribution of P given $X_1, X_2, \ldots, X_n$ is

$$\Pi(\cdot|X_1, X_2, \ldots, X_n) = \int \Pi(\cdot|(h, \theta_1, \theta_2, \ldots, \theta_n, X_1, \ldots, X_n))H(d(h, \underline{\theta})|X_1, X_2, \ldots, X_n) \quad (5.6)$$

Because P and $X_1, X_2, \ldots, X_n$ are conditionally independent given $h, \theta_1, \theta_2, \ldots, \theta_n$,

$$\Pi(\cdot|(h, \theta_1, \theta_2, \ldots, \theta_n, X_1, \ldots, X_n)) = D_\alpha + \sum \delta_{\theta_i}$$

and

$$\Pi(\cdot|X_1, X_2, \ldots, X_n) = \int D_{\alpha + \sum \delta_{\theta_i}} H(d(h, \underline{\theta})|X_1, X_2, \ldots, X_n)$$

As before, if μ and α_h are densities with respect to Lebesgue measure then the posterior density of $(h, \theta_1, \theta_2, \ldots, \theta_n)$ is given by

$$\frac{\mu(h)\tilde{\alpha}(\theta_1, \theta_2, \ldots, \theta_n) \prod_1^n K_h(X_i - \theta_i)}{\int \mu(h)\tilde{\alpha}(\theta_1, \theta_2, \ldots, \theta_n) \prod_1^n K_h(X_i - \theta_i) dh d\underline{\theta}}$$

where $\tilde{\alpha}$ is given by 3.15.

An expression analogous to (5.4) for the Bayes estimate can be written. In the next two sections we look at consistency problems in the case when K gives rise to histograms and when K is the standard normal density.

Ishwaran [103] has used a general polya urn scheme to model θ_is and used these to construct measures analogous to a prior and established consistency of the posterior. These are then applied to a variety of interesting problems.

5.5 Random Histograms

In this section we consider priors that choose at random first a bin of width h and then a histogram with bins $(ih, (i+1)h : h \in \mathcal{N})$ where $\mathcal{N} = \{0 \pm 1 \pm 2 \ldots\}$. Formally, in the hierarchical model we take $\Theta = \mathcal{N}$ and the kernel $K(x) = I_{(0,1]}(x)$.

Thus the model consists of, $h \sim \mu$; given h; choose P on integers with $P \sim D_{\alpha_h}$ and $X_1, X_2, \ldots, X_n$ are, given h, P, i.i.d. $f_{h,P}$ where

$$f_{h,P}(x) = \sum_{i=-\infty}^{\infty} \frac{P\{i\}}{h} I_{(ih,(i+1)h]}(x)$$

One could introduce intermediate latent variables $\theta_1, \theta_2, \ldots, \theta_n$ which are given h, P; i.i.d. P. However, they are not of much use here because X_i completely determines θ_i, namely, $\theta_i = j$ iff $X_i \in (jh, (j+1)h]$.

For each h, let n_{jh} be the number of X_is in the bin $(jh, (j+1)h]$ and $J_h = \{j : n_{jh} > 0\}$.

A bit of reflection shows that the posterior distribution of P given $h, X_1, X_2, \ldots, X_n$ is $D_{\alpha_h + \sum n_{jh}\delta_j}$, where δ_j is the point mass at j.

If μ is a density on $(0, \infty)$ then the joint density of h and $X_1, X_2, \ldots, X_n$ is

$$\frac{\mu(h) \prod_1^\infty [\alpha_h(i)]^{[n_{hi}-1]} h^{-n}}{M_h^{[n]}}$$

where $M_h = \alpha_h(N)$ for any positive real x and positive integer k, $x^{[k]} = x(x+1)\ldots(x+k-1)$. Hence the posterior density $\Pi(h|X_1, X_2, \ldots, X_n)$ is

$$\frac{\mu(h) \prod_1^\infty [\alpha_h(i)]^{[n_{hi}-1]} h^{-n}}{\int_0^\infty \mu(h) \prod_1^\infty [\alpha_h(i)]^{[n_{hi}-1]} h^{-n} dh} \tag{5.7}$$

Thus the posterior is of the same form as the prior, with μ updated to (5.7) and α_h updated to $\alpha_h + \sum n_{hj}\delta_j$.

Since each D_{α_h} leads to the expected density

$$f_{\bar{\alpha}_h}(x) = \sum \frac{\bar{\alpha}_h(j)}{h} I(jh, (j+1)h](x)$$

the prior expectation is given by

$$f_0(x) = \int f_{\bar{\alpha}_h}(x)\mu(h)dh$$

Using the conjugacy of the prior, an expression for the Bayes estimate given the sample can be written.

A choice of μ which is positive in a neighborhood of 0 will allow for wide variability in the choice of histograms and will ensure that the prior has all densities as its support. If the prior belief leads to the density f_0 then an appropriate choice of $\bar{\alpha}_h$ would be

$$\bar{\alpha}_h(j) = \int_{jh}^{(j+1)h} f_0(x)dx$$

Of course, this choice would lead to a prior expected density, which may not be equal to f_0, but it can be viewed as an approximation to f_0.

5.5.1 Weak Consistency

Gasperini introduced these priors in his thesis and under some assumptions on α_h showed that if the true f_0 is not constant on any interval then under the posterior distribution given $X_1, X_2, \ldots, X_n$, h goes to 0, as $n \to \infty$. Thus the posterior stays away from densities that are far from f_0. Under additional assumptions on f_0, he also showed that the Bayes estimate of f converges in L_1 to f_0. In the spirit of Chapter 4 we investigate the consistency properties of the posterior. We confine ourselves to the case when the random histograms all have support on $(0, \infty]$, that is, the case when P is a probability on $\mathcal{N}^+ = \{0, 1, 2, \ldots\}$. This restriction is not required but simplifies the proof of Lemma 5.5.2. Some of the following calculations are taken from Gasperini's thesis, but the main ideas of the proof and the main results are different.

The consistency results in this chapter typically describe a large class of densities where consistency obtains. We saw in Chapter 4 that when we have a prior Π on densities, the Schwartz condition $\Pi(K_{f_0}(\epsilon)) > 0$ for all $\epsilon > 0$ (recall $K_{f_0}(\epsilon)$ is the ϵ Kullback-Leibler neighborhood of f_0) ensures weak consistency at f_0. Thus it seems appropriate, in the context of histogram priors, that we should attempt to describe f_0s which would satisfy Schwartz's condition. This would entail relating the tail behavior of f_0 to the tail behavior of α_hs. This is to be expected but leads to somewhat cumbrous and restrictive conditions. It turns out that histogram priors are amenable to direct calculations that lead to consistency results.

To be more specific, recall that Schwartz's condition (Lemma 4.4.1) was used to show that for all $\beta > 0$,

$$e^{n\beta} \int_{\mathcal{F}} \prod \frac{f(x_i)}{f_0(x_i)} d\Pi(f) \to \infty \text{ a.s. } P_{f_0}^{\infty}$$

Under some assumptions we will establish this result directly. The following proposition indicates the steps involved.

Proposition 5.5.1. *Let $\mathcal{F}$ be a family of densities. For each $h \in H$, Π_h is a prior on $\mathcal{F}$; μ is a prior on H, i.e., $h \sim \mu$; given h; $f \sim \Pi_h$ and given h, f; $X_1, X_2, \ldots, X_n$ are i.i.d. f. If for a density f_0,*

for every $\beta > 0$

$$\mu \left\{ h : e^{n\beta} \int_{\mathcal{F}} \prod \frac{f(x_i)}{f_0(x_i)} d\Pi_h(f) \to \infty \text{ a.s. } P_{f_0}^{\infty} \right\} > 0 \tag{5.8}$$

then the posterior is weakly consistent at f_0.

Proof. Let U be a weak neighborhood of f_0 and let Π be the prior on the space of densities induced by μ, Π_h. Since we have exponentially consistent tests for testing f_0 against U^c, it follows from Lemma 4.4.2 that for some β_0

$$e^{n\beta_0}\int_{U^c}\prod_1^n \frac{f(x_i)}{f_0(x_i)}d\Pi(f) \to 0 \text{ a.s. } P_{f_0}^\infty$$

To establish consistency it is enough to show that

$$\liminf_{n\to\infty} e^{n\beta_0}\int_{\mathcal{F}}\prod_1^n \frac{f(x_i)}{f_0(x_i)}d\Pi(f) = \liminf_{n\to\infty} e^{n\beta_0}\int_{\mathcal{F}}\prod_1^n \frac{f(x_i)}{f_0(x_i)}d\Pi_h(f)d\mu(h)$$
$$\to\infty \text{ a.s. } P_{f_0}^\infty$$

Consider

$$\left\{(h,\underline{x}) : \underline{x}\in\mathbb{R}^\infty, h\in H : e^{n\beta_0}\int_{\mathcal{F}}\prod_1^n \frac{f(x_i)}{f_0(x_i)}d\Pi_h(f)\to\infty\right\}$$

By assumption for h in a set of positive μ measure, the $h-$ section of E has measure 1 under $P_{f_0}^\infty$. By Fubini there is a $F\subset\mathbb{R}^\infty, P_{f_0}^\infty(F) = 1$ and for $\underline{x}\in F$, the $\underline{x}-$ section of E has positive μ measure and for each $\underline{x}\in F$ by Fatou

$$\liminf_{n\to\infty}\int_H\left[e^{n\beta_0}\int_{\mathcal{F}}\prod_1^n \frac{f(x_i)}{f_0(x_i)}d\Pi_h(f)\right]d\mu(h) = \infty$$

□

Assumptions on the Prior (Gasperini)

(i) μ is a prior for h with support $(0,\infty)$.

(ii) For each h, α_h is a probability measure on $\mathcal{N}^+$, and for all h, $\alpha_h(1) > 0$.

(iii) For each h, there is a constant $K_h > 0$ such that

$$\frac{\alpha_h(j)}{\alpha_h(j+1)} < K_h \text{ for } j = 0,1,2\ldots$$

Theorem 5.5.1. *Suppose that the prior satisfies the assumptions just listed. If f_0 is a density such that*

(a) $\int x^2 f_0(x)dx < \infty$ *and*

(b) $\lim_{h\to 0} \int f_0 \log(f_{0,h}/f_0) = 0,$

then the posterior is weakly consistent at f_0.

Proof. Let $I_{nh} = \int_{\mathcal{F}_h} \prod_1^n (f(x_i)/f_0(x_i)) D_{\alpha_h}(df)$

To apply the last proposition it is enough to show that for any $\beta > 0$ there exists h_0 such that for each h in $(0, h_0)$,

$$\exp[n(\beta + \frac{\log I_{nh}}{n})] \to \infty \text{ a.s. } P_{f_0}^{\infty} \tag{5.9}$$

and this follows if for any $\epsilon > 0$,there exists h_0 such that for $h \in (0, h_0)$,

$$\lim_n \frac{\log I_{nh}}{n} > -\epsilon \text{ a.s. } P_{f_0}^{\infty}$$

Then by taking $\epsilon = \beta/2$, (5.9) would be achieved.

$$\frac{\log I_{nh}}{n} = \frac{1}{n} \log \int_{\mathcal{F}_h} \prod_1^n \frac{f(x_i)}{f_{0h}(x_i)} D_{\alpha_h}(df) + \frac{1}{n} \sum_1^n \log \frac{f_{0h}(x_i)}{f_0(x_i)}$$

where $f_{0h}(x) = (1/h) \int_{ih} (i+1)h f_0(y)dy$ for $x \in (ih, (i+1)h]$.

By assumption b and SLLN for some h_0, whenever $h < h_0$,

$$\lim_n \frac{1}{n} \sum_1^n \log \frac{f_{0h}(x_i)}{f_0(x_i)} > \frac{-\epsilon}{2} \text{ a.s. } P_{f_0}^{\infty}$$

Note that whenever $f \in \mathcal{F}_h$, f is a constant on $(ih, (i+1)h] : i \geq 0$. Consequently for $f \in \mathcal{F}_h$,

$$\prod_1^n \frac{f(x_i)}{f_{0h}(x_i)} = \prod_{i \in J_h} \frac{(f_h^*(i))^{n_{ih}}}{(f_{0h}^*(i))^{n_{ih}}}$$

where $n_{ih} = \#\{x_i \in (ih, (i+1)h]\}$, $J_h = \{i : n_{ih} > 0\}$, and for any density f, f_h^* denotes the probability on $\mathcal{N}$ given by $f_h^*(j) = \int_{jh}^{(j+1)h} f(x)dx$. Also let f_h denote the histogram $f_h(x) = f^*(i)/h$ for $x \in (ih, (i+1)h]$.

Since D_{α_h} is Dirichlet and $\alpha_h(\mathcal{N}) = 1$,

$$\frac{1}{n}\int_{\mathcal{F}_h}\prod_{i\in J_h}\frac{(f_h^*(i))^{n_{ih}}}{h^n}D_{\alpha_h}(df) = \frac{1}{n}\frac{1}{\Gamma(n+1)}\prod_{i\in J_h}\frac{1}{h^n}\frac{\Gamma(\alpha_h(i)+n_{ih})}{\Gamma(\alpha_h(i))}$$

Therefore

$$\frac{1}{n}\log\int_{\mathcal{F}_h}\prod_1^n\frac{f(x_i)}{f_{0h}(x_i)}D_{\alpha_h}(df) = \frac{1}{n}\log\frac{1}{\Gamma(n+1)}\prod_{i\in J_h}\frac{1}{h^n}\frac{\Gamma(\alpha_h(i)+n_{ih})}{\Gamma(\alpha_h(i))} - \log\prod_{i\in J_h}\frac{f_{0h}^*(i)}{h^n}$$

It is shown in Lemma 5.5.2 that

$$\frac{1}{n}\log\frac{1}{\Gamma(n+1)}\prod_{i\in J_h}\frac{1}{h^n}\frac{\Gamma(\alpha_h(i)+n_{ih})}{\Gamma(\alpha_h(i))} - \sum_{i\in J_h}n_{ih}\log\frac{n_{ih}}{n} \to 0 \text{ a.s.}P_{f_0}^\infty \tag{5.10}$$

Using (5.10) we have

$$\begin{aligned}
&\lim_{n\to\infty}\frac{1}{n}\int_{\mathcal{F}_h}\prod_{i\in J_h}\frac{(f_h^*(i))^{n_{ih}}}{h^n}D_{\alpha_h}(df) \\
&= \lim_{n\to\infty}\left[\frac{1}{n}\log\frac{1}{\Gamma(n+1)}\prod_{i\in J_h}\frac{1}{h^n}\frac{\Gamma(\alpha_h(i)+n_{ih})}{\Gamma(\alpha_h(i))} - \log h - \sum_{i\in J_h}\log f_{0h}^*(i) + \log h\right] \\
&\qquad = \lim_{n\to\infty}\sum_{i\in J_h}\frac{n_{ih}}{n}\log\frac{n_{ih}}{n} - \log h - \frac{1}{n}\log\prod_{i\in J_h}\frac{(f_{0h}^*(i))^{n_{ih}}}{h^n} \\
&\qquad = -\sum_{i\in J_h}\frac{n_{ih}}{n}\log\frac{n_{ih}}{n} - \log h - \sum_{i\in J_h}\frac{n_{ih}}{n}\log f_{0h}^*(i) + \log h \\
&\qquad\qquad \to 0 \text{ a.s. } P_{f_0}^\infty
\end{aligned} \tag{5.11}$$

□

Lemma 5.5.1. *Under the assumptions of the theorem,*

$$\frac{\max_{i\in J_h} i}{\sqrt{n}} \to 0 \ a.s\ P_{f_0}^\infty$$

Consequently

$$\frac{\#J_h}{\sqrt{n}} \le \frac{\max_{i\in J_h} i}{\sqrt{n}} \to 0 \ a.s\ P_{f_0}^\infty$$

Proof.

$$\max_{i \in J_h} i \leq \{\frac{\max(X_1, X_2, \ldots, X_n)}{h}\} + 1$$

Now $\max(X_1, X_2, \ldots, X_n)/\sqrt{n} \to 0$. This follows from: If $Y_1, Y_2, \ldots, Y_n$ are i.i.d. ($X_i^2 = Y_i$ in our case) then $\max(Y_1, Y_2, \ldots, Y_n)/n \to 0$ iff $EY_1 < \infty$. Recall assumption (a) of Theorem 5.5.1. □

Lemma 5.5.2. *Under the assumptions of the theorem*

$$\frac{1}{n} \log \frac{1}{\Gamma(n+1)} \prod_{i \in J_h} \frac{1}{h^n} \frac{\Gamma(\alpha_h(i) + n_{ih})}{\Gamma(\alpha_h(i))} - \sum_{i \in J_h} n_{ih} \log \frac{n_{ih}}{n} \to 0 \ a.s. \ P_{f_0}^{\infty} \tag{5.12}$$

Proof. Let $l_n(h)$ stand for the first term on the left-hand side. Then

$$l_n(h) = \frac{1}{n} \log \frac{1}{\Gamma(n+1)} \prod_{i \in J_h} \frac{1}{h^n} \frac{\Gamma(\alpha_h(i) + n_{ih})}{\Gamma(\alpha_h(i))}$$

$$\begin{aligned} l_n(h) =& \frac{1}{n} \log \frac{1}{\Gamma(n+1)} \prod_{i \in J_h} \frac{1}{h^n} \frac{\Gamma(\alpha_h(i) + n_{ih})}{\Gamma(\alpha_h(i))} \frac{1}{n} \sum_{i \in J_h} \log \Gamma(\alpha_h(i) + n_{ih}) \\ &- \frac{1}{n} \sum_{i \in J_h} \log \Gamma(\alpha_h(i)) - \log h - \frac{1}{n} \log \Gamma(n+1) \end{aligned}$$

We first show that

$$\frac{1}{n} \sum_{i \in J_h} \log \Gamma(\alpha_h(i)) \to 0 \text{ a.s. } P_{f_0}^{\infty}$$

Since $\Gamma(x) \leq 1/x$ for $0 \leq x \leq 1$, for $h < \epsilon$,

$$0 \leq \frac{1}{n} \sum_{i \in J_h} \log \Gamma(\alpha_h(i)) \leq \frac{1}{n} \sum_{1}^{n} \log \frac{1}{\alpha_h(i)}$$

By using a telescoping argument, the right-hand side of the expression becomes

$$\frac{1}{n}\sum_{i=2}^{N}\sum_{j=2}^{k}\left[\log\frac{1}{\alpha_h(i)} - \log\frac{1}{\alpha_h(i-1)}\right] + \frac{N}{n}\log 1\alpha_h(1)$$

$$= \frac{1}{n}\sum_{2}^{N}(N-j+1)\log\frac{\alpha_h(j-1)}{\alpha_h(j)} + \frac{N}{n}\log 1\alpha_h(1)$$

$$\leq \frac{(N+1)(N+2)}{2n}K_h + \frac{N}{n}\log\frac{1}{\alpha_h(1)} \to 0 \text{ a.s. } P_{f_0}^{\infty} \quad (5.13)$$

By Stirling's approximation for all $x \geq 1$,

$$\log\Gamma(x) = (x-\frac{1}{2})\log x - x + \log\sqrt{2\pi} + R(x) \qquad 0 < R(x) < 1$$

and we can write

$$\frac{1}{n}\log\frac{1}{\Gamma(n+1)}\prod_{i\in J_h}\frac{1}{h^n}\frac{\Gamma(\alpha_h(i)+n_{ih})}{\Gamma(\alpha_h(i))}\frac{1}{n}\sum_{i\in J_h}\log\Gamma(\alpha_h(i)+n_{ih}) - \frac{1}{n}\sum_{i\in J_h}\log\Gamma(\alpha_h(i))$$

$$= \frac{1}{n}\sum_{i\in J_h}\{(\alpha_h(i)+n_{ih}-\frac{1}{2})\log(\alpha_h(i)+n_{ih})\}$$

$$-\frac{1}{n}\sum_{i\in J_h}\left[\alpha_h(i) - n_{ih} - \log\sqrt{2\pi} + R(\alpha_h(i)+n_{ih})\right] - \log h$$

$$-\frac{1}{n}\{(n+\frac{1}{2})\log(n+1) - (n+1) + \log\sqrt{2\pi} + R(n)\} \quad (5.14)$$

Since $\sum_{i\in J_h} n_{ih} = n$ and

$$\frac{1}{n}\sum_{i\in J_h}\left[-\alpha_h(i) + \log\sqrt{2\pi} + R(\alpha_h(i)) + n_{ih} - \log h\right]$$

$$\leq \frac{(\max_{i\in J_h} i)(2+\log\sqrt{2\pi})}{n} \to 0 \text{ a.s. } P_{f_0}^{\infty} \quad (5.15)$$

we get

$$\lim_{n\to\infty} |l_n(h) - \sum_{i\in J_h} \frac{n_{ih}}{nh} \log \frac{n_{ih}}{nh}|$$
$$\leq \frac{1}{n} \sum_{i\in J_h} \{(\alpha_h(i) + n_{ih} - \frac{1}{2}) \log(\alpha_h(i) + n_{ih})\}$$
$$- \sum_{i\in J_h} \frac{n_{ih}}{nh} \log n_{ih} + \log n + \log h \quad (5.16)$$

By adding and subtracting $1/n \sum_{i\in J_h} n_{ih} - 1/2 \log n_{ih}$ we have

$$|\frac{1}{n} \sum_{i\in J_h} \{(\alpha_h(i) + n_{ih} - \frac{1}{2}) \log(\alpha_h(i) + n_{ih}) - \sum_{i\in J_h} \frac{n_{ih}}{nh} \log n_{ih}|$$
$$\leq |\frac{1}{n} \sum_{i\in J_h} \alpha_h(i) \log(\alpha_h(i) + n_{ih})|$$
$$+ \frac{1}{n} \sum_{i\in J_h} (n_{ih} - \frac{1}{2}) \log(1 + \frac{\alpha_h(i)}{n_{ih}}| + \frac{1}{n} \sum_{i\in J_h} \frac{1}{2} \log n_{ih} \quad (5.17)$$

Using $\log(1+x) \leq x$

$$\leq \frac{\log(n+1)}{n} + \frac{1}{n} + \frac{\log n}{2n} \# J_h$$

The last term in this expression goes to 0 by Lemma 5.5.2. □

The condition $\alpha(j-1)/\alpha(j) < K$ essentially requires that the prior does not vanish too rapidly in the tails. If our prior expectation f_0 is unimodal then it is easy to see that the condition holds with $K = \int_{m-h}^{m+h} f_0(x) ds$, where m is the mode of f_0.

5.5.2 L_1-Consistency

We next turn to L_1-consistency. We will use Theorem 4.4.4. Recall that Theorem 4.4.4 required two sets of conditions—one being the Schwartz condition and the other was construction of a sieve $\mathcal{F}_n$ with metric entropy $n\beta$ and such that $\Pi(\mathcal{F}_n^c)$ is exponentially small. A look at the proof of Theorem 4.4.4 shows that the Schwartz condition can be replaced by

$$\text{for all } \beta > 0, \quad \liminf_{n\to\infty} e^{n\beta} \int \prod_1^n \frac{f(X_i)}{f_0(X_i)} \Pi(df) = \infty \text{ a.s } P_{f_0}^\infty$$

Since we have already discussed this aspect in the last section, here we shall concentrate on the construction of a sieve.

To look ahead our sieve will be $\mathcal{F}_n = \cup_{h>h_n}\mathcal{F}_{a_{n,h}}$ where $\mathcal{F}_{a_{n,h}}$ is the set of histograms with support $[-a_n, a_n]$. We will compute the metric entropy of $\mathcal{F}_n$ and show that for a suitable choice of h_n, a_n it is of the order $n\beta$. What is then left is to ensure that the prior gives exponentially small mass to $\mathcal{F}_n^c$

Proposition 5.5.2.

$$\textit{Let } \mathbb{P}_k^\delta = \{(P_1, P_2, \ldots, P_k) : P_i \geq 0, \sum_1^k P_i \geq 1-\delta\}$$

Then

$$J(\mathbb{P}_k^\delta, 2\delta) \leq (\frac{k}{\delta} + \frac{1}{2})\log(1+\delta) + k\log(1+\delta) - \frac{1}{2}\log K + 1$$

Proof. Let K^* be the largest integer less than or equal to k/δ and consider

$$\mathbb{P}^* = \{P \in \mathbb{P}_k^\delta : P_i = j\frac{\delta}{k} \text{ for some integer } j\}$$

We will show that given any $P \in \mathbb{P}_k^\delta$ there is $P^* \in \mathbb{P}^*$ with $\|P - P^*\| < 2\delta$. The logarithm of the cardinality of $\mathbb{P}^*$ then gives an upper bound for $J(\mathbb{P}_k^\delta, 2\delta)$.

Let $P \in \mathbb{P}_k^\delta$. Then since

$$|\frac{P_i}{\sum P_j} - P_i| = \frac{P_i}{\sum P_j}(1 - \sum P_j) \leq \frac{P_i}{\sum P_j}\delta,$$

we have $\|(P_i/\sum P_j) - P_i\| < \delta$.

Given $P \in \mathbb{P}_k^\delta$ with $\sum P_i = 1$, let P^* be such that

$$P_i^* = j\frac{\delta}{k} \text{ for some integer } j \text{ and } P_i - P_i^* < \frac{\delta}{k}$$

Then $P^* = (P_1^*, P_2^*, \ldots, P_k^*) \in \mathbb{P}^*$ and also $\|P - P^*\| < \delta$. Thus we have shown that $\mathbb{P}^*$ is a 2δ net in $\mathbb{P}_k^\delta$.

To compute the number of elements in $\mathbb{P}^*$, consider k^* points $a_1, a_2, \ldots, a_{k^*}$, each endowed with a weight of δ/k. If we place $(k-1)$ sticks among these points, then these divide $a_1, a_2, \ldots, a_{k^*}$ into k parts, those to the left of the first stick, those between the first and second, and so on, the last part being all those $a_i^\prime s$ to the right of the last stick. Adding the weight of each of these parts gives a $(P_1^*, P_2^*, \ldots, P_k^*) \in \mathbb{P}^*$ and any element of $\mathbb{P}^*$ corresponds to a k partition of $a_1, a_2, \ldots, a_{k^*}$. The number of ways of partitioning k^* elements into k parts (some may be empty) is $\binom{k^*+k-1}{k-1}$.

Recall Stirling's approximation

$$x! = \sqrt{2\pi} x^{x+\frac{1}{2}} e^{-x+\frac{\theta}{12x}} \qquad 0 < \theta < 1$$

so that

$$\begin{aligned}\binom{k^*+k-1}{k-1} &= \frac{(k^*+k-1)!}{(k-1)!k^*!} \\ &\leq \frac{(k^*+k)!}{k!k^*!} \\ &\leq \frac{\sqrt{2\pi}(k^*+k)!^{(k^*+k)!+\frac{1}{2}} e^{-(k^*+k)!+\frac{\theta}{12(k^*+k)!}}}{\sqrt{2\pi} k^{k+\frac{1}{2}} e^{-k+\frac{\theta}{12k}} \sqrt{2\pi} (k^*)^{k^*+\frac{1}{2}} e^{-k^*+\frac{\theta}{12k^*}}}\end{aligned}$$

and therefore

$$\log\binom{k^*+k-1}{k-1} \leq \log\frac{(k^*+k)^{k^*+\frac{1}{2}}}{(k^*)^{k^*+\frac{1}{2}}} + \log\frac{(k^*+k)^k}{k^{k^*+\frac{1}{2}}} + \epsilon$$

where

$$\epsilon = \log\frac{1}{\sqrt{2\pi}} + \frac{\theta}{12(k+k^*)} - \frac{\theta}{k^*} - \frac{\theta}{k} < 1$$

so that,

$$\begin{aligned}J(\mathbb{P}_k^\delta, 2\delta) \leq (k^* + \frac{1}{2}) \log(1+\frac{k}{k^*}) &+ k\log(1+\frac{k^*}{k}) \\ &- \frac{1}{2}\log k + 1\end{aligned}$$

substituting $k^* \leq k/\delta$ we get the proposition. □

Lemma 5.5.3. *Suppose*

$$P \in \mathbb{P}_k^\delta = \{(P_1, P_2, \ldots, P_k) : 1 \geq P_i \geq 0, \sum_1^k P_i \geq 1 - \delta\}$$

$\delta < 1, h_0 < h < 1$ *and* $h - h_0 = \epsilon < \delta h_0/2(K+1)$. *If* f_h *is the histogram* $f_h(x) = \sum(P_i/h) I_{(ih,(i+1)h]}(x)$ *and* f_{h_0} *is the histogram* $f_{h_0}(x) = \sum(P_i)h) I_{(ih_0,(i+1)h_0]}(x)$, *then* $\|f_h - f_{h_0}\| < 3\delta$.

Proof. Let

$$I_1 = (0, h], I_2 = (h, 2h], \ldots I_k = ((k-1)h, kh]$$

and

$$J_1 = (0, h_0], J_2 = (h_0, 2h_0], \ldots J_k = ((k-1)h_0, kh_0]$$

Because $k\epsilon < h$, for $i < k$,

$$I_i = (I_i \cap J_i) \cup (I_i \cap J_{i+1}$$

Further,

$$I_i \cap J_{i+1} = ((i+1)h_0, (i+1)h)$$

Since $f_h = f_{h_0}$ on $I_i \cap J_i$, we have

$$\int_{I_i} |f_h - f_{h_0}| dx = |\frac{P_i}{h} - \frac{P_{(i+1)}}{h}|(i+1)(h - h_0)$$

and because $\sum P_i \leq 1$ and $h < h_0$,

$$\begin{aligned}\int_0^{kh} |f_h - f_{h_0}| dx &= \sum_1^{k-1} |P_i - P_{(i+1)}| \frac{(i+1)(h-h_0)}{h} + P_k \frac{(i+1)(h-h_0)}{h} \\ &\leq \sum_1^k P_i \frac{(i+1)(h-h_0)}{h} \\ &\leq 2(k+1)\frac{\epsilon}{h_0} \leq \delta\end{aligned} \tag{5.18}$$

□

A bit of notational clarification: For every h, a_n/h will not be an integer and hence when we write $\mathcal{F}_{a_n,h}$ what we mean is the set of all histograms from 0 to $[a_n/h]$ where $[a_n/h]$is the largest integer less than or equal to a_n/h. In our calculations, to avoid notational mess, we pretend that a_n/h is an integer.

Lemma 5.5.4. *For $a > 0$, let $\mathcal{F}_{a,h}$ be all histograms from $[0, a]$ with bin width h. Then*

$$\cup_{h > h_0} \mathcal{F}_{a,h} = \cup_{2h_0 > h > h_0} \mathcal{F}_{a,h}$$

Proof. For any $h > h_0$, for some integer m, $(h/m) \in (h_0, 2h_0)$. The conclusion follows because any histogram with bin width h can also be viewed as a histogram with bin width h/m. □

We put all the previous steps together in the next proposition Let $\mathcal{F}^{\delta}_{a,h}$ be all histograms f_h in $\mathcal{F}_{a,h}$ such that $P_{f_h}[0,a] > 1 - \delta$.

Proposition 5.5.3.

$$J\left(\cup_{h'>h}\mathcal{F}^{\delta}_{a,h'}, 5\delta\right) \le \log(\frac{2a}{h} + 1) + (\frac{a}{h\delta}\log(1+\delta) + \frac{a}{n}\log(1+\frac{1}{\delta}) + 1$$

Proof. By Lemma 5.5.4

$$\cup_{h'>h}\mathcal{F}_{a,h'} = \cup_{2h>h'>h}\mathcal{F}_{a,h'}$$

Set $k = 2a/h$ and $\epsilon = \delta h^2/(2a+1)$

Let $N^* = [h\epsilon]+1$ where for any a, $[a]$ is the largest integer less than or equal to a, and $h_i = h + i\epsilon, i = 1, 2, \cdots, N^*$. Then by Proposition 5.5.2, given any $f \in \cup_{2h>h'>h}\mathcal{F}_{a,h'}$, there is some h_i such that $\|f - f_{h_i}\| < 3\delta$. Use of Proposition 5.5.1 at each of $\mathcal{F}_{a,h_i}$, and a bit of algebra gives the result. □

Theorem 5.5.2. *Let μ be a probability measure on $(0,\infty)$ such that 0 is in the support of μ. α is a probability measure on $\mathbb{R}$. Our setup is $h \sim \mu$, the prior on $\mathcal{F}_h$ is D_{α_h} where $\alpha_h(i) = \alpha(ih, (i+1)h]$. Let $a_n \to \infty, h_n \to 0$ such that $(a_n/nh_n) \to 0$.*
If

(i) for some $\beta_0, \beta_1, C_1, C_2 > 0$,

$$\alpha(-a_n, a_n] > 1 - C_1 e^{-n\beta_0}$$

(ii) $\mu(0, h_n) < C_2 e^{-n\beta_1}$

then the posterior is strongly consistent at any f_0 satisfying (5.8).

Proof. If $\frac{a_n}{nh_n} \to 0$, it follows from Proposition 5.5.3 that $J(\mathcal{F}_n, \delta) < n\beta$ for large enough n. An easy application of Markov inequality with condition (i), and using (ii) gives $\Pi(\mathcal{F}_n^c) < Ce^{-n\gamma}$ for some C and γ. Theorem 4.4.4 gives the conclusion. □

Thus if $a_n = n^a$ and $h_n = n^{-b}$ then what we need is $a + b < 1$. For example if α is normal then one can take $a_n = n^{-1/2}$. The condition would then be satisfied if $h_n = n^{-b}$ with $b < 1/2$.

5.6 Mixtures of Normal Kernel

Another case of special interest is when K is the normal These priors were introduced by Lo [131], (see also Ghorai and Rubin[72] and West [168] who obtained expressions for the resulting posterior and predictive distributions. These can be further generalized by eliciting the base measure $\alpha = M\alpha_0$ of the Dirichlet up to some parameters and then considering hierarchical priors for these hyperparameters.

5.6.1 Dirichlet Mixtures: Weak Consistency

Returning to the mixture model, let ϕ and ϕ_h denote, respectively the standard normal density and the normal density with mean 0 and standard deviation h. Let $\Theta = \mathbb{R}$ and $\mathcal{M}$ be the set of probability measures on Θ. If P is in $\mathcal{M}$, then $f_{h,P}$ will stand for the density

$$f_{h,P}(x) = \int \phi_h(x - \theta)dP(\theta)$$

Note that $f_{h,P}$ is just the convolution $\phi_h * P$.

To get a feeling for the developments, we first look at the case where $h = h_0$ is fixed and our model is $P \sim \Pi$ and given P, $X_1, X_2, \ldots, X_n$ are i.i.d. f_p. In this case, the induced prior is supported by $\mathcal{F}_{h_0} = \{f_{h_0,P} : P \in \mathcal{M}\}$, and the following facts are easy to establish from Scheffe's theorem:

(i) The map $P \mapsto f_{h_0,P}$ is one-to-one, onto $\mathcal{F}_{h_0}$. Further $P_n \to P_0$ weakly if and only if $\|f_{h_0,P_n} - f_{h_0,P}\| \to 0$.

(ii) $\mathcal{F}_{h_0}$ is a closed subset of $\mathcal{F}$.

Fact (ii) shows that $\mathcal{F}_{h_0}$ is the support of Π, and hence consistency is to be sought only for densities of the form $f_{h_0,P}$. Theorem 5.6.1 implies consistency for such densities. Fact (i) shows that if the interest is in the posterior distribution of P, then weak consistency at P_0 is equivalent to strong consistency of the posterior of the density at $f_{h_0,P}$.

In order to establish weak consistency of the posterior distribution of f we need to verify the Schwartz condition. Following is a proposition that though not useful when Π is D_α is useful in other contexts.

Proposition 5.6.1.

$$K(f_P, f_Q) \leq K(P, Q)$$

Proof. A bit of change of variables and order of integration would show that

$$K(f_P, f_Q) = K(\int P_x\phi(x)dx, \int Q_x\phi(x)dx)$$

where P_x is the measure P shifted by x. Using the convexity of the K-L divergence and observing $K(P_x, Q_x) = K(P, Q)$ for all x, we have

$$K(f_P, fQ) = K(\int P_x\phi(x)dx, \int Q_x\phi(x)dx) \leq \int K(P_x, Q_x)\phi(x)dx = K(P, Q)$$

□

Thus if we have a prior Π such that every P is in K-L support then the posterior is weakly consistent at f_P. In fact the earlier remark shows that we have weak consistency at P and hence strong consistency at f_P. The Dirichlet does not have this property. However, we will show in Chapter 6 that for a suitable choice of parameters the Polya tree satisfies this property. Fixing h severely restricts the class of densities and is thus not of much interest.

We turn next to the model with a prior for h. Our model consists of a prior μ for h and a prior Π on $\mathcal{M}$. The prior $\mu\times\Pi$ through the map $(h, P) \mapsto f_{h,P}$ induces a prior on $\mathcal{F}$. We continue to denote this prior also by Π. Thus $(h, P) \sim \mu \times \Pi$ and given (h, P), $X_1, X_2, \ldots, X_n$ are i.i.d. $f_{h,P}$. This section describes a class of densities in the K-L support of Π. By Schwartz's theorem the posterior will be weakly consistent at these densities. The results in this section are largely from [74]. The next two results look at two simple cases and hold for general priors, but Theorem 5.6.3 makes use of special properties of the Dirichlet.

Theorem 5.6.1. *Let the true density f_0 be of the form $f_0(x) = f_{h_0,P_0}(x) = \int \phi_{h_0}(x - \theta)\, dP_0(\theta)$. If P_0 is compactly supported and belongs to the support of Π, and h_0 is in the support of μ, then $\Pi(K_\epsilon(f_0)) > 0$ for all $\epsilon > 0$.*

Proof. Suppose $P_0[-k, k] = 1$. Since P_0 is in the weak support of Π, it follows that $\Pi\{P : P[-k, k] > 1/2\} > 0$. It is easy to see that f_0 has moments of all orders.

For $\eta > 0$, choose k' such that $\int_{|x|>k'} \max(1, |x|) f_0(x)dx < \eta$. For $h > 0$, we write $\int_{-\infty}^{\infty} f_0 \log\left(f_{h,P_0}/f_{h,P}\right)$ as the sum

$$\int_{-\infty}^{-k'} f_0 \log \frac{f_{h,P_0}}{f_{h,P}} + \int_{-k'}^{k'} f_0 \log \frac{f_{h,P_0}}{f_{h,P}} + \int_{k'}^{\infty} f_0 \log \frac{f_{h,P_0}}{f_{h,P}} \tag{5.19}$$

Now

$$\begin{aligned}
&\int_{-\infty}^{-k'} f_0(x) \log\left(\frac{f_{h,P_0}(x)}{f_{h,P}(x)}\right) dx \\
&\leq \int_{-\infty}^{-k'} f_0(x) \log\left(\frac{\int_{-k}^{k} \phi_h(x-\theta) dP_0(\theta)}{\int_{-k}^{k} \phi_h(x-\theta)\, dP(\theta)}\right) dx \\
&\leq \int_{-\infty}^{-k'} f_0(x) \log\left(\frac{\phi_h(x+k)}{\phi_h(x-k) P[-k,k]}\right) dx \\
&= \int_{-\infty}^{-k'} f_0(x) \frac{2k|x|}{h^2} dx - \log(P[-k,k]) \int_{-\infty}^{-k'} f_0(x) dx \\
&< \left(\frac{2k}{h^2} + \log 2\right) \eta
\end{aligned}$$

provided $P[-k,k] > 1/2$. Similarly, we get a bound for the third term in (5.19).

Clearly,

$$c := \inf_{|x|\leq k'} \inf_{|\theta|\leq k} \phi_h(x-\theta) > 0$$

The family of functions $\{\phi_h(x-\theta) : x \in [-k',k']\}$, viewed as a set of functions of θ in $[-k,k]$, is uniformly equicontinuous. By the Arzela-Ascoli theorem, given $\delta > 0$, there exist finitely many points $x_1, x_2, \ldots, x_m$ such that for any $x \in [-k',k']$, there exists an i with

$$\sup_{\theta\in[-k,k]} |\phi_h(x-\theta) - \phi_h(x_i-\theta)| < c\delta \tag{5.20}$$

Let

$$E = \left\{P : \left|\int \phi_h(x_i-\theta) dP_0(\theta) - \int \phi_h(x_i-\theta) dP(\theta)\right| < c\delta; i = 1,2,\ldots,m\right\}$$

Since E is a weak neighborhood of P_0, $\Pi(E) > 0$. Let $P \in E$. Then for any $x \in [-k',k']$, choosing the appropriate x_i from (5.20), using a simple triangulation argument we get

$$\left|\frac{\int \phi_h(x-\theta) dP(\theta)}{\int \phi_h(x-\theta) dP_0(\theta)} - 1\right| < 3\delta$$

and so

$$\left|\frac{\int \phi_h(x-\theta) dP_0(\theta)}{\int \phi_h(x-\theta) dP(\theta)} - 1\right| < \frac{3\delta}{1-3\delta}$$

(provided $\delta < 1/3$).

Thus for any fixed $h > 0$, for P in a set of positive Π-probability, we have

$$\int f_0 \log\left(f_{h,P_0}/f_{h,P}\right) < 2\left(\frac{2k}{h^2} + \log 2\right)\eta + \frac{3\delta}{1-3\delta} \tag{5.21}$$

Now for any h,

$$\int f_0 \log\left(f_0/f_{h,P}\right) = \int f_0 \log\left(f_0/f_{h,P_0}\right) + \int f_0 \log\left(f_{h,P_0}/f_{h,P}\right) \tag{5.22}$$

The first term on the right-hand side of (5.22) converges to 0 as $h \to h_0$. To see this, observe that

$$\frac{\int \phi_{h_0}(x-\theta)dP_0(\theta)}{\int \phi_h(x-\theta)dP_0(\theta)} \leq \sup_{|\theta|\leq k} \frac{\phi_{h_0}(x-\theta)}{\phi_h(x-\theta)}$$

The rest follows by an application of the dominated convergence theorem.

Given any $\epsilon > 0$, choose a neighborhood N of h_0 (not containing 0) such that if $h \in N$, the first term on the right-hand side of (5.22) is less than $\epsilon/2$. Next choose η and δ so that for any $h \in N$, the right-hand side of (5.21) is less than $\epsilon/2$. Because h_0 is in the support of μ, the result follows. □

Remark 5.6.1. In Theorem 5.6.1, the true density is a compact location mixture of normals with a fixed scale. It is also possible to obtain consistency at true densities which are (compact) location-scale mixtures of the normal, provided we use a mixture prior for h as well. More precisely, if we modify the prior so that $(\theta, h) \sim P$ (a probability on $\mathbb{R} \times (0,\infty)$) and $P \sim \Pi$, then consistency holds at $f_0 = \int \phi_h(x - \theta)P_0(d\theta, dh)$ provided P_0 has compact support and belongs to the support of Π. The proof is similar to that of Theorem 3.

Theorem 5.6.1 covers the case when the true density is normal or a mixture of normal over a compact set of locations. This theorem, however, does not cover the case when the true density itself has compact support, like, say, the uniform. The next theorem takes care of such densities.

Theorem 5.6.2. *Let* 0 *be in the support of* μ *and* f_0 *be a density in the support of* Π. *Let* $f_{0,h} = \phi_h * f_0$. *If*

1. $\lim_{h\to 0} \int f_0 \log(f_0/f_{0,h}) = 0$,

2. *f_0 has compact support,*

then $\Pi(K_\epsilon(f_0)) > 0$ *for all* $\epsilon > 0$.

Proof. Note that, for each h,

$$\int f_0 \log(f_0/f_{h,P}) = \int f_0 \log(f_0/f_{0,h}) + \int f_0 \log(f_{0,h}/f_{h,P})$$

Choose h_0 such that for $h < h_0$, $\int f_0 \log(f_0/f_{0,h}) < \epsilon/2$ so all that is required is to show that for all $h > 0$,

$$\Pi\left\{P : \int f_0 \log\left(f_{0,h}/f_{h,P}\right) < \epsilon/2\right\} > 0$$

If f_0 has support in $[-k, k]$. Then

$$\int f_0 \log(f_{0,h}/f_{h,P}) \leq \int_{-k}^{k} f_0(x) \log\left(\frac{\int_{-k}^{k} \phi_h(x-\theta) f_0(\theta) d\theta}{\int_{-k}^{k} \phi_h(x-\theta) dP(\theta)}\right) dx$$

The rest of the argument proceeds in the same lines as in Theorem 5.6.1. □

While the last two theorems are valid for general priors on $\mathcal{M}$, the next theorem makes strong use of the properties of the Dirichlet process. For any P in $\mathcal{M}$, set $\overline{P}(x) = P(x, \infty)$ and $\underline{P}(x) = P(-\infty, x)$.

Theorem 5.6.3. *Let* D_α *be a Dirichlet process on* $\mathcal{M}$. *Let* l_1, l_2, u_1, u_2 *be functions such that for some* $k > 0$ *for all* P *in a set of* D_α*-probability* 1, *there exists* x_0 *(depending on* P*) such that*

$$\begin{gathered} \overline{P}(x) \geq l_1(x), \bar{P}(x + k\log x) \leq u_1(x)\ \forall x > x_0 \\ \text{and} \\ \underline{P}(x) \geq l_2(x), \underline{P}(x - k\log|x|) \leq u_2(x)\ \forall x < -x_0 \end{gathered} \tag{5.23}$$

For any $h > 0$, *define*

$$L_h(x) = \begin{cases} \phi_h(k\log x)(l_1(x) - u_1(x)), & \text{if } x > 0 \\ \phi_h(k\log|x|)(l_2(x) - u_2(x)), & \text{if } x < 0 \end{cases}$$

and assume that $L_h(x)$ *is positive for sufficiently large* $|x|$. *Let* f_0 *be the "true" density and* $f_{0,h} = \phi_h * f_0$. *Assume that* 0 *is in the support of the prior on* h. *If* f_0 *is in the support of* D_α *(equivalently,* $\text{supp}(f_0) \subset \text{supp}(\alpha)$*) and satisfies*

1. $\lim_{h \downarrow 0} \int f_0 \log(f_0/f_{0,h}) = 0;$,

2. for all h, $\lim_{a \uparrow \infty} \int_{-\infty}^{\infty} f_0(x) \log\left(\frac{f_{0,h}(x)}{\int_{-a}^{a} \phi_h(x-\theta) f_0(\theta) d\theta}\right) dx = 0;$ *and*

3. for all h, $\lim_{M \to \infty} \int_{|x|>M} f_0(x) \log\left(\frac{f_{0,h}(x)}{L_h(x)}\right) dx = 0,$

then $\Pi(K_\epsilon(f_0)) > 0$ *for all* $\epsilon > 0$.

Remark 5.6.2. It follows from Doss and Sellke [55] that if $\alpha = M\alpha_0$, where α_0 is a probability measure, then

$$\begin{aligned}
l_1(x) &= \exp[-2\log|\log \overline{\alpha_0}(x)|/\overline{\alpha_0}(x)] \\
l_2(x) &= \exp[-2\log|\log \underline{\alpha_0}(x)|/\underline{\alpha_0}(x)] \\
u_1(x) &= \exp\left[-\frac{1}{\overline{\alpha_0}(x+k\log x)|\log \overline{\alpha_0}(x-k\log x)|^2}\right] \\
u_2(x) &= \exp\left[-\frac{1}{\underline{\alpha_0}(x-k\log|x|)|\log \underline{\alpha_0}(x-k\log|x|)|^2}\right]
\end{aligned}$$

satisfy the requirements of (5.23). For example, when α_0 is double exponential, we may choose any $k > 2$ and the requirements of the theorem are satisfied if f_0 has finite moment-generating function in an open interval containing $[-1, 1]$.

Remark 5.6.3. The following argument provides a method for the verification of Condition 1 of Theorems 5.6.1 and 5.6.2 for many densities. Suppose that f_0 is continuous a.e., $\int f_0 \log f_0 < \infty$, and further assume that, as for unimodal densities, there exists an interval $[a, b]$ such that $\inf\{f(x) : x \in [a, b]\} = c > 0$ and f_0 is increasing in $(-\infty, a)$ and is decreasing in (b, ∞). Note that $\{x : f_0(x) \geq c\}$ is an interval containing $[a, b]$. Replacing the original $[a, b]$ by this new interval, we may assume that $f_0(x) \leq c$ outside $[a, b]$. Choose h_0 such that $N(0, h_0)$ gives probability $1/3$ to $(0, b-a)$. Let $h < h_0$. Let Φ denote the cumulative distribution function of $N(0, 1)$. If $x \in [a, b]$ then

$$f_{0,h}(\theta) \geq \int_a^b f_0(\theta)\phi_h(x-\theta)\, d\theta \geq c(\Phi((b-x)/h) + \Phi((x-a)/h) \geq c/3$$

If $x > b$ then

$$f_{0,h}(\theta) \geq \int_a^x f_0(\theta)\phi_h(x-\theta)\, d\theta \geq f_0(x)\left(\frac{1}{2} + \Phi((b-a)/h) - 1\right) \geq f_0(x)/3$$

Using a similar argument when $x < a$, we have that the function

$$g(x) = \begin{cases} \log\left(3f_0(x)/c\right), & \text{if } x \in [a,b] \\ \log 3, & \text{otherwise} \end{cases}$$

dominates $\log(f_0/f_{0,h})$ for $h < h_0$ and is P_{f_0}-integrable. Since $f_0(x)/f_{0,h}(x) \to 1$ as $h \to 0$ whenever x is a continuity point of f_0 and $\int f_0 \log(f_0/f_{0,h}) \geq 0$, an application of (a version of) Fatou's lemma shows that $\int f_0 \log(f_0/f_{0,h}) \to 0$ as $h \to 0$.

Proof. Let $\epsilon > 0$ be given and $\delta > 0$, to be chosen later. First find h_0 so that $\int f_0 \log(f_0/f_{0,h}) < \epsilon/2$ for all $h < h_0$. Fix $h < h_0$. Choose k_1 such that

$$\int_{-\infty}^{\infty} f_0(x) \log\left(\frac{f_{0,h}(x)}{\int_{-k_1}^{k_1} \phi_h(x-\theta) f_0(\theta) d\theta}\right) dx < \delta$$

Let $p = P[-k_1, k_1]$ and let p_0 denote the corresponding value under P_0. We may assume that $p_0 > 0$. Let P^* denote the conditional probability under P given $[-k_1, k_1]$, i.e., $P^*(A) = P(A \cap [-k_1, k_1])/p$ (if $p > 0$) and P_0^* denoting the corresponding objects for P_0. Let E be the event $\{P : |p/p_0 - 1| < \delta\}$. Because P_0 is in the support of D_α, $D_\alpha(E) > 0$. Now choose $x_0 > k_1$ such that

(i) $\displaystyle\int_{|x|>x_0} f_0(x) \log\left(f_{0,h}(x)/L_h(x)\right) dx < \delta$

(ii) $D_\alpha(E \cap F) > 0$, where

$$F = \left\{ P : \begin{array}{c} \overline{P}(x) \geq l_1(x), \overline{P}(x + k\log x) \leq u_1(x)\ \forall x > x_0 \\ \text{and} \\ \underline{P}(x) \geq l_2(x), \underline{P}(x - k\log|x|) \leq u_2(x)\ \forall x < -x_0 \end{array} \right\}$$

By Egoroff's theorem, it is indeed possible to meet condition (ii).

Consider the event

$$G = \left\{ P : \sup_{-x_0 < x < x_0} \log\left(\frac{\int_{-k_1}^{k_1} \phi_h(x-\theta) dP_0^*(\theta)}{\int_{-k_1}^{k_1} \phi_h(x-\theta) dP^*(\theta)}\right) < 2\delta \right\}.$$

We shall argue that $D_\alpha(E \cap F \cap G) > 0$ and if $P \in (E \cap F \cap G)$ then $\int f_0 \log(f_0/f_{h,P}) < \epsilon$ for a suitable choice of δ.

The events $E \cap F$ and G are independent under D_α, and hence, to prove the first statement, it is enough to show that $D_\alpha(G) > 0$. By intersecting G with E and using the fact that $\{\phi_h(x-\theta) : -x_0 \le x \le x_0\}$ is uniformly equicontinuous when $\theta \in [-k_1, k_1]$, we can conclude that $D_\alpha(G) \ge D_\alpha(G \cap E) > 0$ (see the proof of Theorem 5.6.1).

Now,

$$\begin{aligned}
&\int f_0 \log(f_0/f_{h,P}) \\
&\quad \le \int_{-\infty}^{\infty} f_0(x)\log(f_0(x)/f_{0,h}(x))dx \\
&\qquad + \int_{|x|\le x_0} f_0(x)\log\left(\frac{f_{0,h}(x)}{\int_{-k_1}^{k_1}\phi_h(x-\theta)f_0(\theta)d\theta}\right)dx \\
&\qquad + \int_{|x|\le x_0} f_0(x)\log\left(\frac{\int_{-k_1}^{k_1}\phi_h(x-\theta)f_0(\theta)d\theta}{\int_{-k_1}^{k_1}\phi_h(x-\theta)dP(\theta)}\right)dx \\
&\qquad + \int_{|x|>x_0} f_0(x)\log\left(\frac{f_{0,h}(x)}{\int\phi_h(x-\theta)dP(\theta)}\right)dx
\end{aligned}$$

If $P \in E \cap F \cap G$, then for $x > x_0$,

$$\begin{aligned}
\int_{-\infty}^{\infty}\phi_h(x-\theta)dP(\theta) &\ge \int_x^{x+k\log x}\phi_h(x-\theta)dP(\theta) \\
&\ge \phi_h(k\log x)[\overline{P}(x) - \overline{P}(x+k\log x)]
\end{aligned}$$

and because $P \in F$, the expression is further greater than or equal to

$$\phi_h(k\log x)[l_1(x) - u_1(x)] = L_h(x)$$

Using a similar argument for $x < -x_0$, we get

$$\int_{|x|>x_0} f_0(x)\log\left(\frac{f_{0,h}(x)}{f_{h,P}(x)}\right)dx \le \int_{|x|>x_0} f_0(x)\log\left(\frac{f_{0,h}(x)}{L_h(x)}\right)dx < \delta$$

Since $P \in E \cap G$, for each x in $[-x_0, x_0]$,

$$\log\left(\frac{\int_{-k_1}^{k_1}\phi_h(x-\theta)f_0(\theta)d\theta}{\int_{-k_1}^{k_1}\phi_h(x-\theta)dP(\theta)}\right) = \log\left(\frac{p_0}{p}\frac{\int_{-k_1}^{k_1}\phi_h(x-\theta)dP_0^*(\theta)}{\int_{-k_1}^{k_1}\phi_h(x-\theta)dP^*(\theta)}\right) < 3\delta$$

All these imply that if δ is sufficiently small, then $P \in E \cap F \cap G$ implies that $\int f_0 \log(f_{0,h}/f_{h,P}) < \epsilon$. □

5.6.2 Dirichlet Mixtures: L_1-Consistency

As before, we consider the prior which picks a random density $\phi_h * P$, where h is distributed according to μ and P is chosen independently of h according to D_α. Since we view h as corresponding to window length, it is only the small values of h that are relevant, and hence we assume that the support of μ is $[0, M]$ for some finite M.

In this model the prior is concentrated on

$$\mathcal{F} = \cup_{0<h<M} \mathcal{F}_h$$

where $\mathcal{F}_h = \{\phi_h * P : P \in \mathrm{M}\}$.

In order to apply Theorem 4.4.4, given $U = \{f : \|f - f_0\| < \epsilon\}$, for some $\delta < \epsilon/4$, we need to construct sieves $\{\mathcal{F}_n : n \geq 1\}$ such that $J(\delta, \mathcal{F}_n) \leq n\beta$ and $\mathcal{F}_n^c$ has exponentially small prior probability. Because, as $a_n \to \infty$, $D_\alpha\{P : P[-a_n, a_n] > 1 - \delta\} \to 1$, a natural candidate for $\mathcal{F}_n$ is

$$\mathcal{F}_n = \cup_{h_n<h<M} \mathcal{F}_h^{a_n}$$

where $h_n \downarrow 0$, a_n increases, and $\mathcal{F}_h^{a_n} = \{\phi_h * P : P[-a_n, a_n] > 1 - \delta\}$. What is then needed is an estimate of $J(\delta, \mathcal{F}_n)$. The next theorem provides such an estimate.

The next lemma shows that the restriction $h < M$ simplifies things a bit.

Lemma 5.6.1. *Let $M > 0$ and let $\mathcal{F}_{h,a,\delta}^M = \cup_{h<h'<M} \mathcal{F}_{h',a,\delta}$. If $a > M/\sqrt{\delta}$, then $\mathcal{F}_{h,a,\delta}^M \subset \mathcal{F}_{h,2a,2\delta}$.*

Proof. By Chebyshev's inequality, if $h' < M$ then the probability of $(-a, a]$ under $N(0, h')$ is greater than $1 - \delta$. If $f = \phi_{h'} * P$, then since $\phi_{h'} = \phi_h * \phi_{h^*}$, where $h^* < M$, $f = \phi_h * \phi_{h^*} * P$ and $(\phi_{h^*} * P)(-a, a] > 1 - 2\delta$. $\square$

Theorem 5.6.4. *Let $\mathcal{F}_{h,a,\delta}^M = \cup_{h<h'<M} \{f_{h,P} : P[-a, a] \geq 1 - \delta\}$. Then*

$$J(\delta, \mathcal{F}_{h,a,\delta}^M) \leq K \frac{a}{h},$$

where K is a constant that does depend on δ and M, but not on a or h.

We prove Theorem 5.6.4 through a sequence of lemmas. Let $\mathcal{F}_{h,a} = \{f_{h,P} : P(-a, a] = 1\}$. Without loss of generality, we shall assume that $a \geq 1$

Lemma 5.6.2. $J(2\delta, \mathcal{F}_{h,a}) \leq \left(\sqrt{\frac{8}{\pi}\frac{a}{h\delta}} + 1\right)\left(1 + \log\left(\frac{1+\delta}{\delta}\right).\right)$

Proof. For any $\theta_1 < \theta_2$,

$$
\begin{aligned}
&\|\phi_{\theta_1,h} - \phi_{\theta_2,h}\| \\
&= \frac{1}{\sqrt{2\pi}h} \int_{x>(\theta_1+\theta_2)/2} \exp[-(x-\theta_2)^2/(2h^2)]dx \\
&\quad - \frac{1}{\sqrt{2\pi}h} \int_{x>(\theta_1+\theta_2)/2} \exp[-(x-\theta_1)^2)/(2h^2)]dx \\
&\quad + \frac{1}{\sqrt{2\pi}h} \int_{x<(\theta_1+\theta_2)/2} \exp[-(x-\theta_1)^2/(2h^2)]dx \\
&\quad - \frac{1}{\sqrt{2\pi}h} \int_{x<(\theta_1+\theta_2)/2} \exp[-(x-\theta_2)^2/(2h^2)]dx \\
&= 4\frac{1}{\sqrt{2\pi}} \int_0^{(\theta_2-\theta_1)/(2h)} \exp[-x^2/2]dx \\
&\le \sqrt{\frac{2}{\pi}} \frac{(\theta_2-\theta_1)}{h}
\end{aligned}
$$

Given δ, let N be the smallest integer greater than $\sqrt{8}a/(\sqrt{\pi}h\delta)$. Divide $(-a, a]$ into N intervals. Let

$$
E_i = \left(-a + \frac{2a(i-1)}{N}, -a + \frac{2ai}{N}\right] : i = 1, 2, \ldots, N
$$

and let θ_i be the midpoint of E_i. Note that if $\theta, \theta' \in E_i$, then $|\theta - \theta'| < 2a/N$, and consequently $\|\phi_{\theta,h} - \phi_{\theta',h}\| < \delta$.

Let $\mathcal{P}_N = \{(P_1, P_2, \ldots, P_N) : P_i \ge 0, \sum_{i=1}^N P_i = 1\}$ be the N-dimensional probability simplex and let $\mathcal{P}_N^*$ be a δ-net in $\mathcal{P}_N$, i.e., given $P \in \mathcal{P}_N$, there is $P^* = (P_1^*, P_2^*, \ldots, P_N^*) \in \mathcal{P}_N^*$ such that $\sum_{i=1}^N |P_i - P_i^*| < \delta$.

Let $\mathcal{F}^* = \{\sum_{i=1}^N P_i^* \phi_{\theta_i,h} : P^* \in \mathcal{P}_N^*\}$. We shall show that $\mathcal{F}^*$ is a 2δ net in $\mathcal{F}_{h,a}$. If $f_{h,P} = \phi_h * P \in \mathcal{F}_{h,a}$, set $P_i = P(E_i)$ and let $P^* \in \mathcal{P}_N^*$ be such that $\sum_{i=1}^N |P_i - P_i^*| < \delta$.

Then

$$\begin{aligned}
&\left\| \int \phi_{\theta,h} dP(\theta) - \sum_{i=1}^{N} P_i^* \phi_{\theta_i,h} \right\| \\
&\leq \left\| \int \phi_{\theta,h} dP(\theta) - \sum_{i=1}^{N} \int I_{E_i}(\theta) \phi_{\theta_i,h} dP(\theta) \right\| + \left\| \sum_{i=1}^{N} P_i \phi_{\theta_i,h} - \sum_{i=1}^{N} P_i^* \phi_{\theta_i,h} \right\| \\
&\leq \int \sum_{i=1}^{N} I_{E_i}(\theta) \| \phi_{\theta,h} - \phi_{\theta_i,h} \| dP(\theta) + \sum_{i=1}^{N} |P_i - P_i^*| \\
&\leq 2\delta
\end{aligned}$$

This shows that $J(2\delta, \mathcal{F}_{h,a}) \leq J(\delta, \mathcal{P}_N)$, and we calculate $J(\delta, \mathcal{P}_N)$ along the lines of Barron, Schervish and Wasserman as follows: Since $|P_i - P_i^*| < \delta/N$ for all i implies that $\sum_{i=1}^{N} |P_i - P_i^*| < \delta$, an upper bound for the cardinality of the minimal δ-net of $\mathcal{P}_N$ is given by

$$\begin{aligned}
&\text{number of cubes of length } \delta/N \text{ covering } [0,1]^N \\
&\qquad \times \text{ volume of } \left\{ (P_1, P_2, \ldots, P_N) : P_i \geq 0, \sum_{i=1}^{N} P_i \leq 1 + \delta \right\} \\
&= (N/\delta)^N (1+\delta)^N \frac{1}{N!}
\end{aligned}$$

So,

$$\begin{aligned}
J(\delta, \mathcal{P}_N) &\leq N \log N - N \log \delta + N \log(1+\delta) - \log N! \\
&\leq N \log N - N \log \delta + N \log(1+\delta) - N \log N + N \\
&= N \left(1 + \log \frac{1+\delta}{\delta} \right) \\
&\leq \left(\sqrt{\frac{8}{\pi}} \frac{a}{h\delta} + 1 \right) \left(1 + \log \frac{1+\delta}{\delta} \right)
\end{aligned}$$

□

Lemma 5.6.3. *Let* $\mathcal{F}_{h,a,\delta} = \{ f_{h,P} : P(-a,a] \geq 1 - \delta \}$. *Then* $J(3\delta, \mathcal{F}_{h,a,\delta}) \leq J(\delta, \mathcal{F}_{h,a})$.

Proof. Let $f = \phi_h * P \in \mathcal{F}_{h,a,\delta}$. Consider the probability measure P^* defined by $P^*(A) = P(A \cap (-a, a])/P(-a, a]$. Then the density $f^* = \phi_h * P^*$ clearly belongs to $\mathcal{F}_{h,a}$ and further satisfies $\|f - f^*\| < 2\delta$. □

Proof. Putting Lemmas 5.6.2 , 5.6.3 and 5.6.1 together, we have Theorem 5.6.4. □

The next theorem formulates the result in terms of strong consistency for Dirichlet-normal mixtures.

Theorem 5.6.5. *Suppose that the prior μ has support in $[0, M]$. If for each $\delta > 0$, $\beta > 0$, there exist sequences a_n,$h_n \downarrow 0$ and constants β_0, β_1 (all depending on δ, β and M) such that*

1. *for some β_0, $D_\alpha\{P : P[-a_n, a_n] < 1 - \delta\} < e^{-n\beta_0}$,*
2. *$\mu\{h < h_n\} \leq e^{-n\beta_1}$, and*
3. *$a_n/h_n < n\beta$*

then f_0 is in the K-L support of the prior implies that the posterior is strongly consistent at f_0.

Remark 5.6.4. What was involved in the preceding is a balance between a_n and h_n. Since δ and M are fixed, the constant K obtained in Theorem 5.6.4 does not play any role. If α has compact support, say $[-a, a]$, then we may trivially choose $a_n = a$ and so h_n may be allowed to take values of the order of n^{-1} or larger. If α is chosen as a normal distribution and h^2 is given a (right truncated) inverse gamma prior, then the conditions of the theorem are satisfied if a_n is of the order $\sqrt{n}$ and $h_n = C/\sqrt{n}$ for a suitable (large) C (depending on δ and β).

5.6.3 *Extensions*

The methods developed in this chapter toward the simple mixture models can be used to study many of the variations used in practice. Some of these are discussed in this section.

1. It is often sensible to let the prior depend on the sample size; see for instance Roeder and Wasserman [141]. A case in point, in our context would be when the precision parameter $M = \alpha(\mathbb{R})$ is allowed to depend on the sample size.

   If Π_n is the prior at stage n, then the results goes through if the assumption $\Pi(K_\epsilon(f_0)) > 0$ is replaced by $\liminf_{n\to\infty} \Pi_n(K_\epsilon(f_0)) > 0$. This follows from the fact the Barron's Theorem (see Chapter 4) goes through with a similar change. The only stage that needs some care is an argument which involves Fubini, but it can be handled easily.

2. Another way the Dirichlet mixtures can be extended is by including a further mixing. Formally, Let $X_1, X_2, \ldots$ be observations from a density f where $f = \phi_h * P$, $P \sim D_{\alpha_\tau}$, $h \sim \pi$, τ is a finite-dimensional mixing parameter, which is also endowed with some prior ρ. Let f_0 be the true density. We are interested in verifying the Schwartz condition at f_0 and conditions for strong consistency. By Fubini's theorem, Schwartz's condition is satisfied for the mixture if

$$\rho\{\tau : \text{ Schwartz condition is satisfied with } \alpha_\tau\} > 0 \tag{5.24}$$

   (a) In particular, if f_0 has compact support, then (5.24) reduces to

$$\rho\{\tau : \text{supp}(f_0) \subset \text{supp}(\alpha_\tau)\} > 0 \tag{5.25}$$

   (b) Suppose f_0 is not of compact support and $\tau = (\mu, \sigma)$ gives a location-scale mixture. So we have to seek the condition so that the Schwartz condition holds with the base measure $\alpha((\cdot - \mu)/\sigma)$. We report results only for $\alpha_0 = \alpha/\alpha(\mathbb{R})$ double exponential or normal.

   When α_0 is double exponential, a sufficient condition is that $f_0(\mu+\sigma x)$ has finite moment-generating function on an open interval containing $[-1, 1]$. When α is normal, we need the integrability of $x \log|x| \exp[x^2/2]$ with respect to the density $f_0(\mu+\sigma x)$. For example, if the true density is $N(\mu_0, \sigma_0)$, then the required condition will be $\sigma < \sigma_0$, so we need

$$\rho\{(\mu, \sigma) : \sigma < \sigma_0\} > 0$$

   We omit proof of these statements. Simulation shows inclusion of location, and scale parameters in the base measure improves convergence of the the Bayes estimates to f_0.

   (c) For strong consistency, we further assume that the support of the prior ρ (for (μ, σ)) is compact. For each (μ, τ), find the corresponding $a_n(\mu, \sigma)$ of Theorem 5.6.5, i.e., satisfying

$$D_{\alpha(\mu,\tau)}\{P : P[-a_n(\mu, \tau), a_n(\mu, \tau)] < 1 - \delta\} < e^{-n\beta_0}$$

   for some $\beta_0 > 0$. Now choose $a_n = \sup_{\mu,\sigma} a_n(\mu, \sigma)$. The order of a_n will then be the same as the individual $a_n(\mu, \sigma)$s.

(d) In some special cases, it is also possible to allow unbounded location mixtures. For example, when the base measure is normal, a normal prior for the location parameter is both natural and convenient. Strong consistency continues to hold in this case as long as σ has a compactly supported prior. To see this, observe that $\rho\{|\mu| > \sqrt{n}\}$ is exponentially small and $\sup_{|\mu|\leq\sqrt{n},\sigma} a_n(\mu,\sigma)$ is again of the order of $\sqrt{n}$.

(e) West et al. put a random prior P' on h, independent of P and a Dirichlet prior for P'. This allows different amounts of smoothing near different sets of X_is. Our methods should apply here also. Such techniques, i.e., dependence of h on X_is or on x in the range of X_is have been introduced in the frequentist literature recently and are also known to improve estimates.

5.7 Gaussian Process Priors

Consider the probabilities $p_1, p_2, \dots p_k$ associated with a multinomial with k- cells. Often, for example, when the cells correspond to the bins of a histogram, it would be evident that a priori that the probabilities of adjacent cells would be highly positively correlated and the correlation would drop off for cells are farther apart. The Dirichlet prior for $p_1, p_2, \dots p_k$ results in negative covariance whereas we want positive covariance. It is thus necessary to model other covariance structures. The difficulty is one of specifying covariances which would ensure that the prior sits on $S_k = \{(p_1, p_2, \dots p_k), p_i \geq 0 \sum p_i = 1\}$. Leonard([126],[127]) suggested choosing real variables $Y_1, Y_2, \dots Y_k$ and setting $p_i = \exp(Y_i)/\sum \exp(Y_i)$. This ensures that $p_i \geq 0$ and $\sum p_i = 1$. Further if the distribution of $Y_1, Y_2, \dots Y_k$ is tractable, say $N(\mu, \Sigma)$, then Leonard shows that one can obtain tractable approximations to the posterior.

The situation is even more striking in the case of smooth random densities where smoothness already implies that the value of the density at two points x, y would be close if x and y are close. If we use the method of Section 5.5 calculations indicate that one gets positive covariance (for fixed h) only for very small values of h. In the spirit of Leonard one could choose a stochastic process $\{Y(x) : x \in \mathbb{R}\}$ with smooth sample paths and for any sample path define $f = \exp(y)/(\int(\exp y(t))dt)$. Leonard [127] suggested using a Gaussian process $\{Y(x) : x \in \mathbb{R}\}$. In this section we present these Gaussian process priors along the lines of Lenk [125]. Lenk considers a larger class of priors which gives a unified appearance to the results. An alternative method is to consider $f = \exp Y$ conditioned on $\int \exp Y(t)dt = 1$. Thorburn[157] has taken this approach. While this method is not discussed here, it would be interesting to see how this method relates to those developed by Leonard and Lenk.

Let $\mu : \mathbb{R} \mapsto \mathbb{R}$ and $\sigma : \mathbb{R} \times \mathbb{R} \mapsto \mathbb{R}^+$ be a symmetric function. σ is said to be positive definite if for any $x_1, x_2, \ldots, x_k$, the $k \times k$ matrix with $\sigma(x_i, x_j)$ as its entries is positive definite.

Definition 5.7.1. Let $\mu : \mathbb{R} \mapsto \mathbb{R}$ and σ be a positive definite function on $\mathbb{R} \times \mathbb{R}$. A process $\{Y(x) : x \in \mathbb{R}\}$ is said to be a Gaussian process with mean μ and covariance kernel σ if for any $x_1, x_2, \ldots, x_k$, $Y(x_1), Y(x_2), \ldots, Y(x_k)$ has a k-dimensional normal distribution with mean $\mu(x_1), \mu(x_2), \ldots, \mu(x_k)$ and covariance matrix whose (i,j)th entry is $\sigma(x_i, x_j)$.

The smoothness of the sample paths of a stochastic process is governed by moment conditions. Extensive results of this kind can be found in [36]. Following are a few that we use.

Theorem 5.7.1. *Let $\{\xi(x) : x \in \mathbb{R}\}$ be a stochastic process. Suppose that for positive constants $p \geq r$,*

$$E|\xi(t+h) - \xi(t)|^p \leq K|h|^{1+r} \text{ for all } t, h$$

Let $0 < a < r/p$. Then there is a process $\{\eta(x) : x \in \mathbb{R}\}$ equivalent to $\{\xi(x) : x \in \mathbb{R}\}$ (i.e. a process with the same finite-dimensional distributions as $\{\xi(x) : x \in \mathbb{R}\}$) such that

$$|\eta(t+h) - \eta(t)| \leq A|h|^a \text{ whenever } |h| < \delta$$

As an example consider the standard Brownian motion. A Gaussian process with $\mu = 0$ and $\sigma(x,y) = x \wedge y$. Let $h > 0$ then

$$E|\xi(t+h) - \xi(t)|^4 = 3\{Var(\xi(t+h) - \xi(t))\}^2 = 3h^2$$

So we can take $p = 4, r = 1$ to conclude that the sample paths are Lipschitz of order at least a, where $0 < a < 1/4$.

More generally, since $\frac{\xi(t+h)-\xi(t)}{h}$ is $N(0,1)$,

$$E|\xi(t+h) - \xi(t)|^{2k} = A_k h^k$$

and we can choose $p = 2k, r = k-1, 0 < a < (k-1)/2k$. Letting $k \to \infty$, we see that the sample functions are Lipshitz of order a for any $0 < a < 1/2$.

Theorem 5.7.2. *If for positive constants $p < r$ and K,*

$$E|\xi(t+h) - \xi(t)|^p \leq \frac{K|h|}{|\log|h||^{1+r}}$$

and

$$E|\xi(t+h) + \xi(t-h) - 2\xi(t)|^p \leq \frac{K|h|^{1+p}}{|\log|h||^{1+r}}$$

Then there is a process $\eta(t)$ equivalent to $\xi(t)$ such that $\eta'(t)$ exists and is continuous almost surely.

To return to Lenk, we consider a Gaussian process $Y(x)$ with mean μ and covariance kernel σ. Lenk appears to assume that

(i) μ is continuous;

(ii) σ is continuous on $\mathbb{R} \times \mathbb{R}$ and positive definite; and

(iii) there exist positive constants c, β, ϵ and nonnegative integer r such that

$$E|Y(x) - Y(y)|^\beta = C|x-y|^{1+r+\epsilon}$$

Condition (iii) guarantees that if $r \geq 1$ then with probability 1, the sample paths are r times continuously differentiable. A useful case is when σ is of the form $\sigma(x,y) = \rho(|x-y|)$ for some function ρ on $\mathbb{R}$. In this case, the process is stationary, and easier sufficient conditions are available for the sample paths to be smooth.

Theorem 5.7.3. *Let $\sigma(x,y) = \rho(|x-y|)$. If*

1. *for some $a > 3$*

$$\rho(h) = 1 - O\{|\log|h||^{-a}\} \text{ as } h \to 0$$

then there is an equivalent process with continuous sample paths

2. *for some $a > 3$ and $\lambda_2 > 0$,*

$$\rho(h) = 1 - \frac{\lambda h^2}{2} + O(\frac{h^2}{|\log|h||^a}) \text{ as } h \to 0$$

then there is an equivalent process whose sample paths are continuously differentiable

Cramér and Leadbetter [36] remark that $a > 3$ may be replaced by $a > 1$ but the proof requires lot more work. Here are some examples used in Lenk [125].

(i) $\rho(x) = e^{-|x|} = 1 - |x| + O(x^2)$ as $x \to 0$;

(ii) $\rho(x) = (1 - |x|)I_{|x|\le 1} = 1 - |x|$ as $x \to 0$;

(iii) $\rho(x) = e^{-x^2} = 1 - x^2 + O(x^4)$ as $x \to 0$; and

(iv) $\rho(x) = \frac{1}{1+x^2} = 1 - x^2 + O(x^4)$ as $x \to 0$.

Cases (i) and (ii) satisfy condition (1) of the theorem and (iii) and (iv) satisfy condition (2).

Let I be a bounded interval and let $\{Z(x) : x \in \mathbb{R}\}$ be a Gaussian process with mean μ and covariance kernel σ. The log-normal process, denoted by $LN(\mu, \sigma)$, is the process $W(x) = \exp(Z(x))$. We will denote the associated measure on $\mathbb{R}^+$ by $\Lambda(\mu, \sigma)$.

Following is a proposition which will be used later.

Proposition 5.7.1. *Fix $x_1, x_2, \ldots, x_k$ in I and constants $a_1, a_2, \ldots, a_k$.*

$$\textit{Let } \mu^*(x) = \mu(x) + \sum_1^k a_i \sigma(x, x_i)$$

Then

$$\frac{d\Lambda(\mu^*, \sigma)}{d\Lambda(\mu, \sigma)} = \frac{\prod_1^k W(x_i)^{a_i}}{E\prod_1^k W(x_i)^{a_i}} = \frac{\prod_1^k W(x_i)^{a_i}}{e^{\underline{a}\mu_{\underline{x}}' + \underline{a}\frac{\sigma_{\underline{x}}}{2}\underline{a}'}}$$

Here $W \in (\mathbb{R}^+)^I$ and the expectation in the right-hand side is with respect to $\Lambda(\mu, \sigma)$;$\mu_{\underline{x}} = (\mu(x_1), \mu(x_2), \ldots, \mu(x_k))$ and $[\sigma_{\underline{x}}]_{i,j} = \sigma(x_i, x_j), \underline{a} = a_1, a_2, \ldots, a_k$.

We will prove the proposition through a series of simple lemmas.

Lemma 5.7.1. *Let $(Z_1, Z_2, \ldots, Z_k)$ be multivariate normal with mean vector $\underline{\mu} = (\mu_1, \mu_2, \ldots, \mu_k)$ and covariance Σ. If $\underline{\mu}^* = (\mu_1^*, \mu_2^*, \ldots, \mu_k^*) = \underline{\mu} + \underline{a}\Sigma$, where $\underline{a}$ is the vector $(a_1, \cdots, a_k)$ then*

$$\frac{dN(\underline{\mu}^*, \Sigma)}{dN(\underline{\mu}, \Sigma)}(Z_1, Z_2, \ldots, Z_k) = Ke^{\sum_i a_i Z_i}$$

where $K = 1/Ee^{\sum a_i Z_i} = 1/e^{\underline{a}\underline{\mu}' + \underline{a}\frac{\Sigma}{2}\underline{a}'}$.

Proof. For any $\underline{\mu}_1$ and $\underline{\mu}_2$,

$$
\begin{aligned}
((\underline{x}-\underline{\mu}_1)\Sigma^{-1}(\underline{x}-\underline{\mu}_1)' - (\underline{x}-\underline{\mu}_2)\Sigma^{-1}(\underline{x}-\underline{\mu}_2)')\\
=2(\underline{\mu}_2 - \underline{\mu}_1)\Sigma^{-1}\underline{x}' + \underline{\mu}_1\Sigma^{-1}\underline{\mu}_1' - \underline{\mu}_2\Sigma^{-1}\underline{\mu}_2'
\end{aligned}
$$

Only the first term depends on $\underline{x}$. Absorbing the other two terms in the constant and taking $\underline{\mu}_1 = \underline{\mu}_*$ and $\underline{\mu}_2 = \underline{\mu}$ the lemma follows.

□

Lemma 5.7.2. *Let $G(\mu, \sigma)$ stand for the Gaussian measure with mean μ and covariance σ. If μ^* is as in Proposition 5.7.1, then*

$$\frac{dG(\mu^*, \sigma)}{dG(\mu, \sigma)}(Z) = Ke^{\Sigma_1^k a_i Z(x_i)} \tag{5.26}$$

Proof. It is enough to show that the finite-dimensional distributions of the measure defined by (5.26) are those arising from $dG(\mu^*, \sigma)$. But that is precisely the conclusion of the lemma 5.7.2. □

Next we state a simple measure theoretic lemma whose proof is routine.

Lemma 5.7.3. *Suppose P, Q are probability measures on $(\Omega, \underline{\mathcal{A}})$ and T is a 1-1 measurable function from $(\Omega', \mathcal{B})$. If $P \ll Q$ then $PT^{-1} \ll QT^{-1}$ and*

$$\frac{dPT^{-1}}{dQT^{-1}}(\omega') = \frac{dP}{dQ}(T^{-1}(\omega'))$$

Proof. To return to the proposition, it easily follows from Lemma 5.7.2 and by taking $T(Z) = e^Z$ in Lemma 5.7.3. □

We next add another real parameter ξ, and following Lenk we define a generalized log-normal process $LN(\mu, \sigma, \xi)$. When $\xi = 0$ the generalized log-normal process is defined to be $LN(\mu, \sigma)$, i.e., $LN(\mu, \sigma, 0) = LN(\mu, \sigma)$.

For any real ξ, $LN(\mu, \sigma, \xi)$ is defined by

$$\frac{dLN(\mu, \sigma, \xi)}{dLN(\mu, \sigma, 0)}(W) = \frac{[\int_I W(x)dx]^\xi}{C(\xi, \mu)} \tag{5.27}$$

where $C(\xi, \mu) = E\int_I W(x)dx]^\xi$ the expectation being taken under $LN(\mu, \sigma, 0)$. Lenk shows that this expectation exists for all real ξ.

We are now ready to define the random density.

Definition 5.7.2. Let $\{W(x).x \in \mathbb{R}\}$ be a generalized log normal process $LN(\mu, \sigma, \xi)$ on $\mathbb{R}^+$. The distribution of

$$f(x) = \frac{W(x)}{\int_I W(x)dx}$$

is called a *logistic normal* process and denoted by $LNS(\mu, \sigma, \xi)$.

Clearly f is a random density. We next show that if f has logistic normal distribution then so does the posterior given $X_1, X_2, \ldots, X_n$.

Theorem 5.7.4. *If $f \sim LNS(\mu, \sigma, \xi)$ then the posterior given $X_1, X_2, \ldots, X_n$ is $LNS(\mu^*, \sigma, \xi^*)$ where $\mu^*(x) = \mu(x) + \sum_1^n \sigma(x, X_i)$ and $\xi^* = \xi - n$.*

Proof. If $W \sim LN(\mu, \sigma, \xi)$ then by the Bayes theorem (for densities) the posterior Λ^* of W given $X_1, X_2, \ldots, X_n$ is

$$\frac{d\Lambda^*}{d\Lambda(\mu, \sigma, 0)}(W) = K \int_I [W(x)dx]^\xi \frac{\prod_1^n W(x_i)}{[\int_I [W(x)dx]^n]} \tag{5.28}$$

$$= K \int_I [W(x)dx]^{\xi - n} \prod_1^n W(x_i) \tag{5.29}$$

and comparison with (5.26) and (5.27) shows that this is $LNS(\mu^*, \sigma, \xi^*)$. The theorem follows because the distribution of f is just the posterior distribution of $W / \int_I W(x)dx$. □

Even though the transformations $\mu \mapsto \mu^*, \sigma \mapsto \sigma, \xi \mapsto \xi^*$ look simple, any interpretation needs to be tempered. First note that μ, σ, ξ do not identify the prior because if $\mu_1 - \mu_2 \equiv C$ then both $\mu_1, \sigma\xi$ and $\mu_2, \sigma\xi$ will lead to the same prior for f. Second μ and σ do not translate separately to $E(f)$ and $cov(f(x), f(y))$. A change in either μ or σ will affect both $E(f)$ and $cov(f(x), f(y))$. As $n \to \infty$ both $\mu^* \to \infty$ and $\xi^* \to -\infty$ indicating that these cannot be used to do simple minded asymptotics.

Since the prior is on densities, the natural tool to study consistency is the Schwartz theorem and Theorem 4.4.4. When the Gaussian process is a standard Brownian motion, with some work it can be shown that if the true distribution f_0 satisfies $\log f_0$ is bounded then the Schwartz condition holds at f_0. Toward L_1-consistency a natural sieve to consider would be to divide $[a, b]$ into $O(\sqrt{n})$ intervals and to look at the class of functions that have oscillation less than δ in all the intervals. These are just preliminary observations; more careful study needs to be done.

It also appears, that in analogy with Dirichlet mixtures, one should introduce a window h in the covariance and have $\rho_h(x) = (1/h)\rho(x/h)$.

In any case a lot of further work is needed to develop this promising method.

It would also be good to have some theoretical or numerical evidence justifying the numerical calculation of the posterior given in Lenk. For instance, one could compare Lenk's algorithms with approximations based on discretization.

6

Inference for Location Parameter

6.1 Introduction

We begin our discussion of semiparametric problems with inference about location parameters. The related problem of regression is taken up in a later chapter.

Our starting point is an important counterexample of Diaconis and Freedman [46, 45]. Since the Dirichlet process is a very flexible and popular prior for many infinite-dimensional examples, it seems natural to use it for estimating a location parameter. Diaconis and Freedman showed that it leads to posterior inconsistency. Barron suggests that the pathology is more fundamental. We present some of their results in Section 2. Doss [50], [51] and [52], showed the existence of similar phenomena when one wants to estimate a parameter θ that is a median.

A common explanation is that inconsistency is due to the Dirichlet sitting on discrete distributions. It is indeed true that the semiparametric likelihood is difficult to handle when a prior sits on discrete distributions. But Diaconis and Freedman [46] argue in their rejoinder to such comments that they expect the same phenomenon for Polya tree priors that sit on densities. We take up this problem in Sections 6.3 and 6.4 and show that under certain conditions symmetrized Polya tree priors have a rich Kullback-Leibler support so that by Schwartz's theorem, one can show posterior consistency for the location parameter for a large class of true densities.

One lesson that emerges from all this is that the tail free property, which is a natural tool for consistency, is destroyed by the addition of a parameter. Hence the Schwartz criterion is an appropriate tool for proving consistency. In particular, if one wants posterior consistency for certain true P_0s, then it is desirable to have a prior whose Kullback-Leibler support contains them.

Another natural prior to consider is the Dirichlet mixture of normals, which has emerged as the currently most popular prior for Bayesian density estimation. We will explore its properties in the next chapter and return briefly to the location parameter in Chapter 7.

Much of this chapter is based on Diaconis and Freedman [46] and Ghosal et.al. [78].

6.2 The Diaconis-Freedman Example

Suppose we have the model

$$X_i = Y_i + \theta, \qquad i = 1, 2, \ldots, n$$

where given P and θ, Y_is are i.i.d. P. Finally P and θ are independent with Dirichlet process prior D_α for P and a prior density $\mu(\theta)$ for θ. The probability measure $\bar{\alpha}$ has a density g.

Suppose the true value of θ is θ_0 and the true distribution of the Ys is P_0 with density f_0. The densities μ, g, f_0 are all with respect to Lebesgue measure on appropriate spaces.

The main interest is in the location parameter θ and the behavior of the posterior for θ under P_0. Since the random distributions P are not symmetrized around 0, the location parameter has an identifiability problem. For the time being, we ignore this. We will rectify this later by symmetrizing P.

To calculate the posterior, note that the random distribution P' of Xs is a mixture of Dirichlet, i.e., given θ, $P' \sim D_{\alpha_\theta}$, where $\alpha_\theta(\cdot) = \alpha(\mathbb{R})\bar{\alpha}(\cdot - \theta)$. Because P_0 has a density X_is may be assumed to be distinct. Hence by expression (3.17) the posterior density $\Pi(\theta|X_1, X_2, \ldots, X_n)$ is proportional to

$$\mu(\theta) \prod_1^n g(X_i - \theta)$$

As Barron pointed out in his discussion of [46] the Dirichlet is a pathological prior for a parameter in a semiparametric problem. The posterior is the same as if one assumed that X_is are i.i.d. with the parametrized density $g(X_i - \theta)$.

Diaconis and Freedman point out that consequences of choosing g can be serious. If g is a normal density, then one gets consistency, but not when g is Cauchy. An intuitive interpretation of this is that a normal likelihood for θ provides a robust model. For example, the MLE is $\bar{X}$, which is consistent for $E(X) = \theta$ even without normality. On the other hand, a Cauchy likelihood for θ, unlike a Cauchy prior, does not provide robustness. In fact, Diaconis and Freedman provide the following counterexample. They construct an f_0, which has compact support, is symmetric around 0, and infinitely differentiable. Under θ_0 and P_0, nearly half the samples the posterior concentrates around $\theta_0+\delta$ and for nearly another half it concentrates around $\theta_0 - \delta$. The true model P_0 can be chosen to make δ as large as we please. Because we are now essentially dealing with a misspecified model g, when actually f_0 is true, some insight into this phenomenon as well as the argument in [46] can be achieved by studying the asymptotic behavior of the posterior under misspecified models; see [17] and Bunke and Milhaud [28].

We now indicate why the same phenomenon holds even if we symmetrize P to $P^s(A) = (1/2)(P(A) + P(-A))$.

Given P we first generate $Z_1, Z_2, \ldots, Z_n$, i.i.d. P. Then define $Y_i = |Z_i|\delta_i$, where δ_i are i.i.d. and $\delta_i = \pm 1$ with probability 1/2. Then $Y_1, Y_2, \ldots, Y_n$ are i.i.d. P^s. Given Y's and θ; $X_i = Y_i + \theta$ as before. We will provide a heuristic computation of the posterior distribution of θ.

Assume without loss generality that $X_1, X_2, \ldots, X_n$ and $(X_i+X_j)/2, 1 \le i < j \le n$ are all distinct. The variables $(\theta, \underline{X})$, $(\theta, \underline{Z}, \underline{\delta})$, and $(\theta, \underline{Y})$ may be related in two ways. If $\theta \neq (X_i + X_j)/2$ for all pairs i, j then

$$Y_i = |Z_i|\delta_i = X_i - \theta$$

are all distinct. Moreover, all the $|Z_i|$s are also distinct. For, if $|Z_i| = |Z_j|$, then δ_i and δ_j must be of opposite sign and θ must be $(X_i + X_j)/2$, a case we have excluded for the time being. Hence, given $\theta, |Z_1|, |Z_2|, \ldots, |Z_n|$ are n distinct values in a sample of size n from the distribution $P^{|Z|} = P^{s,|Z|}$, where P is D_{α_θ}. Hence one can write down the joint density of $|Z_1|, |Z_2|, \ldots, |Z_n|$ by equation (3.17). Finally, δ_is are independent given θ and $|Z_i|$. Since there is a 1-1 correspondence between Y_i and (Z_i, δ_i), the density of Y_is given θ is

$$C \prod_1^n \left(g^{|z|}(|y_i|)\frac{1}{2} \right) = C \prod_1^n g(y_i) = C \prod_1^n g(X_i - \theta) \tag{6.1}$$

where $C = \{\alpha(\mathbb{R})^{[n]}\}^{-1}\{\alpha(\mathbb{R})\}^n$.

There is a second way in which the Y_is can be related to X_is. Suppose $\theta = (X_i + X_j)/2$. Then $|Z_i| = |Z_j|$ and δ_i and δ_j are of opposite sign. The remaining $|Z|$s—all $(n-2)$ of them—are all distinct and different from the common value of $|Z_i|$ and $|Z_j|$. Hence, given $\theta = (X_i + X_j)/2$, the density of Zs (with respect to $(n-1)$-dimensional Lebesgue measure) is

$$D\left(\prod_{k\neq i,j} g^{|Z|}(|Y_k|)\right) g^{|Z|}(|Y_i|) = C\frac{\prod_1^n g^{|Z|}(|Y_k|)}{g^{|Z|}(|Y_j|)}$$

where $D = C/\alpha(\mathbb{R})$. Finally, given $\theta = (X_i + X_j)/2$, the density of $Y_1, Y_2, \ldots, Y_n$ is

$$C\frac{\prod_1^n g^{|Z|}(|Y_k)}{g^{|Z|}(|Y_j|)}\frac{1}{2^n} = \prod \frac{g(X_i - \theta)}{2g(X_i - X_j)} \tag{6.2}$$

because $|Y_i| = |Y_j| = |X_i - X_j|$ and $g(|X_i - X_j|) = g(X_i - X_j)$.

The density (6.1) multiplied by $\mu(\theta)$ leads to the absolutely continuous part of the posterior for θ, while (6.2) leads to its discrete part. Formally, the discrete part is

$$\Pi_d(\theta|X_1, X_2, \ldots, X_n) = \sum_{i<j} \mu\left(\frac{X_i + X_j}{2}\right) \prod \frac{g(X_i - \theta)}{2g(X_i - X_j)}$$

and the absolutely continuous part has the density

$$\Pi_c(\theta|X_1, X_2, \ldots, X_n) = \mu(\theta) C \prod_1^n g(y_i) = C\mu(\theta)\prod_1^n g(X_i - \theta)$$

Hence the posterior is

$$\Pi(\theta|X_1, X_2, \ldots, X_n) = \frac{C\mu_c(\theta, \underline{X}) + D\mu_d(\theta, \underline{X})}{C_N}$$

where C_N is the norming constant

$$C_N = C\int \mu_c(\theta, \underline{X})d\theta + \sum_{\theta=(X_i+X_j)/2: i<j} \mu_d(\theta, \underline{X})$$

A detailed, rigorous proof appears in lemma 3.1 of and Freedman[45]. The posterior is still pathological and leads to inconsistency.

Diaconis and Freedman give examples of P_0 where one of the two terms in the posterior dominate. In case the first term dominates, the posterior for the symmetrized Dirichlet is similar to the posterior for the Dirichlet, and the proof for consistency in that case applies here.

6.3 Consistency of the Posterior

When P has a symmetrized Dirichlet prior distribution and g is log concave, as for normal, then Diaconis and Freedman [45] show that the posterior is consistent for all θ_0 for essentially "all" true P_0. On the other hand without such assumptions consistency fails, as indicated in the previous section. One explanation is the pathological form of the posterior. A somewhat deeper explanation is the fact that the Dirichlet and the symmetrized Dirichlet live on discrete distributions.

Diaconis and Freedman reacted to this as follows. They argued that discreteness is not the main issue. They construct a class of Polya tree priors, supported by densities and remark "Now consider the location problem; we guess this prior is consistent when expectation is the normal and and inconsistent when it is Cauchy. The real mathematical issue, it seems to us, is to find computable Bayes procedures and figure out when they are consistent."

We believe that Diaconis and Freedman are correct in thinking that existence of density for random P is not enough. What one needs is a stronger notion of support and a prior that has a support rich enough to contain one's favorite P_0s. the weak support is not good enough except for tail free priors. Since addition of a parameter destroys the tail free property, neither tail free priors nor the assumption that P_0 is in the weak support of the prior helps in ensuring consistency. Schwartz's theorem shows that a sufficient condition for consistency is that P_0 is in the Kulback-Leibler support of the prior. Schwartz's theorem is stated next in the form in which we need it.

Our parameter space is $\Theta \times \mathcal{F}^s$ where Θ is the real line and $\mathcal{F}^s$ is the set of all symmetric densities on $\mathbf{R}$. On $\Theta \times \mathcal{F}^s$, we consider a prior $\mu \times \mathbf{P}$ and given (θ, f), $X_1, X_2, \ldots, X_n$ are independent identically distributed as $P_{\theta,f}$, where $P_{\theta,f}$ is the probability measure corresponding to the density $f(x-\theta)$. We denote by f_θ the density $f(x-\theta)$. Given $X_1, X_2, \ldots, X_n$, we consider the posterior distribution $(\mu \times \mathbf{P})(\cdots|X_1, X_2, \ldots, X_n)$ on $\Theta \times \mathcal{F}^s$ given by the density $\prod f_\theta(X_i)/\int \prod f_\theta(X_i) d(\mu \times \mathbf{P})(\theta, f)$. The posterior $(\mu \times \mathbf{P})(\cdots|X_1, X_2, \ldots, X_n)$ is said to be consistent at (θ_0, f_0) if, as $n \to \infty$, $(\mu \times \mathbf{P})(\cdots|X_1, X_2, \ldots, X_n)$ converges weakly to the degenerate measure δ_{θ_0, f_0} almost surely P_{θ_0, f_0}. Clearly, if the posterior is consistent at (θ_0, f_0), the marginal distribution of $(\mu \times \mathbf{P})(\cdots|X_1, X_2, \ldots, X_n)$ on Θ converges to δ_{θ_0} almost surely P_{θ_0, f_0}.

Theorem 6.3.1. *If for all* $\delta > 0$,

$$(\mu \times \mathbf{P})\{(\theta, f) : K(f_{\theta_0}, f_\theta) < \delta\} > 0, \tag{6.3}$$

then the posterior $(\mu \times \mathbf{P})(\cdots|X_1, X_2, \ldots, X_n)$ *is consistent at* (θ_0, f_0).

A naive way to ensure (6.3) is to require that θ_0 and f_0 belong respectively, to the Euclidean and Kullback-Leibler supports of μ and $\mathbf{P}$. The flaw in this argument is that the Kullback-Leibler divergence is not a metric. So even if θ is close to θ_0 and $K(f_0, f)$ is small, we cannot draw any conclusion about $K(f_{0\theta_0}, f_\theta)$ or $K(f, f_\theta)$. A way out is indicated below.

Definition 6.3.1. The map $(\theta, f) \mapsto f_\theta$ is said to be *KL-continuous* at $(0, f_0)$ if

$$K(f_0, f_{0,\theta}) = \int_{-\infty}^{\infty} f_0(x)\log(f_0(x)/f_0(x-\theta))dx \to 0 \quad \text{as} \quad \theta \to 0.$$

We would then call $(0, f_0)$ a *KL-continuity point.*

Let $f^*_{0,\theta}$ be the density defined by $f^*_{0,\theta}(x) = \left(f_{0,\theta}(x) + f_{0,\theta}(-x)\right)/2$, the symmetrization of $f_{0,\theta}$ where $f_{0,\theta}$ stands for $f_0(.-\theta)$. For later convenience we write $\mathbf{P}^*$ instead of $\mathbf{P}$ for a prior on $\mathcal{F}^s$.

Assumption A: Support of μ is $\mathbb{R}$ and for all θ sufficiently small, $f^*_{0,\theta}$ is in the K-L support of P^*.

It is easy to check that this condition holds for many common densities, e.g., for normal or Cauchy. However, it fails for densities like uniform on an interval. For such cases a different method is discussed later.

Theorem 6.3.2. *If μ and $\mathbf{P}^*$ satisfy Assumption A and if $(0, f_0)$ is a KL-continuity point, then the posterior $(\mu \times \mathbf{P}^*)(\cdots|X_1, X_2, \ldots, X_n)$ is consistent at $(0, f_0)$.*

Proof. We first prove it when $\theta = 0$. By Theorem 6.3.1, it is enough to verify that $\mu \times \mathbf{P}^*$ satisfies the Schwartz condition (6.3). For any θ,

$$\begin{aligned} K(f_0, f_\theta) &= \int_{-\infty}^{\infty} f_0 \log(f_0/f_{-\theta}) \\ &= \int_{-\infty}^{\infty} f_{0,\theta}\log f_{0,\theta} - \int_{-\infty}^{\infty} f_{0,\theta}\log f \end{aligned} \tag{6.4}$$

Since

$$\int_{-\infty}^{\infty} f_{0,\theta}\log f^*_{0,\theta} = \int_{-\infty}^{\infty} f^*_{0,\theta}\log f^*_{0,\theta} \tag{6.5}$$

and

$$\int_{-\infty}^{\infty} f_{0,\theta}\log f = \int_{-\infty}^{\infty} f^*_{0,\theta}\log f, \tag{6.6}$$

we have, by the concavity of $\log x$

$$\begin{aligned} K(f_0, f_\theta) &= \int_{-\infty}^{\infty} f_{0,\theta} \log(f_{0,\theta}/f^*_{0,\theta}) + \int_{-\infty}^{\infty} f^*_{0,\theta} \log(f^*_{0,\theta}/f) \\ &\le \frac{1}{2}\int_{-\infty}^{\infty} f_{0,\theta} \log\left(\frac{f_{0,\theta}}{f_{0,\theta}}\right) + \frac{1}{2}\int_{-\infty}^{\infty} f_{0,\theta} \log\left(\frac{f_{0,\theta}}{f_{0,-\theta}}\right) + K(f^*_{0,\theta}, f) \\ &= \frac{1}{2}K(f_0, f_{0,-2\theta}) + K(f^*_{0,\theta}, f) \end{aligned} \tag{6.7}$$

By the KL-continuity assumption there is an ε such that for $|\theta| < \varepsilon$, the first term is less than $\delta/2$. For any θ, by Assumption A, $\{f : K(f^*_{0,\theta}, f) < \delta/2\}$ has positive $\mathbf{P}^*$ measure. Thus we have, for each $\theta \in [-\varepsilon, \varepsilon], \{f : K(f^*_{0,\theta}, f) < \delta/2\}$ is contained in $\{f : K(f_0, f_\theta) < \delta\}$. Since $\mu[-\varepsilon, \varepsilon] > 0$ this completes the proof for $\theta = 0$.

For a general θ_0, $K(f_{0,\theta_0}, f_{\theta_0+\theta}) = K(f_0, f_\theta)$ which by the previous argument is less than δ with positive probability, if f is chosen as before and θ is in $[\theta_0 - \epsilon, \theta_0 + \epsilon]$. □

Assumption A of Theorem 6.3.2 can be verified if $\mathbf{P}^*$ arises as follows. Let P^* be a symmetrization of P obtained by one of the following two methods.

Method 1. Let $\mathbf{P}$ be a prior on $\mathcal{F}$. The map $f \mapsto (f(x) + f(-x))/2$ from $\mathcal{F}$ to $\mathcal{F}^s$ induces a measure on $\mathcal{F}^s$.

Method 2. Let $\mathbf{P}$ be a prior on $\mathcal{F}(\mathbb{R}^+)$—the space of densities on $\mathbb{R}^+$. The map $f \mapsto f^*$, where, $f^*(x) = f^*(-x) = f(x)/2$, gives rise to a measure on $\mathcal{F}^s$.

Lemma 6.3.1. *Let* $\mathbf{P}$ *be a prior on* $\mathcal{F}$ *or on* $\mathcal{F}(\mathbf{R}^+)$ *with a given symmetric* f_0 *in its K-L support. Let* $\mathbf{P}^*$ *be the prior obtained on* $\mathcal{F}^s$ *by Method* 1 *or Method* 2. *If* $f_0 \in \mathcal{F}^s$, *then*

$$\mathbf{P}^*\{f \in \mathcal{F}^s : K(f_0, f) < \delta\} > 0 \tag{6.8}$$

Proof. For Method 1, the result follows from Jensen's inequality; the conclusion is immediate for method 2 because, setting $g_0(x) = 2f_0(x)$ and $g(x) = 2f(x)$ for x in $\mathbf{R}^+$, both g_0, g belong to $\mathcal{F}(\mathbf{R}^+)$ and $K(f_0, f) = K(g_0, g)$. □

The K-L continuity assumptions fails if f_0 has support in a finite interval. However, our next result in this section shows that consistency continues to hold even when f_0 has support in a finite interval, provided f_0 is continuous. The proof consists in approximating f_0 by an f_1 satisfying conditions of Theorem 6.3.2. We first need a lemma to bound a K-L number. It is a slight improvement over a lemma in [78].

Lemma 6.3.2. *Let* f_0 *and* f_1 *be densities so that* $f_0 \le Cf_1$. *Then for any* f,

$$K(f_0, f) \le C \log C + [K(f_1, f) + \sqrt{K(f_1, f)}]$$

Proof. First note that $C \geq 1$. Also

$$\begin{aligned} K(f_0, f) &\leq \int f_0[\log(f_0/f_1)]^+ \leq \int C f_1[\log(C f_1/f)]^+ \\ &\leq C \log C + C \int f_1[\log(f_1/f)]^+ \end{aligned} \tag{6.9}$$

But

$$\int f_1[\log(f_1/f)]^+ \leq K(f_1, f) + \int f_1[\log(f_1/f)]^- \tag{6.10}$$

$$\begin{aligned} f_1[\log(f_1/f)]^- &= \int f_1[\log(f/f_1)]^+ \leq \int f_1 \left(\frac{f}{f_1} - 1\right)^+ \\ &= \frac{\|f - f_1\|}{2} \leq \sqrt{K(f_1, f)} \end{aligned} \tag{6.11}$$

The last inequality follows from Proposition 1.2.2. Combining (6.9), (6.10) and (6.11), one gets the lemma. □

Theorem 6.3.3. *If μ and $\mathbf{P}^*$ satisfy Assumption A, f_0 is continuous and has support in a finite interval $[-a, a]$, and $\log \alpha(x)$ is integrable with respect to $N(\mu, \sigma^2)$ for all (μ, σ), then the posterior $\mathbf{P}(\cdots | X_1, X_2, \ldots, X_n)$ is consistent at (θ, f_0) for all θ.*

Proof. We consider two cases.

Case 1. $\inf_{[-a,a]} f_0(x) = \alpha > 0$.

Let

$$f_1(x) = \begin{cases} (1 - \eta) f_0(x), & \text{for } -a < x < a \\ (\eta/2)\phi_{-a,\sigma^2}, & \text{for } x \leq -a \\ (\eta/2)\phi_{a,\sigma^2}, & \text{for } x \geq a \end{cases} \tag{6.12}$$

where ϕ_{-a,σ^2} and ϕ_{a,σ^2} are, respectively, the densities of $N(-a, \sigma^2)$ and $N(a, \sigma^2)$ and σ^2 is chosen to ensure that f_1 is continuous at a.

We first show that f_1 is KL-continuous, i.e.,

$$\lim_{\theta \to 0} \int_{-\infty}^{\infty} f_1 \log(f_1/f_{1,\theta}) = \int_{-\infty}^{\infty} \lim_{\theta \to 0} f_1 \log(f_1/f_{1,\theta}) = 0 \tag{6.13}$$

It is enough to establish that for some $\varepsilon > 0$, the family $\{\log(f_1/f_{1,\theta}) : |\theta| < \varepsilon\}$ is uniformly integrable with respect to f_1. This follows because for any M,

$$\sup_{|\theta|<\varepsilon} \sup_{|x|<M} |\log(f_1(x)/f_{1,\theta}(x))| < C_M$$

and when M is large, for $|x| > M$, $f_{1,\theta}(x) = (\eta/2)(\sigma\sqrt{2\pi})^{-1}\exp[-(x-a-\theta)^2/(2\sigma^2)]$ for all $|\theta| < \varepsilon$, implying

$$\sup_{|\theta|<\varepsilon} \int_{|x|>M} f_1(x)\log(f_1(x)/f_{1,\theta}(x))dx \to 0 \quad \text{as } M \to \infty$$

It now follows from Lemma 6.3.2 that, by setting $C = (1-\eta)^{-1}$ and choosing η close to 1 so that $(C+1)\log C < \delta/2$, we can choose a δ^* such that $K(f_1, f) < \delta^*$ implies $K(f_0, f) < \delta$; consequently $\{(\theta, f) : K(f_1, f_\theta) < \delta^*\} \subset \{(\theta, f) : K(f_0, f_\theta) < \delta\}$. Theorem 6.3.2 shows that the set on the left hand side has positive $\mu \times \mathbf{P}^*$ measure.

Case 2. $\inf_{[-a,a]} f_0(x) = 0$.

By the continuity of f_0, we can, given any $\eta > 0$, choose a C such that $\int_{-a}^{a}(f_0 \vee C) = 1+\eta$, where $a \vee b = \max(a, b)$. Set $f_1 = (1+\eta)^{-1}(f_0 \vee C)$. Then $f_0 \leq (1+\eta)f_1$ and using Lemma 6.3.2, we can choose η and δ^* small such that $\{f : K(f_1, f) < \delta^*\} \subset \{f : K(f_0, f) < \delta\}$. Since f_1 is covered by Case 1, the theorem follows. □

In the remaining section we concentrate on constructing Polya tree priors which satisfy conditions of Theorem 6.3.2 for many f_0s.

6.4 Polya Tree Priors

The main result in this section is Theorem 6.4.1. It implies that Assumption A is true if P^* is a symmetrization of the Polya tree prior in this theorem and $K(f_{0,\theta_0}, \alpha) < \infty$ for all θ_0.

We already discussed the basic properties of Polya trees in Chapter 3. They are recalled below. Let $E = \{0, 1\}$ and E^m be the m-fold Cartesian product $E \times \cdots \times E$ where $E^0 = \emptyset$. Further, set $E^* = \cup_{m=0}^{\infty} E^m$. Let $\pi_0 = \{\mathbf{R}\}$ and for each $m = 1, 2, \ldots$, let $\pi_m = \{B_\varepsilon : \varepsilon \in E^m\}$ be a partition of $\mathbf{R}$ so that sets of π_{m+1} are obtained from a binary split of the sets of π_m and $\cup_{m=0}^{\infty}\pi_m$ is a generator for the Borel σ-field on $\mathbf{R}$. Let $\Pi = \{\pi_m : m = 0, 1, \ldots\}$.

A random probability measure P on $\mathbf{R}$ is said to possess a Polya tree distribution with parameters $(\Pi, \mathcal{A})$; we write $\mathsf{P} \sim \mathrm{PT}(\Pi, \mathcal{A})$, if there exist a collection of non-negative numbers $\mathcal{A} = \{\alpha_\varepsilon : \varepsilon \in E^*\}$ and a collection $\mathcal{Y} = \{Y_\varepsilon : \varepsilon \in E^*\}$ of random variables such that the following hold:

(i) the collection $\mathcal{Y}$ consists of mutually independent random variables;

(ii) for each $\varepsilon \in E^*$, Y_ε has a beta distribution with parameters $\alpha_{\varepsilon 0}$ and $\alpha_{\varepsilon 1}$;

(iii) the random probability measure P is related to $\mathcal{Y}$ through the relations

$$\mathsf{P}(B_{\varepsilon_1\cdots\varepsilon_m}) = \left(\prod_{j=1;\varepsilon_j=0}^{m} Y_{\varepsilon_1\cdots\varepsilon_{j-1}}\right)\left(\prod_{j=1;\varepsilon_j=1}^{m}(1-Y_{\varepsilon_1\cdots\varepsilon_{j-1}})\right) \quad m=1,2,\ldots,$$

where the factors are Y_0 or $1-Y_0$ if $j=1$.

We restrict ourselves to partitions $\Pi = \{\pi_m : m = 0,1,\ldots\}$ that are determined by a strictly positive continuous density α on $\mathbf{R}$ in the following manner: The sets in π_m are intervals of the form $\{x : (k-1)/2^m < \int_{-\infty}^{x}\alpha(t)dt \le k/2^m\}$, $k = 1,2,\ldots,2^m$. We term the measure (corresponding to) α as the base measure because its role is similar to the base measure of Dirichlet process.

Our next theorem refines theorem 2 of Lavine [119] by providing an explicit condition on the parameters.

Theorem 6.4.1. *Let f_0 be a density and $\mathbf{P}$ denote the prior* $\mathrm{PT}(\Pi,\mathcal{A})$*, where $\alpha_\varepsilon = r_m$ for all $\varepsilon \in E^m$ and $\sum_{m=1}^{\infty} r_m^{-1/2} < \infty$. Further assume that $K(f_0,\alpha) < \infty$. Then for every $\delta > 0$,*

$$\mathbf{P}\{\mathsf{P} : K(f_0,f) < \delta\} > 0 \tag{6.14}$$

Proof. By Theorem 3.3.7, the weaker condition $\sum_{m=0}^{\infty} r_m^{-1} < \infty$ implies the existence of a density of the random probability measure. Considering the transformation $x \mapsto \int_{-\infty}^{x}\alpha(t)dt$, assume that f and f_0 are densities on $[0,1]$. Moreover, Π is then the canonical binary partition. By the martingale convergence theorem, there exists a collection of numbers $\{y_\varepsilon : \varepsilon \in E^*\}$ from $[0,1]$ such that, with probability one

$$f_0(x) = \lim_{m\to\infty}\left(\prod_{j=1;\varepsilon_j=0}^{m} 2y_{\varepsilon_1\cdots\varepsilon_{j-1}}\right)\left(\prod_{j=1;\varepsilon_j=1}^{m} 2(1-y_{\varepsilon_1\cdots\varepsilon_{j-1}})\right) \quad . \tag{6.15}$$

where the limit is taken through a sequence $\varepsilon_1\varepsilon_2\cdots$ which corresponds to the dyadic expansion of x. It similarly follows that

$$f(x) = \lim_{m\to\infty}\left(\prod_{j=1;\varepsilon_j=0}^{m} 2Y_{\varepsilon_1\cdots\varepsilon_{j-1}}\right)\left(\prod_{j=1;\varepsilon_j=1}^{m} 2(1-Y_{\varepsilon_1\cdots\varepsilon_{j-1}})\right) \tag{6.16}$$

for almost every realization of f. Now for any $N \ge 1$,

$$K(f_0,f) = M_N + R_{1N} - R_{2N} \tag{6.17}$$

where

$$M_N = \mathsf{E}\left[\log\left(\prod_{j=1;\varepsilon_j=0}^{N}\left(\frac{y_{\varepsilon_1\cdots\varepsilon_{j-1}}}{Y_{\varepsilon_1\cdots\varepsilon_{j-1}}}\right)\prod_{j=1;\varepsilon_j=1}^{N}\left(\frac{1-y_{\varepsilon_1\cdots\varepsilon_{j-1}}}{1-Y_{\varepsilon_1\cdots\varepsilon_{j-1}}}\right)\right)\right] \tag{6.18}$$

$$R_{1N} = \mathsf{E}[\log(\prod_{j=N+1;\varepsilon_j=0}^{\infty} 2y_{\varepsilon_1\cdots\varepsilon_{j-1}} \prod_{j=N+1;\varepsilon_j=1}^{\infty} 2(1-y_{\varepsilon_1\cdots\varepsilon_{j-1}}))] \tag{6.19}$$

and

$$R_{2N} = \mathsf{E}[\log(\prod_{j=N+1;\varepsilon_j=0}^{\infty} 2Y_{\varepsilon_1\cdots\varepsilon_{j-1}} \prod_{j=N+1;\varepsilon_j=1}^{\infty} 2(1-Y_{\varepsilon_1\cdots\varepsilon_{j-1}}))] \tag{6.20}$$

with E standing for the expectation with respect to the distribution of $(\varepsilon_1, \varepsilon_2, \ldots)$ for a fixed realization of the Ys. The εs come from the binary expansion of x, and x is distributed according to the density f_0.

By the definition of a Polya tree, M_N and R_{2N} are independent for all $N \geq 1$. To prove (6.14), we show that for any $\delta > 0$, there is some $N \geq 1$ such that

$$\mathbf{P}\{M_N < \delta\} > 0 \tag{6.21}$$

$$|R_{1N}| < \delta \tag{6.22}$$

and

$$\mathbf{P}\{|R_{2N}| < \delta\} > 0 \tag{6.23}$$

The set $\{(Y_\varepsilon : \varepsilon \in E^m, m = 0, \ldots, N-1) : M_N < \delta\}$ is a nonempty open set in $\mathbf{R}^{2^N-1}$; it is open by the continuity of the relevant map and it is nonempty as $(y_\varepsilon : \varepsilon \in E^m, m = 0, \ldots, N-1)$ belongs to this set. Thus (6.21) follows by the nonsingularity of the beta distribution. Relation (6.22) follows from lemma 2 of Barron [6]. To complete the proof, it remains to show (6.23) for some $N \geq 1$. We actually prove the stronger fact

$$\lim_{N\to\infty} \mathbf{P}\{|R_{2N}| \geq \delta\} = 0 \tag{6.24}$$

Let $\mathbf{E}$ stand for the expectation with respect to the prior distribution.i.e., the distribution of the Ys and E, as before, the expectation with respect to the distribution of

$(\varepsilon_1, \varepsilon_2, \ldots)$. Now

$$
\begin{aligned}
&\mathbf{P}\{|R_{2N}| \geq \delta\} \\
&\quad \leq \delta^{-1}\mathbf{E}|R_{2N}| \\
&\quad \leq \delta^{-1}\mathbf{E}\,\mathsf{E}[\sum_{j=N+1;\varepsilon_j=0}^{\infty} |\log(2Y_{\varepsilon_1\cdots\varepsilon_{j-1}})| + \sum_{j=N+1;\varepsilon_j=1}^{\infty} |\log(2(1-Y_{\varepsilon_1\cdots\varepsilon_{j-1}}))|] \\
&\quad = \delta^{-1}\mathsf{E}[\sum_{j=N+1;\varepsilon_j=0}^{\infty} \mathbf{E}|\log(2Y_{\varepsilon_1\cdots\varepsilon_{j-1}})| + \sum_{j=N+1;\varepsilon_j=1}^{\infty} \mathbf{E}|\log(2(1-Y_{\varepsilon_1\cdots\varepsilon_{j-1}}))|] \quad (6.25) \\
&\quad \leq \delta^{-1}\mathsf{E}[\sum_{j=N+1}^{\infty} \max\{\mathbf{E}|\log(2Y_{\varepsilon_1\cdots\varepsilon_{j-1}})|, \mathbf{E}|\log(2(1-Y_{\varepsilon_1\cdots\varepsilon_{j-1}}))|] \\
&\quad \leq \delta^{-1}\sum_{j=N+1}^{\infty} \max_{(\varepsilon_1\cdots\varepsilon_{j-1})\in E^{j-1}} \max\{\mathbf{E}|\log(2Y_{\varepsilon_1\cdots\varepsilon_{j-1}})|, \mathbf{E}|\log(2(1-Y_{\varepsilon_1\cdots\varepsilon_{j-1}}))|] \\
&\quad = \delta^{-1}\sum_{j=N+1}^{\infty} \eta(r_{j-1})
\end{aligned}
$$

where $\eta(k) = E|\log(2U_k)|$ with $U_k \sim$Beta(k,k). By Lemma 6.4.1, $\eta(k) = O(k^{-1/2})$ as $k \to \infty$. Since $\sum_{m=1}^{\infty} r_m^{-1/2} < \infty$ by assumption, the right-hand side of (6.25) is the tail of a convergent series. This completes the proof of (6.24) and hence of the theorem as well. □

Remark 6.4.1. Essentially the same proof shows that the Kullback-Leibler neighborhoods would continue to have positive measure when the prior is modified as follows: Divide $\mathbf{R}$ into $k+1$ intervals $I_1, \ldots, I_{k+1}$ and assume that $(P(I_1), \ldots, P(I_k))$ have a joint density which is positive everywhere on the k-dimensional set $\{(a_1, \ldots, a_k) : a_i > 0, j = 1, \ldots, k, \sum_{j=1}^{k} a_i < 1\}$. For each I_j, the conditional distribution given $P(I_j)$ has a Polya tree prior satisfying the assumptions of the theorem. These priors are special cases of the priors constructed by Diaconis and Freedman. Moreover, it follows from theorem 1 of Lavine [119] that such priors can approximate any prior belief up to any desired degree of accuracy in a strong sense.

Remark 6.4.2. It is not necessary that for each m, $\alpha_{\varepsilon_1\cdots\varepsilon_m}$ be the same for all $(\varepsilon_1, \ldots, \varepsilon_m) \in E^m$. The proof goes through even when only $\alpha_{\varepsilon_1\cdots\varepsilon_{m-1}0} = \alpha_{\varepsilon_1\cdots\varepsilon_{m-1}1}$ for all $(\varepsilon_1, \ldots, \varepsilon_{m-1}) \in E^{m-1}$, $m \geq 1$, and $r_m := \min\{\alpha_{\varepsilon_1\cdots\varepsilon_m} : (\varepsilon_1, \ldots, \varepsilon_m) \in E^m\}$ satisfies the condition $\sum_{m=1}^{\infty} r_m^{-1/2} < \infty$.

Lemma 6.4.1. *If $U_k \sim$beta(k,k), then $E|\log(2U_k)| = O(k^{-1/2})$ as $k \to \infty$.*

Proof. The proof uses Laplace's method with a rigorous control of the error term. Let $\eta_k = E|\log(2U_k)|$, i.e.,

$$\eta_k = \frac{1}{B(k,k)} \int_0^1 |\log(2u)| u^{k-1}(1-u)^{k-1} du \tag{6.26}$$

$$= \frac{1}{B(k,k)} \int_0^1 |\log(2(1-u))| u^{k-1}(1-u)^{k-1} du \tag{6.27}$$

Adding (6.26) and (6.27) and observing that $\log(2u)$ and $\log(2(1-u))$ are always of the opposite sign,

$$2\eta_k = \frac{1}{B(k,k)} \int_0^1 |\log(u/(1-u))| u^{k-1}(1-u)^{k-1} du \tag{6.28}$$

This implies by Jensen's inequality that

$$\begin{aligned} 4\eta_k^2 &\le \frac{1}{B(k,k)} \int_0^1 (\log(u/(1-u)))^2 u^{k-1}(1-u)^{k-1} du \\ &= \frac{1}{B(k,k)} \int_0^1 \{1 + (\log(u/(1-u)))^2\} u^{k-1}(1-u)^{k-1} du - 1 \end{aligned} \tag{6.29}$$

We approximate the integral by Laplace's method. Let

$$\{1 + (\log(u/(1-u)))^2\} u^{k-1}(1-u)^{k-1} = \exp(g_k(u)) \tag{6.30}$$

where

$$g_k(u) = (k-1)\log u + (k-1)\log(1-u) + h(u)$$

and

$$h(u) = \log\{1 + (\log(u/(1-u)))^2\}$$

Clearly, $g_k(1/2) = -2(k-1)\log 2$, $g_k'(1/2) = 0$ and $g_k'(u)$ is decreasing in u so that $g_k(u)$ has a unique maximum at $1/2$. Fix $\delta > 0$ and let $\lambda = \sup\{h''(u) : |u - 1/2| < \delta\}$. Then on $u \in (1/2-\delta, 1/2+\delta)$, we have

$$g_k(u) \le -2(k-1)\log 2 - \frac{(u-\frac{1}{2})^2}{2}(8(k-1) - \lambda) \tag{6.31}$$

Thus

$$
\begin{aligned}
4\eta_k^2 & \\
&\leq \frac{1}{B(k,k)}\int_{1/2-\delta}^{1/2+\delta} \exp[-2(k-1)\log 2 - 4(k-1)\left(1-\frac{\lambda}{8(k-1)}\right)(u-\frac{1}{2})^2]du \\
&\quad + \frac{1}{B(k,k)}\int_{|u-\frac{1}{2}|>\delta} \{1+(\log(u/(1-u)))^2\}u^{k-1}(1-u)^{k-1}du - 1 \qquad (6.32)\\
&\leq \frac{\Gamma(2k)}{(\Gamma(k))^2}2^{-2(k-1)}\int_{-\infty}^{\infty} \exp[-4(k-1)\left(1-\frac{\lambda}{8(k-1)}\right)(u-\frac{1}{2})^2]du \\
&\quad + \frac{1}{B(k,k)}\int_{|u-\frac{1}{2}|>\delta} \{1+(\log(u/(1-u)))^2\}u^{k-1}(1-u)^{k-1}du - 1
\end{aligned}
$$

Since the function $u(1-u)\{1+(\log(u/(1-u))^2\}$ is bounded on $(0,1)$ by, say, M, the second term on the right-hand side of (6.32) is dominated by

$$
\begin{aligned}
&\frac{M}{B(k,k)}\int_{|u-1/2|>\delta} u^{k-2}(1-u)^{k-2}du \\
&= M\frac{(2k-1)(2k-2)}{(k-1)^2}P\{|U_{k-1}-\frac{1}{2}|>\delta\} \\
&\leq M\frac{(2k-1)(2k-2)}{(k-1)^2}E|U_{k-1}-\frac{1}{2}|^2/\delta^2 \\
&= O(k^{-1})
\end{aligned} \qquad (6.33)
$$

The first term on the right-hand side of (6.32) is

$$
\frac{\Gamma(2k)}{(\Gamma(k))^2}2^{-2k+2}(2\pi)^{1/2}(8(k-1)-\lambda)^{-1/2} \qquad (6.34)
$$

which, by an application of Stirling's inequalities [[171] p. 253], is less than

$$
\begin{aligned}
&\frac{(2k)^{2k-1/2}e^{-2k}(2\pi)^{1/2}\exp[(24k)^{-1}]}{(k^{k-1/2}e^{-k}(2\pi)^{1/2})^2}2^{-2k+2}(2\pi)^{1/2} \\
&\times 2^{-3/2}(k-1)^{-1/2}\left(1-\frac{\lambda}{8(k-1)}\right)^{-1/2} \\
&= \left(\frac{k}{k-1}\right)^{1/2}\exp[(24k)^{-1}]\left(1-\frac{\lambda}{8(k-1)}\right)^{-1/2} \\
&= 1+O(k^{-1})
\end{aligned} \qquad (6.35)
$$

Thus $\eta_k^2 = O(k^{-1})$, completing the proof. □

Remark 6.4.3. While we have discussed consistency issues, it would be interesting to explore how the robustness calculations in Section 4 of Lavine [119] can be made in the context of a location parameter.

We have argued that the Schwartz theorem is the best available tool for handling consistency issues in semiparametric problems. We have also exhibited a Polya tree priors which have a rich K-L support. However, there are caveats. The consistency theorem notwithstanding, computation of the posterior for θ for a density f_0 of the kind used by Diaconis-Freedman shows that convergence for Cauchy base measure is very slow. Even for $n = 500$, one notices the tendency to converge to a wrong value, as in the case of the Dirichlet prior with Cauchy base measure. Rapid convergence does take place if we replace the Cauchy by the normal.

A second fact is that the condition $\sum r_m^{-1/2} < \infty$ implies that the tail of the random P^* is close in some sense to the tail of the prior expected density. This in turn implies that the posterior for f converges to δ_{f_0} rather slowly, which might imply relatively slow convergence also of the posterior for θ. Both these questions can be better understood if one can get rates of convergence of the posterior and see how they depend on the base measure and the r_ms. These are delicate issues.

What happens if $\sum r_m^{-1/2} = \infty$? We have conjectured earlier that then, the Schwartz condition would not hold. If so, it seems likely that in all such cases consistency would depend dramatically on the base measure.

7

Regression Problems

7.1 Introduction

An important semiparametric problem is to make inference about the constants in the regression equation when the error in the regression model

$$Y_i = \alpha + \beta x_i + \epsilon_i, \qquad i = 1, 2, \ldots \tag{7.1}$$

has an unknown, symmetric distribution. This is similar to the location parameter problem, so it is natural to try a symmetrized Polya tree prior for the error distribution. Another prior that suggests itself is a symmetrized version of Dirichlet mixtures of normals of Chapter 5. We explore both priors in this chapter with a focus on posterior consistency. The covariate may arise as fixed nonrandom constants or as i.i.d. observations of a random variable.

Because this is a semiparametric problem, it is natural to try to use Schwartz's theorem. However since the observations are not identically distributed, major changes are needed. We begin with a variant of Schwartz's theorem in Section 7.2. In two of the subsequent sections we discuss how the conditions of the theorem can be verified. Lack of i.i.d. structure for the Y_is necessitates assumptions on the x_is to ensure that the exponentially consistent tests required by Schwartz's theorem exist in the current context. Also certain conditions have to be imposed on f_0 to verify conditions relating to K-L support and variance in the Schwartz theorem. Among other things it is shown that Polya tree priors of the sort considered in the Chapter 6 fulfill the required conditions on the prior.

We then turn to the Dirichlet mixtures of normal. It turns out that the random densities are sufficiently well behaved that the proof for results similar to that outlined in the previous paragraph can be simplified to some extent.

It may be observed that as in the Chapter 6 it may be tempting to use a Dirichlet prior on $\mathcal{F}$. It is easy to show that the posterior for α, β would be pathological in exactly the same way, namely, it would be identical with the posterior arising from assigning a parametric prior on $\mathcal{F}$. The proof is quite similar.

In the literature, the regression problem has been handled by putting a Dirichlet mixture of normals but without symmetrization. This means that there is an identifiability problem for the constant but not for the regression coefficient β. Of course, the posterior for α cannot be consistent, but one can show posterior consistency for β. In many examples, one would want consistency for both α and β, so symmetrization seems desirable. See , Burr et al.[29] for an interesting application.

The final section discusses binary response regression with nonparametric link functions. This chapter is based heavily on [134] and unpublished work of Messan.

7.2 Schwartz Theorem

Fix f_0, α_0, β_0. Let

$$f_{\alpha,\beta,i} = f_{\alpha+\beta x_i}(y) = f(y - (\alpha + \beta x_i)) \tag{7.2}$$

and put $f_{0i} = f_{0,\alpha_0,\beta_0,i}$.

For any two densities f and g, let

$$K(f,g) = \int f \log \frac{f}{g}, \quad V(f,g) = \int f \left(\log \frac{f}{g}\right)^2 \tag{7.3}$$

and put

$$K_i(f,\alpha,\beta) = K(f_{0i}, f_{\alpha,\beta,i}), \quad V_i(f,\alpha,\beta) = V(f_{0i}, f_{\alpha,\beta,i}) \tag{7.4}$$

As mentioned in the introduction, the main tool we use is a variant of Schwartz's theorem. The following theorem is an adaptation to the case when the Y_is are independent but not identically distributed. Here the x_is are nonrandom.

Definition 7.2.1. Let $\mathcal{W} \subset \mathcal{F} \times \mathbb{R} \times \mathbb{R}$. A sequence of test functions $\Phi_n(Y_1, \ldots, Y_n)$ is said to be *exponentially consistent* for testing

$$H_0 : (f, \alpha, \beta) = (f_0, \alpha_0, \beta_0) \qquad \text{against} \qquad H_1 : (f, \alpha, \beta) \in \mathcal{W} \tag{7.5}$$

if there exist constants C_1, C_2, $C > 0$ such that

(a) $E_{\prod_1^n f_{0i}} \Phi_n \leq C_1 e^{-nC}$, and

(b) $\inf_{(f,\alpha,\beta)\in\mathcal{W}} E_{\prod_1^n f_{\alpha,\beta,i}} (\Phi_n) \geq 1 - C_2 e^{-nC}$.

Theorem 7.2.1. *Suppose $\tilde{\Pi}$ is a prior on $\mathcal{F}$ and μ is a prior for (α, β). Let $\mathcal{W} \subset \mathcal{F} \times \mathbb{R} \times \mathbb{R}$. If*

(i) *there is an exponentially consistent sequence of tests for*

$$H_0 : (f, \alpha, \beta) = (f_0, \alpha_0, \beta_0) \quad \textit{against} \quad H_1 : (f, \alpha, \beta) \subset \mathcal{W}$$

(ii) *for all $\delta > 0$,*

$$\Pi\left\{(f, \alpha, \beta) : K_i(f, \alpha, \beta) < \delta \textit{ for all } i, \quad \sum_{i=1}^{\infty} \frac{V_i(f, \alpha, \beta)}{i^2} < \infty\right\} > 0$$

then with $\prod_{i=1}^{\infty} P_{f_{0i}}$ probability 1, the posterior probability

$$\Pi(\mathcal{W}|Y_1, \ldots, Y_n) = \frac{\int_{\mathcal{W}} \prod_{i=1}^{n} \frac{f_{\alpha,\beta\, i}(Y_i)}{f_{0i}(Y_i)} d\Pi(f, \alpha, \beta)}{\int_{\mathcal{F}\times\mathbb{R}\times\mathbb{R}} \prod_{i=1}^{n} \frac{f_{\alpha,\beta\, i}(Y_i)}{f_{0i}(Y_i)} d\Pi(f, \alpha, \beta)} \to 0 \tag{7.6}$$

Note that $V_i(f, \alpha, \beta)$ bounded above in i is sufficient to ensure the summability of $\sum_{i=1}^{\infty} V_i(f, \alpha, \beta)/i^2$.

Proof. The proof is similar to the proof of Schwartz's theorem. If we write (7.6) as

$$\Pi(\mathcal{W}|Y_1, \ldots, Y_n) = \frac{I_{1n}(Y_1, \ldots, Y_n)}{I_{2n}(Y_1, \ldots, Y_n)} \tag{7.7}$$

it can be shown, as in the proof of Schwartz's theorem (Chapter 4), that condition (i) implies that " there exists a $d > 0$ such that $e^{nd} I_{1n}(Y_1, \ldots, Y_n) \to 0$ a.s. "

The denominator can be handled similarly, using Kolomogorov's strong law of large numbers for independent but not identically distributed random variables. Yet, with a later application in mind, we give an argument here with a somewhat weaker assumption than (ii). For any two densities f and g, let

$$V_+(f,g) = \int f \left(\log_+ \frac{f}{g} \right)^2 \tag{7.8}$$

and put

$$V_{+i}(f,\alpha,\beta) = V_+(f_{0i}, f_{\alpha,\beta,i}) \tag{7.9}$$

We will show that " for all $d > 0$, $e^{nd} I_{2n}(Y_1, ..., Y_n) \to \infty$ a.s." under the assumption, (ii)$'$ For all $\delta > 0$,

$$\Pi \left\{ (f,\alpha,\beta) : K_i(f,\alpha,\beta) < \delta \text{ for all } i, \ \sum_{i=1}^{\infty} \frac{V_{+i}(f,\alpha,\beta)}{i^2} < \infty \right\} > 0$$

Because $V_+(f,g) \le V(f,g)$ it is easy to see that (ii) implies (ii)$'$.
Let $\mathcal{V}$ be the set

$$\left\{ (f,\alpha,\beta) : K_i(f,\alpha,\beta) < \delta \text{ for all } i, \ \sum_{i=1}^{\infty} \frac{V_{+i}(f,\alpha,\beta)}{i^2} < \infty \right\}$$

and $W_i = \log_+(f_{0i}/f_{\alpha,\beta,i})(Y_i)$. Applying Kolmogorov's strong law of large numbers for independent non-identical variables to the sequence $W_i - E(W_i)$, it follows that for each $f \in \mathcal{V}$, a.s. $\prod_{i=1}^{\infty} P_{f_{0i}}$,

$$\begin{aligned}
&\liminf_{n\to\infty} \left(\frac{1}{n} \sum_{i=1}^{n} \log \frac{f_{\alpha,\beta,i}(Y_i)}{f_{0i}(Y_i)} \right) \\
&\quad \ge - \limsup_{n\to\infty} \left(\frac{1}{n} \sum_{i=1}^{n} \log_+ \frac{f_{0i}(Y_i)}{f_{\alpha,\beta,i}(Y_i)} \right) \\
&\quad = - \limsup_{n\to\infty} \frac{1}{n} \sum_{i=1}^{n} K_i^+(f,\alpha,\beta) \qquad (7.10) \\
&\quad \ge - \limsup_{n\to\infty} \left(\frac{1}{n} \sum_{i=1}^{n} K_i(f,\alpha,\beta) + \frac{1}{n} \sum_{i=1}^{n} \sqrt{K_i(f,\alpha,\beta)/2} \right) \\
&\quad \ge - \limsup_{n\to\infty} \left(\frac{1}{n} \sum_{i=1}^{n} K_i(f,\alpha,\beta) + \sqrt{\frac{1}{n} \sum_{i=1}^{n} K_i(f,\alpha,\beta)/2} \right)
\end{aligned}$$

Since for $f \in \mathcal{V}$, $n^{-1}\sum_{i=1}^n K_i(f,\alpha,\beta) < \delta$, we have for each $f \in \mathcal{V}$,

$$\liminf_{n\to\infty} \frac{1}{n}\sum_{i=1}^n \log\frac{f_{\alpha,\beta,i}(Y_i)}{f_{0i}(Y_i)} \geq -(\delta + \sqrt{\delta/2}) \tag{7.11}$$

Choosing C so that $\delta + \sqrt{\delta/2} \leq C/8$ and noting that

$$I_{2n} \geq \int_{\mathcal{V}} \prod_{i=1}^n \frac{f_{\alpha,\beta,i}(Y_i)}{f_{0i}(Y_i)} d\Pi(f,\alpha,\beta)$$

it follows from Fatou's lemma that

$$e^{nC/4} I_{2n} \to \infty \tag{7.12}$$

a.s. $\prod_{i=1}^\infty P_{f_{0i}}$. □

Remark 7.2.1. Condition (ii) of the theorem can be weakened. It can be seen from the proof that if the prior assigns positive probability to the following set

$$\frac{1}{n}\sum_{i=1}^n K_i(f,\alpha,\beta) < \delta \text{ for all } n, \quad \sum_{i=1}^\infty \frac{V_i(f,\alpha,\beta) + K_i^2(f,\alpha,\beta)}{i^2} < \infty$$

then also the posterior is consistent.

7.3 Exponentially Consistent Tests

Our goal is to establish consistency for (f,α,β) or for (α,β) at (f_0,α_0,β_0), and thus the sets $\mathcal{W}$ of interest to us are of the type $\mathcal{W} = \mathcal{U}^c$, where $\mathcal{U}$ is a neighborhood of β_0 or α_0 alone or of (f_0,α_0,β_0). In the first case we write $\mathcal{W}$ of this type as a finite union of $\mathcal{W}_i$s and show that condition (i) of Theorem 7.2.1 holds for each of these $\mathcal{W}_i$s.

We begin with a couple of lemmas.

Lemma 7.3.1. *For $i = 1, 2$, let g_{0i} and g_i be densities on $\mathbb{R}$. If for each i there exists a function Φ_i, $0 \leq \Phi_i \leq 1$ such that*

$$E_{g_{0i}}(\Phi_i) = \alpha_i \leq \gamma_i = E_{g_i}(\Phi_i) \tag{7.13}$$

and if

$$\liminf_{n\to\infty} \frac{1}{n}\sum_{i=1}^n (\gamma_i - \alpha_i) > 0 \tag{7.14}$$

then there exists a constant C, sets $B_n \subset \mathbb{R}^n$, $n = 1, 2, \ldots$, and n_0— all depending only on (γ_i, α_i), such that for $n > n_0$

$\left[\prod_{i=1}^n P_{g_{0i}}\right](B_n) < e^{-nC}$, *and*

$\left[\prod_{i=1}^n P_{g_i}\right](B_n) > 1 - e^{-nC}$.

We refer to [134] for a proof. For a density g and $\theta \in \mathbb{R}$, let g_θ stand for the density $g_\theta(y) = g(y - \theta)$.

Lemma 7.3.2. *Let g_0 be a continuous symmetric density on $\mathbb{R}$, with $g_0(0) > 0$. Let η be such that $\inf_{|y|<\eta} g_0(y) = C > 0$.*

(i) *For any $\Delta > 0$, there exists a set B_Δ such that*

$$P_{g_0}(B_\Delta) \leq \frac{1}{2} - C(\Delta \wedge \eta)$$

and for any symmetric density g

$$P_{g_\theta}(B_\Delta) \geq \frac{1}{2} \qquad \text{for all } \theta \geq \Delta$$

(ii) *For any $\Delta < 0$, there exists a set $\tilde{B}_\Delta$ such that*

$$P_{g_0}(\tilde{B}_\Delta) \leq \frac{1}{2} - C(\Delta \wedge \eta)$$

and for any symmetric density g

$$P_{g_\theta}(\tilde{B}_\Delta) \geq \frac{1}{2} \qquad \text{for all } \theta \leq \Delta$$

Proof. (i) Take $B_\Delta = (\Delta, \infty)$. Since $\theta \geq \Delta$ and g_θ is symmetric around θ, $P_{g_\theta}(B_\Delta) \geq \frac{1}{2}$. On the other hand

$$P_{g_0}(B_\Delta) = \frac{1}{2} - \int_0^\Delta g_0(y)dy \leq \frac{1}{2} - \int_0^{\Delta\wedge\eta} g_0(y)dy \leq \frac{1}{2} - C(\Delta \wedge \eta) \tag{7.15}$$

Similarly $\tilde{B}_\Delta = (-\infty, \Delta)$ would satisfy condition (ii). □

Remark 7.3.1. By considering $I_{B_\Delta}(y - \theta_0)$, it is easy to see that Lemma 7.3.2 holds if we replace g_0 by g_{0,θ_0} and require $\theta - \theta_0 > \Delta$ or $\theta - \theta_0 < \Delta$.

Assumption A. There exists $\varepsilon_0 > 0$ such that the covariate values x_i satisfy

$$\liminf_{n\to\infty} \frac{1}{n}\sum_{i=1}^{n} I\{x_i < -\varepsilon_0\} > 0, \qquad \liminf_{n\to\infty} \frac{1}{n}\sum_{i=1}^{n} I\{x_i > \varepsilon_0\} > 0$$

Remark 7.3.2. Assumption A forces the covariate x to take both positive and negative values, i.e., values on both sides of 0. If the condition is satisfied around any point, then by a simple location shift, we can bring it to the present form.

Proposition 7.3.1. *If Assumption A holds, f_0 is continuous at 0 and $f_0(0) > 0$, then there is an exponentially consistent sequence of tests for*

$$H_0 : (f, \alpha, \beta) = (f_0, \alpha_0, \beta_0) \qquad \textit{against} \qquad H_1 : (f, \alpha, \beta) \in \mathcal{W}$$

in each of the following cases:

(i) $\mathcal{W} = \{(f, \alpha, \beta) : \ \alpha > \alpha_0, \ \beta - \beta_0 > \Delta\}$;

(ii) $\mathcal{W} = \{(f, \alpha, \beta) : \ \alpha < \alpha_0, \ \beta - \beta_0 > \Delta\}$;

(iii) $\mathcal{W} = \{(f, \alpha, \beta) : \ \alpha > \alpha_0, \ \beta - \beta_0 < -\Delta\}$; *and*

(iv) $\mathcal{W} = \{(f, \alpha, \beta) : \ \alpha < \alpha_0, \ \beta - \beta_0 < -\Delta\}$.

Proof. (i) Let $K_n = \{i : 1 \le i \le n, \quad x_i > \varepsilon_0\}$ and $\#K_n$ stand for the cardinality of K_n. We will construct a test using only those Y_is for which the corresponding i is in K_n.

If $i \in K_n$, then $(\alpha + \beta x_i) - (\alpha_0 + \beta_0 x_i) > \Delta x_i$, and by Lemma 7.3.2 for each $i \in K_n$, there exists a set A_i such that

$$\alpha_i := P_{f_{0i}}(A_i) < \frac{1}{2} - C(\eta \wedge \Delta x_i)$$

and

$$\gamma_i := \inf_{(f,\alpha,\beta)\in\mathcal{W}} P_{f_{\alpha,\beta,i}}(A_i) \ge \frac{1}{2}$$

where ":=" stands for equality by definition.

If $i \leq n$ and $i \notin K_n$, set $A_i = \mathbb{R}$, so that $\alpha_i = \gamma_i = 1$. Thus

$$\begin{aligned}
&\liminf_{n\to\infty}\left(n^{-1}\sum_{i=1}^{n}(\gamma_i - \alpha_i)\right) \\
&\quad \geq \liminf_{n\to\infty}\left(n^{-1}\sum_{i\in K_n} C(\eta\wedge\Delta x_i)\right) \qquad (7.16)\\
&\quad \geq C(\eta\wedge\Delta\epsilon)\liminf_{n\to\infty}\#K_n/n > 0
\end{aligned}$$

With $\Phi_i = I_{A_i}$, the result follows from Lemma 7.3.1.

(ii) In this case we construct tests using Y_i such that $i \in M_n := \{1 \leq i \leq n : x_i < -\varepsilon_0\}$. If $i \in M_n$, then

$$(\alpha + \beta x_i) - (\alpha_0 + \beta_0 x_i) < \Delta x_i < -\Delta\varepsilon_0$$

Now using condition (ii) of Lemma 7.3.2, we get sets $\tilde{B}_i$ and then obtain exponentially consistent tests using Lemma 7.3.1 as in part (i). The other two cases follow similarly.
□

The union of the $\mathcal{W}$'s in Proposition 7.3.1 is the set $\{(f,\alpha,\beta) : |\beta - \beta_0| > \Delta\}$. The case for α alone can be proved in exactly the same way. Combining all eight exponentially consistent tests for α and β one can get an exponentially consistent test for $\alpha = \alpha_0, \beta = \beta_0$.

If random fs are not symmetrized around zero, α is not identifiable. So the posterior distribution for α will not be consistent. Consistency for β will continue to hold under appropriate conditions. To prove the existence of uniformly consistent tests for β in the nonsymmetric case, we pair Y_is and consider the difference $Y_i - Y_j$, which has a density that is symmetric around $\beta(x_i - x_j)$. We can now handle the problem in essentially the same way as in Proposition 7.3.1 to construct strictly unbiased tests. The verification of the other conditions in Sections 7.4, 7.5 and 7.6 is along similar lines.

The next proposition considers neighborhoods of f_0 to get posterior consistency for the true density rather than only the parametric part. We need an additional assumption.

Assumption B. For some L, $|x_i| < L$ for all i.

Proposition 7.3.2. *Suppose that Assumption* B *holds. Let $\mathcal{U}$ be a weak neighborhood of f_0 and let $\mathcal{W} = \mathcal{U}^c \times \{(\alpha,\beta) : |\alpha - \alpha_0| < \Delta,\ |\beta - \beta_0| < \Delta\}$. Then there exists* *an exponentially consistent sequence of tests for testing*

$$H_0 : (f, \alpha, \beta) = (f_0, \alpha_0, \beta_0) \quad \textit{against} \quad H_1 : (f, \alpha, \beta) \in \mathcal{W}$$

Proof. Without loss of generality take

$$\mathcal{U} = \left\{ f : \int \Phi(y) f(y) - \int \Phi(y) f_0(y) < \varepsilon \right\} \tag{7.17}$$

where $0 \leq \Phi \leq 1$ and Φ is uniformly continuous.

Since Φ is uniformly continuous, given $\varepsilon > 0$, there exists $\delta > 0$ such that $|y_1 - y_2| < \delta$ implies $|\Phi(y_1) - \Phi(y_2)| < \varepsilon/2$.

Let Δ be such that

$$|(\alpha - \alpha_0) + (\beta - \beta_0) x_i| < \delta$$

for $\alpha, \beta \in \mathcal{W}$ and all x_i. Set $\tilde{\Phi}_i(y) = \Phi(y - (\alpha_0 + \beta_0 x_i))$. Then

$$E_{f_{0i}} \tilde{\Phi}_i = E_{f_0} \Phi_i, \qquad E_{f_{\alpha,\beta,i}} \tilde{\Phi}_i = E_{f_{(\alpha-\alpha_0),(\beta-\beta_0),i}} \Phi \tag{7.18}$$

Noting that

$$\begin{aligned} &\int \Phi(y - ((\alpha - \alpha_0) + (\beta - \beta_0) x_i)) f_{(\alpha-\alpha_0)+(\beta-\beta_0)x_i}(y) dy \\ &= \int \Phi(y) f(y) dy \end{aligned}$$

we have

$$\begin{aligned} &\int \tilde{\Phi}_i(y) f_{\alpha,\beta,i}(y) dy \\ &\quad \geq \int \Phi(y) f(y) dy - \int |\Phi(y) - \Phi(y - ((\alpha - \alpha_0) + (\beta - \beta_0) x_i))| \\ &\qquad\qquad \times f_{(\alpha-\alpha_0)+(\beta-\beta_0)x_i}(y) dy \\ &\quad \geq \int \Phi(y) f(y) dy - \frac{\varepsilon}{2} \end{aligned}$$

in the last step, we used the uniform continuity of Φ. An application of Lemma 7.3.1 completes the proof. □

If one is interested in showing posterior probability of $f \in U$, $|\alpha - \alpha_0| < \Delta$, $|\beta - \beta_0| < \delta$ goes to 1 a.s. (f_0, α_0, β_0), then it is necessary to get an exponential sequence of tests for $H_0 : (f, \alpha, \beta) = (f_0, \alpha_0, \beta_0)$ against $H_1 : f \in U^c$ or $|\alpha - \alpha_0| > A$ or $|\beta - \beta_0| > \delta$. For this, one has only to combine Propositions 7.3.1, its analogoue for α, and Proposition 7.3.2.

7.4 Prior Positivity of Neighborhoods

In this section we develop sufficient conditions to verify condition (ii) of Theorem 7.2.1. A similar problem in the context of location parameter was studied in Chapter 6. There, we managed with Kullback-Leibler continuity of f_0 at θ_0—the true value of the location parameter, and the requirement that $\Pi\{K(f^*_{0,\theta}, f) < \delta\} > 0$ for all θ in a neighborhood of θ_0 and where $f^*_{0,\theta}$ is close to but different from $f_{0,\theta}$. However, this approach does not carry over to the regression context because, even though the true parameter remains (α_0, β_0), for each i we encounter different parameters $\theta_i = \alpha_0 + \beta_0 x_i$. Here we take a different approach. Since we have no assumptions on the structure of the random density f, the assumption on f_0 is somewhat strong. This condition is weakened in Section 7.7, where we consider Dirichlet mixture of normals. In that case, the random f is better behaved.

Lemma 7.4.1. *Suppose $f_0 \in \mathcal{F}$ satisfies the following condition: There exists $\eta > 0$, C_η and a symmetric density g_η such that, for $|\eta'| < \eta$,*

$$f_0(y - \eta') < C_\eta g_\eta(y) \qquad \textit{for all } y \tag{7.19}$$

Then

(a) *for any $f \in \mathcal{F}$ and $|\theta| < \eta$*

$$K(f_0, f_\theta) \le C_\eta \log C_\eta + \left[K(g_\eta, f) + \sqrt{K(g_\eta, f)}\right]$$

(b) *if, in addition,* $\mathrm{var}_{g_\eta}(\log(g_\eta/f)) < \infty$, *then*

$$\sup_{|\theta|<\eta} \mathrm{var}_{f_0}\left(\log_+ \frac{f_0}{f_\theta}\right) < \infty$$

Proof. Part (a) is an immediate consequence of Lemma 6.3.2 and the fact that $K(f_{0,\theta}, f) = K(f_0, f_\theta)$, which follows from the symmetry of f_0 and f.

For (b), note that

$$\int f_0 \left[\log_+ \frac{f_0}{f_\theta}\right]^2 = \int f_{0,\theta} \left[\log_+ \frac{f_{0,\theta}}{f}\right]^2 \le C_\eta \int g_\eta \left[\log_+ \frac{C_\eta g_\eta}{f}\right]^2 \tag{7.20}$$

A remark here: We work with $\mathrm{var}_{f_0}\left(\log_+ f_0/f_\theta\right)$ rather than $\mathrm{var}_{f_0}\left(\log f_0/f_\theta\right)$ because the condition $f_\theta < C_\eta g_\eta$ does not imply $\left[\log f_{0,\theta}/f\right]^2 \le C_\eta g_\eta \left[\log C_\eta g_\eta/f\right]^2$. $\square$

We write the assumption of Lemma 7.4.1 as follows:

Assumption C. For $\eta > 0$, sufficiently small, there is $g_\eta \in \mathcal{F}$ and constant $C_\eta > 0$ such that for $|\eta'| < \eta$,

$$f_0(y - \eta') < C_\eta g_\eta(y) \quad \text{for all } y$$

and

$$C_\eta \to 1 \quad \text{as } \eta \to 0$$

Proposition 7.4.1. *Suppose Assumptions B and C hold. Let $\tilde{\Pi}$ be a prior for f and μ be a prior for (α, β). If (α_0, β_0) is in the support of μ and if for all η sufficiently small and for all $\delta > 0$*

$$\tilde{\Pi}\left\{K(g_\eta, f) < \delta,\ \mathrm{var}_{g_\eta}\left(\log \frac{g_\eta}{f}\right) < \infty\right\} > 0 \tag{7.21}$$

then for all $\delta > 0$ and some $M > 0$,

$$(\tilde{\Pi} \times \mu)\left\{(f, \alpha, \beta) : K_i(f, \alpha, \beta) < \delta,\ V_i(f, \alpha, \beta) < M \text{ for all } i\right\} > 0 \tag{7.22}$$

Proof. Choose η, δ_0 such that (7.21) holds with $\delta = \delta_0$ and

$$(C_\eta + 1)\log C_\eta + C_\eta\left[\delta_0 + \sqrt{\delta_0}\right] < \delta$$

Let

$$V = \left\{(\alpha, \beta) : |\alpha - \alpha_0| < \frac{\eta}{2},\quad |\beta - \beta_0| < \frac{\eta}{2L}\right\}$$

Note that

$$K_i(f_0, \alpha, \beta) = K(f_0, f_{(\alpha-\alpha_0)+(\beta-\beta_0)x_i})$$

and

$$V_i(f_0, \alpha, \beta) = V(f_0, f_{(\alpha-\alpha_0)+(\beta-\beta_0)x_i})$$

and $(\alpha, \beta) \in V$ implies that $|(\alpha - \alpha_0) + (\beta - \beta_0)x_i| < \eta$ for all x_i. An application of Lemma 7.19 immediately gives the result. □

Theorem 7.4.1. *Suppose that*

(i) *the covariates $x_1, x_2, \ldots$ satisfy Assumptions A and B;*

(ii) *f_0 is continuous, $f_0(0) > 0$, and f_0 satisfies Assumption C;*

(iii) *for all sufficiently small* η *and for all* $\delta > 0$,

$$\tilde{\Pi}\{K(g_\eta, f) < \delta, \quad V(g_\eta, f) < \infty\} > 0$$

where g_η *is as in Assumption C.*

Then for any neighborhood $\mathcal{U}$ *of* f_0,

$$\Pi\{(f,\alpha,\beta) : f \in \mathcal{U}, |\alpha - \alpha_0| < \delta, |\beta - \beta_0| < \delta | Y_1, Y_2, \ldots, Y_n\} \to 1 \tag{7.23}$$

a.s. $\prod_{i=1}^\infty P_{f_{0i}}$.
In other words, the posterior distribution is weakly consistent at (f_0, α_0, β_0).

Proof. The proof follows from the remarks after Proposition 7.3.2. □

Remark 7.4.1. Assumption (ii) of Theorem 7.4.1 is satisfied if f_0 is Cauchy or normal. If f_0 is Cauchy, then $g_\eta = f_0$ satisfies Assumption C. If f_0 is normal, then Assumption C holds with $g_\eta = f^s_{0,\eta}$, where

$$f^s_{0,\eta} = \frac{1}{2}\{f_0(y - \eta) + f_0(-y - \eta)\} \tag{7.24}$$

Remark 7.4.2. Assumption B is used in two places: Propositions 7.3.2 and 7.4.1. For specific f_0s one may be able to obtain the conclusion of Proposition 7.4.1 without Assumption B. In such cases one would be able to get consistency at (α_0, β_0) without having to establish consistency at (f_0, α_0, β_0).

7.5 Polya Tree Priors

In this section we note that Polya tree priors, with a suitable choice of parameters, satisfy condition (iii) of Theorem 7.19 and hence the posterior distribution is weakly consistent. To obtain a prior on symmetric densities, we consider Polya tree priors on densities f on the positive half-line and then considering the symmetrization $f^s(y) = \frac{1}{2}f(|y|)$. Since $K(f,g) = K(f^s, g^s)$ and $V(f,g) = V(f^s, g^s)$, this symmetrization presents no problems.

We briefly recall Polya tree priors from Chapter 3. Let $E = \{0,1\}$, $E^m = \{0,1\}^m$ and $E^* = \bigcup_{m=1}^\infty E^m$. For each m, $\{B_{\underline{\epsilon}} : \underline{\epsilon} \in E^m\}$ is a partition of $\mathbb{R}^+$ and for each $\underline{\epsilon}$, $\{B_{\underline{\epsilon}0}, B_{\underline{\epsilon}1}\}$ is a partition of $B_{\underline{\epsilon}}$. Further $\{B_{\underline{\epsilon}} : \underline{\epsilon} \in E^*\}$ generates the Borel σ-algebra.

A random probability measure P on $\mathbb{R}^+$ is said to be distributed as a Polya tree with parameters $(\Pi, \mathcal{A})$, where Π is a sequence of partitions as described in the last paragraph, and $\mathcal{A} = \{\alpha_{\underline{\epsilon}} : \underline{\epsilon} \in E^*\}$ is a collection of nonnegative numbers, if there exists a collection $\{Y_{\underline{\epsilon}} : \underline{\epsilon} \in E^*\}$ of mutually independent random variables such that

(i) each $Y_{\underline{\epsilon}}$ has a beta distribution with parameters $\alpha_{\underline{\epsilon}0}$; and $\alpha_{\underline{\epsilon}1}$

(ii) the random measure P is given by

$$P(B_{\epsilon_1\cdots\epsilon_m}) = \left[\prod_{j=1,\ \epsilon_j=0}^{m} Y_{\epsilon_1\cdots\epsilon_{j-1}}\right]\left[\prod_{j=1,\ \epsilon_j=1}^{m} (1 - Y_{\epsilon_1\cdots\epsilon_j})\right]$$

We restrict ourselves to partitions $\Pi = \{\Pi_m : m = 0, 1, \ldots\}$ that are determined by a strictly positive, continuous density α on $\mathbb{R}^+$ in the following sense: The sets in Π_m are intervals of the form

$$\left\{y : \frac{k-1}{2^m} < \int_{-\infty}^{y} \alpha(t)dt \le \frac{k}{2^m}\right\}$$

Theorem 7.5.1. *Let $\tilde{\Pi}$ be a Polya tree prior on densities on $\mathbb{R}^+$ with $\alpha_{\underline{\epsilon}} = r_m$ for all $\underline{\epsilon} \in E^m$. If $\sum_{m=1}^{\infty} r_m^{-1/2} < \infty$, then for any density g such that $K(g, \alpha) < \infty$ and $\text{var}_g(\log g) < \infty$ for all $\delta > 0$,*

$$\lim_{M\to\infty} \tilde{\Pi}\{f : K(g,f) < \delta,\ V(g,f) < M\} > 0 \tag{7.25}$$

The proof is along similar lines as that of Theorem 6.4.1. We refer to [134] for details.

Although Polya trees give rise to naturally interpretable priors on densities and leads to consistent posterior, sample paths of Polya trees are, however, very rough and have discontinuities everywhere. Such a drawback can be easily overcome by considering a mixture of Polya trees. Posterior consistency continues to hold this case, because by Fubini's theorem, prior positivity holds under mild uniformity conditions. Such priors are worth further study.

7.6 Dirichlet Mixture of Normals

In this section, we look at random densities that arise as mixtures of normal densities. Let ϕ_h denote the normal density with mean 0 and standard deviation h. For any probability P on $\mathbb{R}$, $f_{h,P}$ will stand for the density

$$f_{h,P}(y) = \int \phi_h(y-t)dP(t) \tag{7.26}$$

Our model consists of prior μ for h and a prior $\tilde{\Pi}$ for P. Consistency issues related to these priors, in the context of density estimation, based on [74], were discussed in Chapter 5. Here we look at similar issues when the error density f in the regression model is endowed with these priors.

To ensure that the prior sits on symmetric densities, we let P be a random probability on $\mathbb{R}^+$ and set

$$f_{h,P}(y) = \frac{1}{2}\int \phi_h(y-t)dP(t) + \frac{1}{2}\int \phi_h(y+t)dP(t) \tag{7.27}$$

We will denote by $\tilde{\Pi}$ both the prior for P and the prior for $f_{h,P}$.

The following lemma shows that the random f generated by the prior under consideration is more regular than those generated by Polya tree priors, and hence the conditions on f_0 are more transparent than those in Section 7.5 or those in Ghosal, Ghosh, and Ramamoorthi [78].

Lemma 7.6.1. *Let f_0 be a density such that*

$$\int y^2 f_0(y)dy < \infty \quad \text{and} \quad \int f_0(y)\log f_0(y)dy < \infty \tag{7.28}$$

If $f(y) = \int \phi_h(y-t)dP(t)$ and $\int t^2 dP(t) < \infty$, then

(i) $\displaystyle\lim_{\theta\to 0}\int f_0(y)\log\frac{f_0(y)}{f_\theta(y)}dy = \int f_0(y)\log\frac{f_0(y)}{f(y)}dy$, *and*

(ii) $\displaystyle\lim_{\theta\to 0}\int f_0(y)\left[\log\frac{f_0(y)}{f_\theta(y)}\right]^2 dy = \int f_0(y)\left[\log\frac{f_0(y)}{f(y)}\right]^2 dy.$

Proof. We have

$$\log f_\theta(y) = \log\int \phi_h(y-(t+\theta))dP(t)$$

and hence

$$|\log f_\theta(y)| \le |\log\sqrt{2\pi}h| + \left|\log\int e^{-(y-\theta-t)^2/(2h^2)}dP(t)\right| \tag{7.29}$$

Since $\log \int e^{-(y-\theta-t)^2/(2h^2)} dP(t) < 0$, by Jensen's inequality applied to $-\log x$, the last expression is bounded by

$$|\log \sqrt{2\pi} h| + \int \frac{(y-\theta-t)^2}{h^2} dP(t)$$

Hence

$$\begin{aligned}
&\left| f_0(y) \log \frac{f_0(y)}{f_\theta(y)} \right| \\
&\quad \leq |f_0(y) \log f_0(y)| + f_0(y) |\log f_\theta(y)| \\
&\quad \leq |f_0(y) \log f_0(y)| + |\log \sqrt{2\pi} h| + f_0(y) \int \frac{(y-\theta-t)^2}{h^2} dP(t)
\end{aligned}$$

The dominated Convergence Theorem now yields the result. □

We now return to the regression model.

Theorem 7.6.1. *Suppose* $\tilde{\Pi}$ *is a normal mixture prior for* f. *If*

(i) *Assumptions A and B hold,*

(ii) $\tilde{\Pi}\{f : K(f_0, f) < \delta, \quad V(f_0, f) < \infty\} > 0$ *for all* $\delta > 0$,

(iii) $E_{f_0}(\log f_0)^2 < \infty$, *and*

(iv) $\int \int t^2 dP(t) d\tilde{\Pi}(P) < \infty$,

then the posterior $\Pi(\cdot|Y_1, \ldots, Y_n)$ *is weakly consistent for* (f, α, β) *at* (f_0, α_0, β_0) *provided* (α_0, β_0) *is in the support of the prior for* (α, β).

Proof. By condition (iv), $\left\{P : \int t^2 dP(t) < \infty\right\}$ has $\tilde{\Pi}$ probability 1. So we may assume that

$$\tilde{\Pi}\left\{f : f = f_P, \text{ (ii) holds}, \quad \int t^2 dP(t) < \infty\right\} > 0 \tag{7.30}$$

Let $\mathcal{U} = \left\{f : f = f_P, \text{ (ii) holds}, \int t^2 dP(t) < \infty\right\}$.

For every $f \in \mathcal{U}$, using Lemma 7.6.1, choose δ_f such that, for $\theta < \delta_f$

$$\left| \int f_0 \log \frac{f_0}{f} - \int f_0 \log \frac{f_0}{f_\theta} \right| < \delta \tag{7.31}$$

Now choose ε_f such that $|\alpha-\alpha_0+(\beta-\beta_0)x_i| < \delta_f$ whenever $|\alpha-\alpha_0| < \varepsilon_f, \quad |\beta-\beta_0| < \varepsilon_f/L$.

Clearly, if $f \in \mathcal{U}$ and $|\alpha-\alpha_0| < \varepsilon_f$ and $|\beta-\beta_0| < \varepsilon_f/L$, we have

$$K_i(f,\alpha,\beta) < 2\delta \quad \text{and} \quad V_i(f,\alpha,\beta) < V(f_0,f)+\delta \tag{7.32}$$

Since

$$\tilde{\Pi}\{(f,\alpha,\beta): f \in \mathcal{U}, \quad |\alpha-\alpha_0| < \varepsilon_f, \quad |\beta-\beta_0| < \varepsilon_f/L\} > 0 \tag{7.33}$$

we have

$$\Pi\left\{(f,\alpha,\beta): K_i(f_0,\alpha,\beta) < \delta \text{ for all } i, \quad \sum_{i=1}^{\infty} \frac{V_i(f,\alpha,\beta)}{i^2} < \infty\right\} > 0 \tag{7.34}$$

An application of Theorem 7.2.1 completes the proof. □

It was shown in Chapter 5 that if f_0 has compact support or if $f_0 = f_P$ with P having compact support, then $\tilde{\Pi}\{f: K(f_0,f) < \delta\} > 0$ for all $\delta > 0$. The argument given there also shows that in these cases, (ii) of Theorem 7.6.1 holds when $\tilde{\Pi}$ is Dirichlet with base measure γ. In Chapter 5 we also described f_0s whose tail behavior is related to that of γ such that $\tilde{\Pi}\{f: K(f_0,f) < \delta\} > 0$. In the case when the prior is Dirichlet, the double-integral in (iv) is finite if and only if $\int t^2 d\gamma(t) < \infty$. While normal f_0 is covered by these results, the case of Cauchy f_0 cannot be resolved by the methods in that chapter. However, Dirichlet mixtures of both location and scale parameters of normal may be able to handle Cauchy, which is a scale mixture of normal. Results of Chapter 5 may need to be generalized to prove posterior consistency for these priors. .

7.7 Binary Response Regression with Unknown Link

One of the most popular models in bioassay involves regression of the probability of some event on a covariate x. The regression is taken to be linear in logit or probit scale. In this section we consider the same problem with a nonparametric link function, instead of a logit or probit model. We indicate, without going into details, how posterior consistency can be established.

Consider k levels of a drug on a suitable scale, say, $x_1,\ldots,x_k$, with probability of a response (which may be death or some other specified event) p_i, $i=1,\ldots,k$. The ith level of the drug is given to n_i subjects and the number of responses r_i noted.

We thus get k independent binomial variables $B(n_i, p_i)$. The object is often to find x such that $p = 0.5$. Often, p_i is modeled as

$$p_i = F(\alpha + \beta x_i) = H(x_i) \tag{7.35}$$

where F is a response distribution and $\alpha + \beta x_i$ is a linear representation of $F^{-1}(p_i) = y_i$. Here p_i may be estimated by r_i/n_i, but if the n_is are small, the estimates will have large variances, so the model provides a way of combining all the data. In a logit model, F is taken as a logistic distribution function. In a probit model the link function is the normal distribution function. The choice of the functional form of the link function is somewhat arbitrary, and this may substantially affect inference, particularly at the two ends where data are sparse. In recent years, there has been a lot of interest in link functions with unknown functional form. In nonparametric problems of this kind, one puts a prior on F or H. Such an approach was taken by Albert and Chib ([1]) , Chen and Dey ([31]), Basu and Mukhopadhyay ([11, 12]) and some other authors. If one puts a prior on F, one has to put conditions on F like specifying two values of two quantiles to make (F, α, β) identifiable. In this case, one can develop sufficient conditions for posterior consistency at (F_0, α_0, β_0) using our variant of Schwartz's theorem. However, in practice, one often puts a Dirichlet process or some other prior on F and independently of this, a prior on (α, β). Due to the discreteness of Dirichlet selections, many authors actually prefer the use of other priors such as Dirichlet scale mixtures of normals, see Basu and Mukhopadhyay ([11, 12]) and the references therein. Because of the lack of identifiability, the posterior for (α, β) is not consistent. On the other hand, a Dirichlet process prior and a prior on (α, β) provides a prior on H and one can ask for posterior consistency of $H^{-1}(1/2)$ at, say, $H_0^{-1}(1/2)$. This problem can be solved by the methods developed earlier in this chapter.

Without loss of generality, one may take $n_i = 1$ for all i. To verify condition (ii) of Theorem 7.2.1, consider

$$Z_i = \log \frac{(H_0(x_i))^{r_i}(1 - H_0(x_i))^{1-r_i}}{(H(x_i))^{r_i}(1 - H(x_i))^{1-r_i}} \tag{7.36}$$

where r_i is 1 or 0 with probability $H(x_i)$ and $1 - H(x_i)$, respectively, and the true H is denoted by H_0. Then it is easily found that

$$E_{H_0}(Z_i) = H_0(x_i) \log \frac{H_0(x_i)}{H(x_i)} + (1 - H_0(x_i)) \log \frac{1 - H_0(x_i)}{1 - H(x_i)} \tag{7.37}$$

and

$$E_{H_0}(Z_i^2) \le 2H_0(x_i)\left(\log\frac{H_0(x_i)}{H(x_i)}\right)^2 + 2(1-H_0(x_i))\log\left(\frac{1-H_0(x_i)}{1-H(x_i)}\right)^2 \tag{7.38}$$

Assume that x_is lie in a bounded interval containing $H_0^{-1}(1/2)$, and the support of H_0 contains a bigger interval. Since the range of x_is is bounded, the sequence of formal empirical distributions $n^{-1}\sum_{i=1}^n \delta_{x_i}$ of $x_1,\ldots,x_n$ is relatively compact. Assume that all limits of subsequences converge to distributions which give positive measure to all nondegenerate intervals, provided they lie in a certain interval containing $H_0^{-1}(1/2)$. Therefore, a positive fraction of x_is lie in an interval of positive length if the interval is close to the point $H_0^{-1}(1/2)$. Also assume that H_0 is continuous and the support of the prior for H contains H_0. For example, if the prior is Dirichlet with a base measure whose support contains the support of H_0, then the above condition is satisfied. Mixture priors often have large supports also. For instance, the Dirichlet scale mixture of normal prior used by Basu and Mukhopadhyay ([11, 12]) will have this property if the true link function is also a scale mixture of normal cumulative distribution functions.

If H_ν is a sequence converging weakly to H_0, then by Polya's theorem, the convergence is uniform. Note that for $0 < p < 1$, the functions $p\log(p/q) + (1-p)\log((1-p)/(1-q))$ and $p(\log(p/q))^2 + (1-p)(\log((1-p)/(1-q)))^2$ in q converge to 0 as $q \to p$, uniformly in p lying in a compact subinterval of $(0,1)$. Thus given $\delta > 0$, we can choose a weak neighborhood $\mathcal{U}$ of H_0 such that if $H \in \mathcal{U}$, then $E_{H_0}(Z_i) < \delta$ and $E_{H_0}(Z_i^2)$'s are bounded. By the assumption on the support of the prior, condition (ii) of Theorem 7.2.1 holds.

For existence of exponentially consistent tests in condition (i) of Theorem 7.2.1, consider, without loss of generality, testing $H^{-1}(1/2) = H_0^{-1}(1/2)$ against $H^{-1}(1/2) > H_0^{-1}(1/2) + \varepsilon$ for small $\varepsilon > 0$. Let

$$K_n = \left\{i : H_0^{-1}(1/2) + \varepsilon/2 \le x_i \le H_0^{-1}(1/2) + \varepsilon\right\}$$

Since

$$E_H(r_i) = H(x_i) \le H(H_0^{-1}(1/2) + \varepsilon) \le \frac{1}{2} \tag{7.39}$$

and

$$E_{H_0}(r_i) = H_0(x_i) \ge H_0(H_0^{-1}(1/2) + \varepsilon/2) > \frac{1}{2} \tag{7.40}$$

the test

$$\frac{1}{\#K_n} \sum_{i \in K_n} r_i < \frac{1}{2} + \eta \tag{7.41}$$

for $\eta = (H_0(H_0^{-1}(1/2) + \varepsilon/2) - 1/2)/2$ is exponentially consistent by Hoeffeding's inequality and the fact that $\#K_n/n$ converge to positive limits along subsequences. Therefore Theorem 7.2.1 applies and the posterior distribution of $H^{-1}(1/2)$ is consistent at $H_0^{-1}(1/2)$.

7.8 Stochastic Regressor

In this section, we consider the case that the independent variable X is stochastic. We assume that the X observations $X_1, X_2, \ldots$ are i.i.d. with a probability density function $g(x)$ and are independent of the errors $\epsilon_1, \epsilon_2, \ldots$. We will argue that all the results on consistency hold under appropriate conditions.

Let $G(x) = \int_{-\infty}^{x} g(u)du$, denote the cumulative distribution function of X. We shall assume that the following condition holds.

Assumption D. The independent variable X is compactly supported and $0 < G(0-) \leq G(0) < 1$.

Under these assumptions, results follow from a conditionality argument and the corresponding results for the nonstochastic case, conditioned on a sequence $x_1, x_2, \ldots$ such that Assumptions A and B hold. Note that if g satisfies Assumption D, under P_g^{∞}, almost all sequences $x_1, x_2, \ldots$ satisfy Assumptions A and B. For details see [134]. Thus if X is stochastic and Assumption D replaces Assumptions A and B in Theorems 7.5.1 and 7.6.1, posterior consistency holds.

7.9 Simulations

Additional insight can often be obtained by carrying out simulations. In the mixture model that we have discussed, one can study the effect on the posterior of β by varying the ingredients in the mixture model. There is an additional issue of symmetrization. After fixing the prior, one can generate observations from carefully chosen parameters and error density and in each case examine the behavior of the posterior. Extensive simulations of this kind have been done by Charles Messan using WINBUGS, and we present a few of these.

First we look at two cases for the kernel: normal and Cauchy. The base measure for the Dirichlet process is $N(0, 1)$. Figure 7.1 displays the simulated posterior when observations were generated from (true f_0 is) normal. The value of β is 3.0., and the random densities are not symmetrized. It is clear from the graphs that, in this case, the posterior behaves well, and in addition to consistency also shows asymptotic normality.

In figure 7.2, the setup for priors is the same as that just considered, but the posterior is evaluated when the true f_0 is Cauchy. Clearly, things do not seem to go well. Both consistency and asymptotic normality seem to be in doubt.

One could see if the introduction of a hyperparameter for the base measure of the Dirichlet process would lead to amelioration of the situation. Figures 7.3 and 7.4 show the result of simulations with a hyperparameter for the base measure. There seems to be some improvement. The estimates are closer to the true value of $\beta = 3$, and there is a suggestion of asymptotic normality.

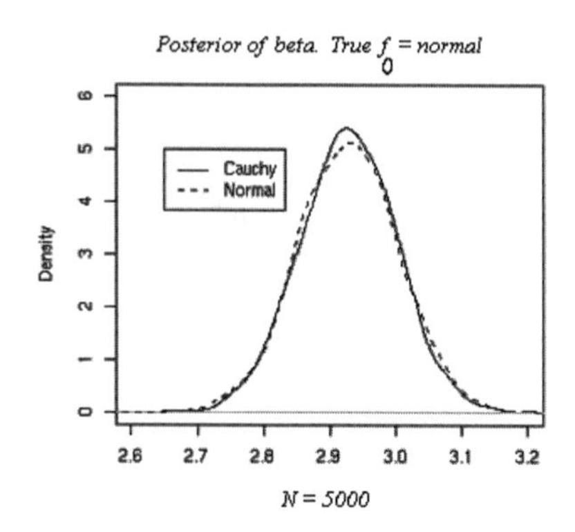

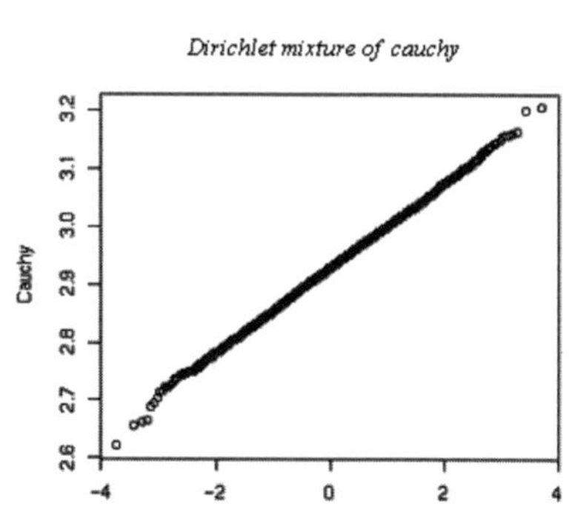

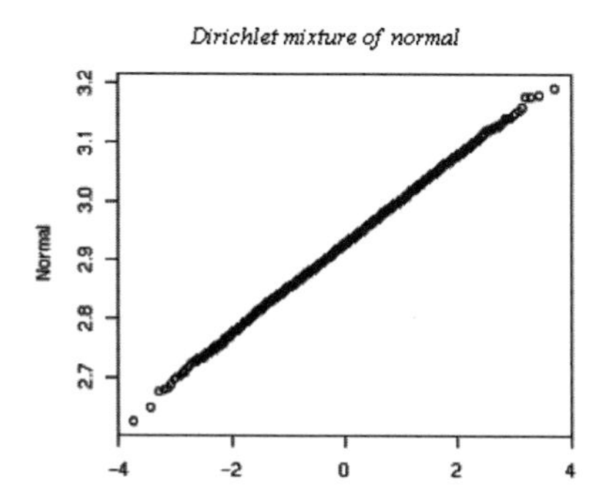

Figure 7.1: Sample size $n = 50$ true $f_0 = N(0,1)$. **Priors:** base measure of the two Dirichlet mixtures $G_0 = N(0,1)$

Classical estimate of beta:

$\hat{\beta} = 2.9248$, Var($\hat{\beta}$) = 0.0040 Est.Var ($\hat{\beta}$) = 0.0052

MCMC estimates of beta: Hyperparameter of Dirichlet $M = 100$

Dirichlet mixture of cauchy: $\hat{\beta}_C = 2.928$ Var($\hat{\beta}_C$) = 0.0053 Skewness = - 0.0295 Kurtosis = 0.0507

Dirichlet mixture of normal: $\hat{\beta}_N = 2.928$ Var($\hat{\beta}_N$) = 0.0058 Skewness = - 0.0220 Kurtosis = 0.0183

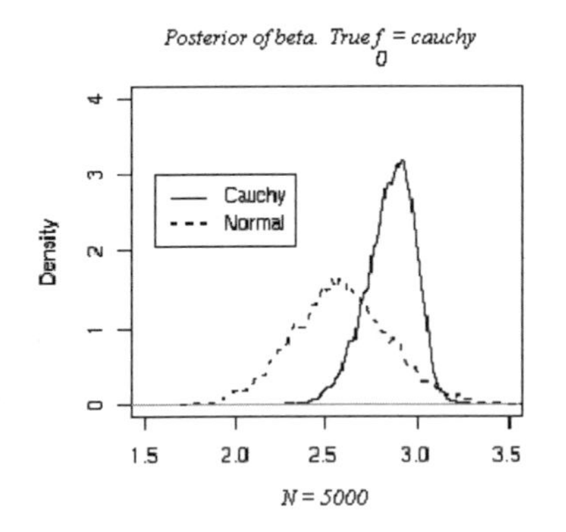

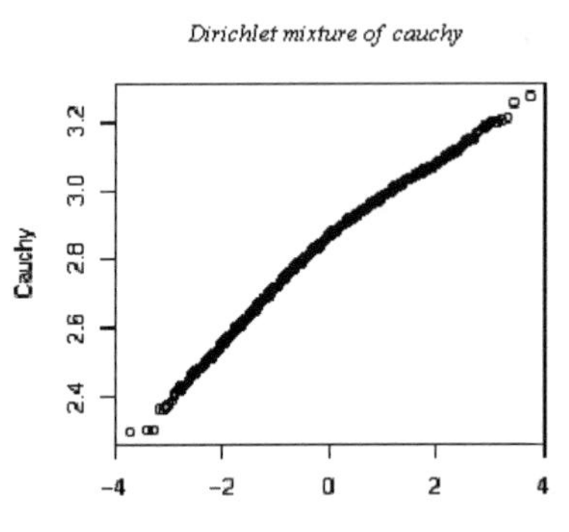

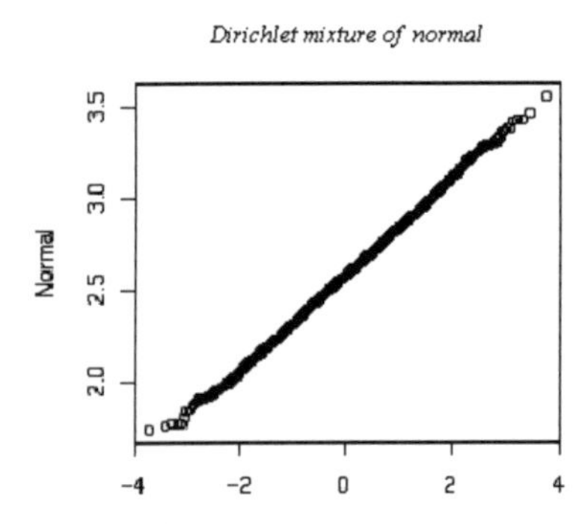

Figure 7.2: Sample size $n = 50$ true $f_0 = Cauchy(0,1)$ **Priors:** base measure of the two Dirichlet mixtures $G_0 = N(0,1)$

Classical estimate of beta:

$\hat{\beta} = 2.5682$, $\text{Var}(\hat{\beta}) = \text{infinite}$

MCMC estimates of beta: Hyperparameter of Dirichlet $M = 100$

Dirichlet mixture of cauchy: $\hat{\beta}_C = 2.855$ $\text{Var}(\hat{\beta}_C) = 0.0177$ Skewness = - 0.5098 Kurtosis = 0.2774

Dirichlet mixture of normal: $\hat{\beta}_N = 2.578$ $\text{Var}(\hat{\beta}_N) = 0.0694$ Skewness = 0.0753 Kurtosis = - 0.0061

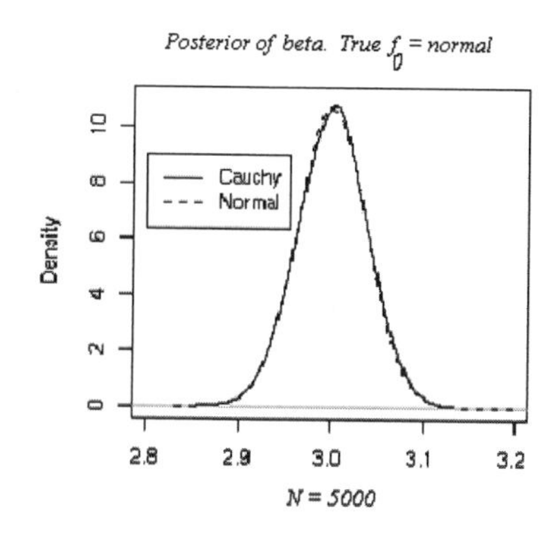

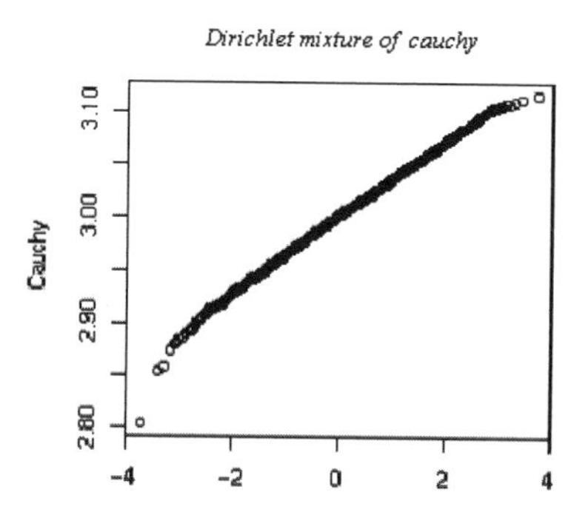

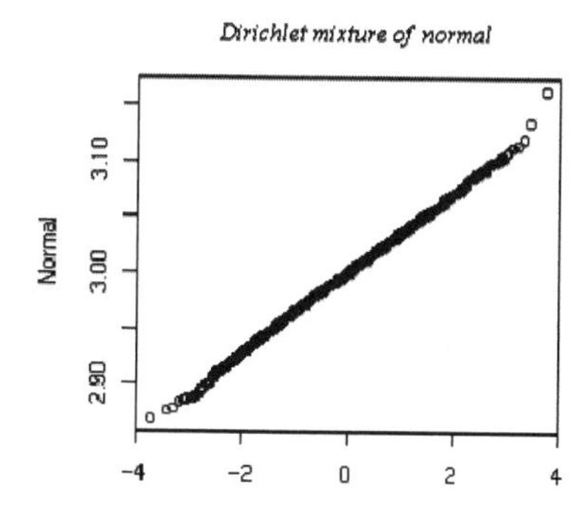

Figure 7.3: Sample size $n = 50$ True $f_0 = N(0, 0.5)$ **Priors:** base measure of Dirichlet: $N(\mu, \sigma)$

$\mu|\sigma \sim N(0, 2\sigma)$

$\sigma \sim Unif(0,10)$

Classical estimate of beta:

$\hat{\beta} = 2.9982$, $\text{Var}(\hat{\beta}) = 0.0012$ $\text{Est.Var}(\hat{\beta}) = 0.0012$ Bandwidth h: $h \sim Unif(0,4)$

MCMC estimates of beta: Hyperparameter of Dirichlet $M = 100$

Dirichlet mixture of cauchy: $\hat{\beta}_C = 3.002$ $\text{Var}(\hat{\beta}_C) = 0.0013$ Skewness = - 0.0938 Kurtosis = 0.2093

Dirichlet mixture of normal: $\hat{\beta}_N = 3.0$ $\text{Var}(\hat{\beta}_N) = 0.0013$ Skewness = - 0.0210 Kurtosis = 0.1751

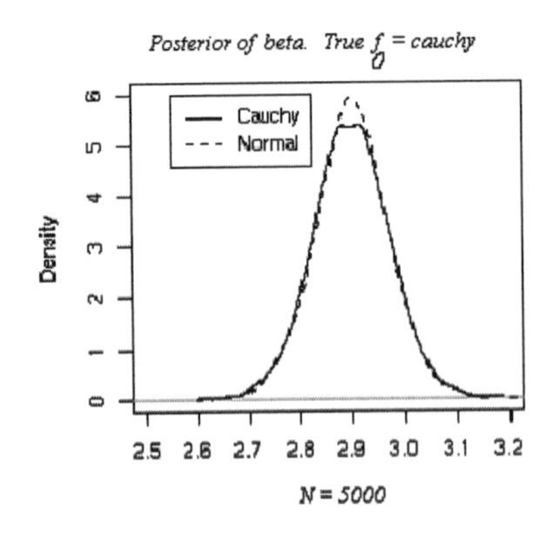

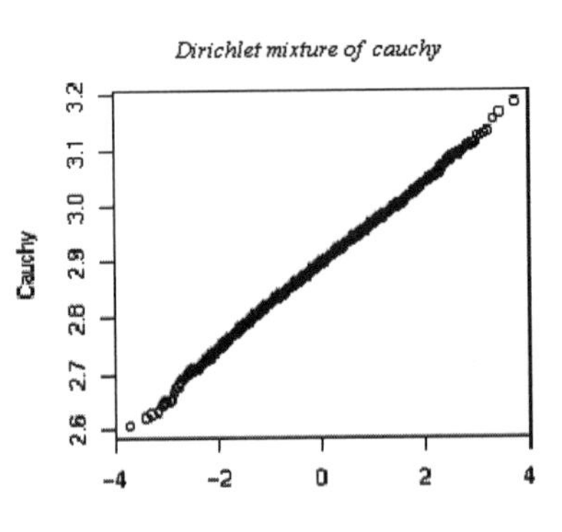

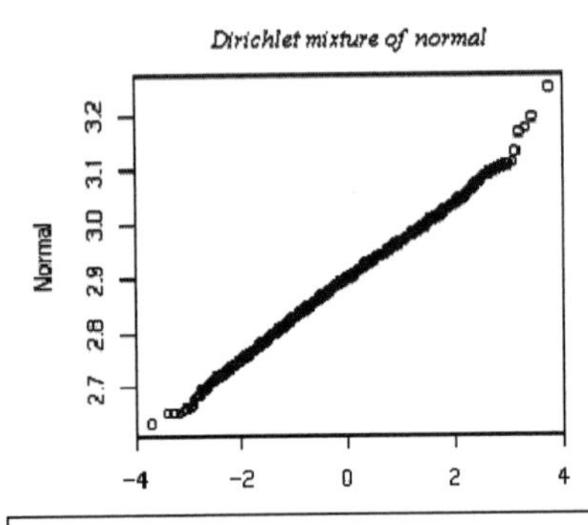

Figure 7.4: Sample size $n = 50$ True $f_0 =$ *cauchy*(0, 0.5) **Priors:** base measure of Dirichlet: $N(\mu, \sigma)$

$\mu|\sigma \sim N(0,2\sigma)$

$\sigma \sim Unif(0,10)$

Classical estimate of beta:

$\hat{\beta} = 2.4641$, $\text{Var}(\hat{\beta}) =$ infinite Bandwidth h: $h \sim Unif(0,4)$

MCMC estimates of beta: Hyperparameter of Dirichlet $M = 100$

Dirichlet mixture of cauchy: $\hat{\beta}_C = 2.898$ $\text{Var}(\hat{\beta}_C) = 0.0053$ Skewness = - 0.0753 Kurtosis = 0.2729

Dirichlet mixture of normal: $\hat{\beta}_N = 2.899$ $\text{Var}(\hat{\beta}_N) = 0.0050$ Skewness = - 0.0623 Kurtosis = 0.3620

8
Uniform Distribution on Infinite-Dimensional Spaces

8.1 Introduction

Except for a noninformative choice of the base measure α for a Dirichlet very little is known about noninformative priors in nonparametric or infinite-dimensional problems. In this chapter we explore how one may construct a prior that is noninformative, i.e., completely nonsubjective in the sense of Chapter 1, for nonparametric problems. One way of thinking of them is as a uniform distribution over an infinite-dimensional space. Our approach has some similarities with that of Dembski [40], as well as many differences.

Several new approaches to construction of such a prior are discussed in Section 8.2. The remaining sections attempt some validation. In Section 8.3 we show that one of our methods would lead to the Jeffreys prior for parametric models under regularity conditions. We also briefly discuss what would be reference priors from this point of view. Section 8.4 contains an application of our ideas to a density estimation problem of Wong and Shen [172]. We show that for our hierarchical noninformative prior, the posterior is consistent–a sort of weak frequentist validation. The proof of consistency is interesting in that the Schwartz condition is not assumed. We also show that the rate of convergence of the posterior is optimal. In particular, this implies that the Bayes estimate of the density corresponding to this prior achieves the optimal frequentist rate–a strong frequentist validation. We offer these tentative ideas to be tried out on different problems. Computational or other considerations may require replacing $\mathcal{P}_i$ by other sieves, which need not be finite, changing an index i to h, which may take values in a continuum, and distributions on $\mathcal{P}_i$ which are not uniform. These relaxations will create a very large class of priors that are nonsubjective in some sense and from which it may be convenient to elicit a prior. This approach includes some of the priors in Chapter 5, namely, the random histograms and the Dirichlet mixture of normals with standard deviation h. The parameter h can be viewed as indexing a sieve. This chapter is almost entirely based on [73] and [80]

8.2 Towards a Uniform Distribution

8.2.1 *The Jeffreys Prior*

By way of motivation we begin with a regular parametric model. Let $\Theta \subset \mathbb{R}^p$. A uniform distribution on Θ should be associated with the geometry on Θ induced by the statistical problem. To do this, let $I(\underline{\theta}) = [I_{i,j}(\underline{\theta})]$ be the $p \times p$ Fisher information (positive definite) matrix. As shown by Rao [2], the matrix induces a Riemannian metric on Θ through the integration of

$$\rho(d\underline{\theta}) = \sum_i \sum_j I_{i,j}(\underline{\theta}) d\theta_i d\theta_j$$

over all curves connecting θ to θ' and minimizing over curves. The minimizing curve is a geodesic. If the model is $N(\underline{\theta}, \Sigma)$, then $I_{i,j} = \Sigma^{-1}$ and we get the famous Mahalanobis distance. Cencov [30] has shown the Riemannian geometry induced by Rao's metric is the unique Riemannian metric that changes in a natural way under 1-1 smooth transformations of Θ onto itself. The Jeffreys prior $\{\det I(\underline{\theta})\}^{1/2}$ can be motivated as follows.

Fix a $\underline{\theta}$ and consider a 1-1 smooth transformation

$$\underline{\theta} \mapsto \psi(\underline{\theta}) = \underline{\psi}$$

such that the information matrix I^{ψ} with the new parametrization $\underline{\psi}$ is identity at $\underline{\psi}(\underline{\theta}_0)$. This implies that the local geometry in the $\underline{\psi}$-space is Euclidean near $\underline{\psi}(\underline{\theta}_0)$ and hence the Lebesgue measure $d\underline{\psi}$ is a suitable uniform distribution near $\underline{\psi}(\underline{\theta}_0)$. If we lift this back to the θ-space making use of the Jacobian and the elementary fact

$$[\frac{\partial \theta_j}{\partial \psi_i}][I_{i,j}(\underline{\theta})][\frac{\partial \theta_j}{\partial \psi_i}]' = I^{\psi} = I$$

we get Jeffreys prior in the θ-space, namely,

$$d\underline{\psi} == \{det[\frac{\partial \theta_i}{\partial \psi_j}]\}^{-1} d\underline{\theta} = \{det[I_{i,j}(\underline{\theta})]\}^{1/2} d\underline{\theta}$$

Another way of deriving the Jeffreys prior in a similar spirit is given in Hartigan ([93] pp. 48, 49). The basic paper for the Jeffreys prior is Jeffreys [106]. These references are relevant for Section 8.3 especially Remark 8.4.1.

8.2.2 *Uniform Distribution via Sieves and Packing Numbers*

Suppose we have a model $\mathcal{P}$ which is equipped with a metric ρ and is compact. In applications we use the Hellinger metric. The compactness assumption can then be relaxed in at least some σ compact cases in a standard way. Our starting point is a sequence ϵ_i diminishing to zero and sieves $\mathcal{P}_i$ where $\mathcal{P}_i$ is a finite set whose elements are separated from each other by at least ϵ_i and has cardinality $D(\epsilon_i, \mathcal{P})$, the largest m for which there are $P_1, P_2, \ldots, P_m \in \mathcal{P}$ with $\rho(P_j, P_{j'}) > \epsilon_i, j \neq j', j, j' = 1, 2, \ldots, m$. Clearly, given any $P \in \mathcal{P}$ there exists $P' \in \mathcal{P}_i$ such that $\rho(P, P') \leq \epsilon_i$. Thus $\mathcal{P}_i$ approximates $\mathcal{P}$ within ϵ_i and no subset of it will have this property.

In the first method we choose $\epsilon_{i(n)}$, tending to 0 in some suitable way. It is then convenient to think of $\mathcal{P}_{i(n)}$ as a finite approximation to $\mathcal{P}$ with the approximation depending on the sample size n. The idea is that the approximating finite model is made more and more accurate by increasing its cardinality with sample size. In the first method our noninformative prior is just the uniform distribution on $\mathcal{F}_{i(n)}$.

This seems to accord well with Basu's [9] recommendation in the parametric case to approximate the parameter space Θ by a finite set and then put a uniform distribution. It is also intuitively plausible that the complexity or richness of a model $\mathcal{P}_{i(n)}$ may be allowed to depend on the sample size. Since this prior depends on the sample size, we consider two other approaches that are more complicated but do not depend on sample size.

In the second approach, we consider the sequence of uniform distributions Π_i on $\mathcal{P}_i$ and consider any weak limit Π^* of $\{\Pi_i\}$ as a noninformative prior. If Π^* is unique, it is simply the uniform distribution defined and studied by Dembski [40].

In the infinite-dimensional case, evaluation of the limit points may prove to be impossible. However, the first approach may be used, and $\Pi_{i(n)}$ may be treated as an approximation to a limit point Π^*.

We now come to the third approach. Here, instead of a limit, we consider the index as a hyperparameter and consider a hierarchical prior which picks up the index i with probability λ_i and then uses Π_i.

8.3 Technical Preliminaries

Let K be a compact metric space with a metric ρ. A finite subset S of K is called *ϵ-dispersed* if $\rho(x, y) \geq \epsilon$ for all $x, y \in S, x \neq y$. A maximal ϵ-dispersed set is called an *ϵ-net* and an ϵ-net with maximum possible cardinality is said to be an *ϵ-lattice*. The cardinality of an ϵ-lattice is called the *packing number* (or *ϵ-capacity*) of K and is denoted by $D(\epsilon, K) = D(\epsilon, K, \rho)$. As K is totally bounded, $D(\epsilon, K)$ is finite. Closely related to packing numbers are *covering numbers* $N(\epsilon, K, \rho)$–the maximum number of balls of radius ϵ needed to cover K. Clearly,

$$N(\epsilon, K, \rho) \leq D(\epsilon, K, \rho) \leq N(\epsilon/2, K, \rho)$$

In view of this, our arguments could also be stated in terms of covering numbers.

Define the ϵ-probability P_ϵ by

$$P_\epsilon(X) = \frac{D(\epsilon, X)}{D(\epsilon, K)}, \qquad X \subset K$$

It follows that $0 \leq P_\epsilon(\cdot) \leq 1, P_\epsilon(\emptyset) = 0, P_\epsilon(K) = 1$. P_ϵ is subadditive and for $X, Y \subset K$. Because K is compact, subsequences of μ_ϵ will have weak limits. If all the subsequences have the same limits, then K is called *uniformizable* and the common limit point is called the *uniform probability* on K.

The following result of Dembski [40]) will be used in the next section.

Theorem 8.3.1 (Dembski). *Let (K, ρ) be a compact metric space. Then the following assertions hold.*

(a) If K is uniformizable with uniform probability μ, then $\lim_{\epsilon\to 0} P_\epsilon(X) = \mu(X)$ for all $X \subset K$ with $\mu(\partial X) = 0$.

(b) If $\lim_{\epsilon\to 0} P_\epsilon(X)$ exists on some convergence-determining class in K, then K is uniformizable.

To extend these ideas to noncompact σ-compact spaces, one can take a sequence of compact sets $K_n \uparrow K$ having uniform probability μ_n. Any positive Borel measure μ satisfying

$$\mu(\cdot \cap K_n) = \frac{\mu_n(\cdot \cap K_n)}{\mu_n(K_1)}$$

may be thought of as an (improper) uniform distribution on K. Such a measure would be unique up to a multiplicative constant by lemma 2 of Dembski [40].

8.4 The Jeffreys Prior Revisited

Let X_is be i.i.d. with density $f(.;\theta)$(with respect to a σ-finite measure ν), and Θ is an open subset of $\mathbb{R}^d$. Assume that $\{f(.;\theta) : \theta \in \Theta\}$ is a regular parametric family, i.e., there exist $\{\psi(.;\theta) \in (L_2(\nu))^d$ such that for any compact $K \subset \Theta$

$$\sup_{\theta \in K} \int |f^{1/2}(x;\theta + \mathbf{h}) - f^{1/2}(x;\theta) - \mathbf{h}^T\psi(x;\theta)|^2 \nu(dx) = o(\|h\|^2) \tag{8.1}$$

as $\|h\| \to 0$. Define the Fisher information by the relation

$$\mathbf{I}(\theta) = 4 \int \psi(x;\theta)(\psi(x;\theta))^T \nu(dx) \tag{8.2}$$

Assume that $\mathbf{I}(\theta)$ is positive definite and the map $\theta \mapsto \mathbf{I}(\theta)$ is continuous. Further, assume the following stronger form of identifiability: On every compact set $K \subset \Theta$,

$$\inf\{\int \left(f^{1/2}(x;\theta_1) - f^{1/2}(x;\theta_2)\right)^2 \nu(dx) : \theta_1, \theta_2 \in K, \|\theta_1 - \theta_2\| \geq \epsilon\} > 0, \quad \epsilon > 0$$

For i.i.d. observations equip Θ with the Hellinger distance, as defined in Chapter 1, namely,

$$H(\theta_1, \theta_2) = \left(\int |f^{1/2}(x;\theta_1) - f^{1/2}(x;\theta_2)|^2 \nu(dx)\right)^{1/2} \tag{8.3}$$

The following result is the main theorem of this section.

Theorem 8.4.1. *Fix a compact subset K of Θ. Then for all $Q \subset K$ with vol $(\partial Q) = 0$, we have*

$$\lim_{\epsilon \to 0} \frac{D(\epsilon, Q)}{D(\epsilon, K)} = \frac{\int_Q \sqrt{det\mathbf{I}(\theta)}d\theta}{\int_K \sqrt{det\mathbf{I}(\theta)}d\theta} \tag{8.4}$$

By using Theorem 8.3.1 we conclude that the Jeffreys measure μ on Θ defined by

$$\mu(Q) \propto \int_K \sqrt{det\mathbf{I}(\theta)}d\theta \qquad Q \subset \Theta \tag{8.5}$$

is the (possibly improper) noninformative prior on Θ in the sense of the second approach described in the introduction.

The idea is to approximate the packing number of relatively small sets by the Jeffreys prior measure for those sets (see 8.13, 8.14) and then fit these small sets into a given set Q or K. One has to check that the approximation remains good at this higher scale [vide 8.16].

Proof. Fix $0 < \eta < 1$. Cover K by J cubes of length η. In each cube fix an interior cube with length $\eta - \eta^2$. The interior cube will provide an approximation from below.

Since by continuity, the eigenvalues of $\mathbf{I}(\theta)$ are uniformly bounded away from zero and infinity on K, by standard arguments [see theorem I.7.6. in [102]], it follows from (8.1) that there exist $M > m > 0$ such that

$$m\|\theta_1 - \theta_2\| \le H(\theta_1, \theta_2) \le M\|\theta_1 - \theta_2\|, \qquad \theta_1, \theta_2 \in K \tag{8.6}$$

Given $\eta > 0$ choose $\epsilon > 0$ so that $\epsilon/(2m) < \eta^2$. Any two interior cubes are separated by at least η/m in terms of Euclidean distance and by ϵ in terms of the Hellinger distance.

For $Q \subset K$, let Q_j be the intersection of Q with the jth cube and Q'_j be the intersection with the jth interior cube, $j = 1, 2 \ldots, J$. Then

$$Q_1 \cup Q_2 \cup \ldots \cup Q_J = Q'_1 \cup Q'_2 \cup \ldots \cup Q'_J \tag{8.7}$$

Hence

$$\sum_{j=1}^{J} D(\epsilon, Q'_j, H) \le D(\epsilon, Q, H) \le \sum_{j=1}^{J} D(\epsilon, Q_j, H) \tag{8.8}$$

In particular, with $Q = K$, we obtain

$$\sum_{j=1}^{J} D(\epsilon, K'_j, H) \le D(\epsilon, K, H) \le \sum_{j=1}^{J} D(\epsilon, K_j, H) \tag{8.9}$$

where K_j and K'_j are defined in the same way.

For the jth cube, choose $\theta_j \in K$. By an argument similar to that for (8.6), for all θ, θ' in the jth cube,

$$\frac{\underline{\lambda}(\eta)}{2}\sqrt{(\theta - \theta')^T \mathbf{I}(\theta_j)(\theta - \theta')} \le H(\theta, \theta') \le \frac{\bar{\lambda}(\eta)}{2}\sqrt{(\theta - \theta')^T \mathbf{I}(\theta_j)(\theta - \theta')} \tag{8.10}$$

where $\bar{\lambda}(\eta)$ and $\underline{\lambda}(\eta)$tend to 1 as $\eta \to 0$.

Let

$$\underline{H}_j(\theta, \theta') = \frac{\underline{\lambda}(\eta)}{2}\sqrt{(\theta - \theta')^T \mathbf{I}(\theta_j)(\theta - \theta')}$$

and

$$\bar{H}_j(\theta, \theta') = \frac{\bar{\lambda}(\eta)}{2}\sqrt{(\theta - \theta')^T \mathbf{I}(\theta_j)(\theta - \theta')}$$

Then from (8.10),

$$D(\epsilon, Q_j, H) \leq D(\epsilon, Q_j, \underline{H}) \tag{8.11}$$

$$D(\epsilon, Q'_j, H) \leq D(\epsilon, Q'_j, \bar{H}) \tag{8.12}$$

By the second part of theorem IX of Kolmogorov and Tihomirov [115], for some constants τ_j, τ'_j and absolute constants A_d (depending only on the dimension d),

$$D(\epsilon, Q_j, \underline{H}) \sim A_d vol(Q_j) \sqrt{det \mathbf{I}(\theta_j)} (\underline{\lambda}(\eta))^{-d} \epsilon^{-d} \tag{8.13}$$

and

$$D(\epsilon, Q_j, \bar{H}) \sim A_d vol(Q'_j) \sqrt{det \mathbf{I}(\theta_j)} (\bar{\lambda}(\eta))^{-d} \epsilon^{-d} \tag{8.14}$$

where the symbol $\sim$ means that the limit of the ratio of the two sides is 1 as $\epsilon \to 0$. As all metrics, $\underline{H}_j$ and $\bar{H}_j; j = 1, 2, \ldots, J$ arise from elliptic norms, it can be easily concluded by making a suitable linear transformation that $\tau_j = \tau'_j = \tau$ (say) for all $j = 1, 2, \ldots, J$. Thus we obtain from (8.7)–(8.14) that

$$\limsup_{\epsilon \to 0} \frac{D(\epsilon, Q, H)}{D(\epsilon, K, H)} \leq \frac{\sum_{j=1}^J vol(Q_j) \sqrt{det \mathbf{I}(\theta_j)}}{\sum_{j=1}^J vol(K_j) \sqrt{det \mathbf{I}(\theta_j)}} \left(\frac{\bar{\lambda}(\eta)}{\underline{\lambda}(\eta)} \right)^{-d} \tag{8.15}$$

and

$$\limsup_{\epsilon \to 0} \frac{D(\epsilon, Q, H)}{D(\epsilon, K, H)} \leq \frac{\sum_{j=1}^J vol(Q'_j) \sqrt{det \mathbf{I}(\theta_j)}}{\sum_{j=1}^J vol(K_j) \sqrt{det \mathbf{I}(\theta_j)}} \left(\frac{\underline{\lambda}(\eta)}{\bar{\lambda}(\eta)} \right)^{-d} \tag{8.16}$$

Now let $\eta \to 0$. By the convergence of sums $\sum_{j=1}^J vol(Q_j) \sqrt{det \mathbf{I}(\theta_j)}$ to $\int_Q \sqrt{\mathbf{I}(\theta)} d\theta$ and $\sum_{j=1}^J vol(Q'_j) \sqrt{det \mathbf{I}(\theta_j)} \to \int_Q \sqrt{\mathbf{I}(\theta)} d\theta$ and similarly for sums involving K_js and K'_js. Also $\underline{\lambda}(\eta) \to 1$ and $\bar{\lambda}(\eta) \to 1$, so the desired result follows. □

Remark 8.4.1. It has been pointed out to us by Prof.Hartigan that Jeffreys had envisaged constructing noninformative priors by approximating Θ with Kullback-Leibler neighborhoods . He asked us if the construction in this section can be carried out using the Kullback-Leibler neighborhoods . Because the Kullback-Leibler divergence is not a metric there would be obvious difficulties in formalizing the notion of an ϵ-net. However, if the family of densities $\{f_\theta : \theta \in \Theta\}$ have well-behaved tails such that, for any $\theta, \theta', K(\theta, \theta') \leq \phi(H(\theta, \theta'))$, where $\phi(\epsilon)$ goes to 0 as ϵ goes to 0, then any ϵ-net $\{\theta_1, \ldots, \theta_k\}$ in the Hellinger metric can be thought of as a Kullback-Leibler net in the sense that

1. $K(\theta_i, \theta_j) > \epsilon$ for $i, j, = 1, 2, \ldots k$; and

2. for any θ there exists an i such that $K(\theta_i, \theta) < \phi(\epsilon)$.

In such situations, the above theorems allow us to view the Jeffreys prioras a limit of uniform distributions arising out of Kullback-Leibler neighborhoods. Wong and Shen [172] show that a suitable tail behavior is that for all θ, θ',

$$\int_{f_\theta/f_{\theta'} \geq exp\frac{1}{\delta}} f_\theta(\frac{f_\theta}{\mathcal{F}_{\theta'}})^\delta < M$$

We now consider the case when there is a nuisance parameter. Let θ be the parameter of interest and ϕ be the nuisance parameter. We can write the information matrix as

$$\begin{pmatrix} I_{11}(\theta,\phi) & I_{12}(\theta,\phi) \\ I_{12}(\theta,\phi) & I_{22}(\theta,\phi) \end{pmatrix} \tag{8.17}$$

In view of Theorem 8.4.1, and in the spirit of reference priors of Bernardo [18], the prior for ϕ given θ is specified as $\Pi(\phi|\theta) = \sqrt{I_{11}(\theta, \phi)}$. So it is only necessary to construct a noninformative marginal prior for θ. Assume, as before, that the parameter space is compact. With n i.i.d. observations, the joint density of the observations given θ only is given by

$$g(\mathbf{x}^n, \theta) = (c(\theta))^{-1} \int \prod_1^n f(x_i, \theta, \phi) \sqrt{I_{22}(\theta, \phi)} d\phi \tag{8.18}$$

where $c(\theta) = \int \prod_1^n f(x_i, \theta, \phi) \sqrt{I_{22}(\theta, \phi)} d\phi$ is the constant of normalization. Let $I_n(\theta, g)$ denote the information in the family $\{g(\mathbf{x}^n, \theta) : \theta \in \Theta\}$. Under appropriate regularity conditions, it can be shown that the information per observation $I_n(\theta, g)/n$ satisfies

$$\lim_{n\to\infty} I_n(\theta, g)/n = (c(\theta))^{-1} \int I_{11.2}(\theta, \phi) \sqrt{I_{22}(\theta, \phi)} d\phi = J(\theta) \text{ (say)} \tag{8.19}$$

where $I_{11.2} = I_{11} - I_{12}^2/I_{22}$ is the (11) element in the inverse of the information matrix. Let $H_n(\theta, \theta + h)$ be the Hellinger distance between $g(\mathbf{x}^n, \theta)$ and $g(\mathbf{x}^n, \theta + h)$. Locally, as $h \to 0, H_n^2(\theta, \theta + h)$ behaves like $h^2 I_n(\theta, g)$. Hence by Theorem 8.4.1, the noninformative (marginal) prior for θ would be proportional to $\sqrt{I_n(\theta, g)}$. In view of (8.19), passing to the limit as $n \to \infty$, the (sample size–independent) marginal noninformative prior for θ should be taken to be proportional to $(J(\theta))^{1/2}$, and so the prior for (θ, ϕ) is proportional to $J(\theta)\pi(\phi|\theta)$. Generally, for noncompact parameter space, one can proceed like Berger and Bernardo [14]. Informally, we can sum up as follows. The prior for θ based on the current approach is obtained by taking the average of $I^{11}(\theta, \phi)$ with respect to $\sqrt{I_{22}(\theta, \phi)}$ and then taking the square root. The reference prior of Berger and Bernardo or the probability matching prior takes average geometric and harmonic means of other functions of $\sqrt{I^{11}(\theta, \phi)}$ and then transforms back. In the examples of Datta and Ghosh [38], we believe that they reduce to the same prior.

8.5 Posterior Consistency for Noninformative Priors for Infinite-Dimensional Problems

In this section, we show that in a certain class of infinite dimensional families, the third approach mentioned in the introduction leads to consistent posterior.

Theorem 8.5.1. *Let $\mathcal{P}$ be a family of densities where $\mathcal{P}$, metrized by the Hellinger distance, is compact. Let ε_n be a positive sequence satisfying*

$$\sum_{n=1}^{\infty} n^{1/2}\varepsilon_n < \infty$$

Let $\mathcal{P}_n$ be a ε_n-net in $\mathcal{P}$, μ_n be the uniform distribution on $\mathcal{P}_n$, and μ be the probability on $\mathcal{P}$ defined by $\mu = \sum_{n=1}^{\infty} \lambda_n \mu_n$, where λ_ns are positive numbers adding up to unity. If for any $\beta > 0$,

$$\lim_{n \to \infty} e^{\beta n} \frac{\lambda_n}{D(\varepsilon_n, \mathcal{P}_n)} = \infty \tag{8.20}$$

then the posterior distribution based on the prior μ and i.i.d. observations $X_1, X_2, \ldots$ is strongly consistent at every $p_0 \in \mathcal{P}$.

Proof. Since $\mathcal{P}$ is compact under the Hellinger metric, the weak topology and the Hellinger topology coincide on $\mathcal{P}$. Consequently weak neighborhoods and strong neighborhoods coincide and so do the notions of weak and strong consistency.

To prove consistency, by Remark 4.5.1, it is enough to show that for every δ, if $U_n^\delta = \{P : H(P_0, P) < \delta/n\}$ then for all $\beta > 0$,

$$e^{n\beta}\Pi(U_n^\delta) \to \infty$$

Because $\sum_{n=1}^{\infty} n^{1/2}\varepsilon_n < \infty$, given δ, there is a n_0 such that for $n > n_0, \varepsilon_n < \delta/n$; so that for $n > n_0$, there is a $P_n \in \mathcal{P}_n$ such that $H(P_0, P_n) < \delta/n$.

Since $\Pi\{P_n\} = \lambda_n / D(\varepsilon_n, \mathcal{P}_n)$ and by assumption, for all $\beta > 0$,

$$\lim_{n\to\infty} e^{\beta n} \frac{\lambda_n}{D(\varepsilon_n, \mathcal{P}_n)} = \infty$$

and $\Pi(U_n^\delta) > \Pi\{P_n\}$; consistency follows. □

Remark 8.5.1. Consistency is obtained in the Theorem 8.5.1 by requiring (8.20) for sieves whose width ε_n was chosen carefully. However, it is clear from the proof that consistency would follow for sieves with width $\varepsilon_n \downarrow 0$ by imposing (8.20) for a carefully chosen subsequence.

Precisely, if $\varepsilon_n \downarrow 0, \mathcal{P}_n$ an ε_n-net, μ is the probability on $\mathcal{P}$ defined by $\mu = \sum_1^\infty \lambda_n \mu_n$ and δ_n is a positive summable sequence, then by choosing $j(n)$ with

$$\varepsilon_{j(n)} \leq \sqrt{\frac{2}{n}\delta_n} \tag{8.21}$$

the posterior is consistent, if

$$\exp[n\beta] \frac{\lambda_{j(n)}}{D(\varepsilon_{j(n)}, \mathcal{P}_n)} \to \infty \tag{8.22}$$

A useful case corresponds to

$$D(\varepsilon, \mathcal{P}) \leq A \exp[c\epsilon^{-}\alpha] \tag{8.23}$$

where $0 < \alpha < 2/3$ and A and c are positive constants, $\delta_n = n^{-\gamma}$ for some $\gamma > 1$. If in this case $j(n)$ is the smallest integer satisfying (8.21), then (8.22) becomes

$$\exp[n\beta - c\varepsilon_{j(n)}^{-\alpha}]\lambda_{j(n)} \to \infty \tag{8.24}$$

If $\varepsilon_n = \varepsilon/2^n$ for some $\varepsilon > 0$ and λ_n decays no faster than n^{-s} for some $s > 0$ then (8.24) holds. Moreover, the condition $0 < \alpha < 2$ in (8.23) is enough for posterior consistency in probability.

We can apply this in the following example [see Wong and Shen [172]] the following.

Example 8.5.1. Let

$$\begin{aligned} \mathcal{P} = \{g^2 : g \in C^r[0,1], \int_0^1 g^2(x)dx = 1, \\ \|g^{(j)}\|_{\text{sup}} \leq L_j, j = 1, 2, \ldots r \\ |g^{(r)}(x_1) - g^{(r)}(x_2)| \leq L_{r+1}|x_1 - x_2\|^m\} \end{aligned}$$

where r is a positive integer and $0 \leq m \leq 1$ and L's are fixed constants. By theorem 15 of Kolomogorov and Tihomirov [115] $D(\varepsilon, \mathcal{P}, h) \leq \exp[c\varepsilon^{-1/r+m}]$.

8.6 Convergence of Posterior at Optimal Rate

This section is based on Ghosal, Ghosh and van der Vaart ([80]).

We present a result concerning rate of convergence of the posterior relative to L_1, L_2, and Hellinger metrics. The two main elements controlling the rate of convergence are the size of the model (measured by packing or covering numbers) and the amount of prior mass given to a shrinking ball around the true measure. It is the latter quantity that is easy to estimate for the hierarchical noninformative priors introduced in Section 8.1. and appearing in Theorem 8.5.1 of the preceding section. See also Shen and Wasserman [150]

Theorem 8.6.1. *Suppose for a sequence ϵ_n with $\epsilon_n \to 0$ and $n\epsilon_n^2 \to \infty$, a constant $C > 0$ and sets $\mathcal{P}_n \subset \mathcal{P}$ we have*

$$\log D(\epsilon_n, \mathcal{P}_n, d) \leq n\epsilon_n^2 \tag{8.25}$$

$$\Pi_n(\mathcal{P}\backslash\mathcal{P}_n) \leq \exp(-n\epsilon_n^2(C+4)) \tag{8.26}$$

$$\Pi_n\left(P : -E_0(\log\frac{p}{p_0}) \leq \epsilon_n^2, E_0(\log\frac{p}{p_0})^2 \leq \epsilon_n^2\right) \geq \exp(-n\epsilon_n^2 C). \tag{8.27}$$

Then for sufficiently large M, we have that

$$\Pi_n(P : d(P, P_0) \geq M\epsilon_n | X_1, X_2, \ldots, X_n) \to 0 \text{ in } P_0^n \text{ probability}$$

See [80] for a proof.

Condition (8.25) requires that the "model" $\mathcal{P}_n$ is not too big and (8.26) ensures that its complement will not alter too much. It is true for every $\epsilon_n' \geq \epsilon_n$ as soon at it is true for ϵ_n and thus can be seen as defining a minimal possible value of ϵ_n. Condition (8.25) ensures the existence of certain tests and could be replaced by a testing condition in the spirit of LeCam [120]. Note that the metric d used here reappears in the assertion of the theorem. Since the total variation metric is bounded above by twice the Hellinger metric, the assertion of the theorem using the Hellinger metric is stronger, but also condition (8.25) will be more restrictive, so that we really have two theorems. In the case that the densities are uniformly bounded, one can have a third theorem, when using the L_2-distance, which in that case will be bounded above by a multiple of the Hellinger distance. If the densities are also uniformly bounded and uniformly bounded away from zero, then these three distances are equivalent and are also equivalent to the Kullback-Leibler number and L_2-norm appearing in condition (8.27).

A rate ϵ_n satisfying (8.25) for $\mathcal{P} = \mathcal{P}_n$ and d the Hellinger metric is often viewed as giving the "optimal" rate of convergence for estimators of P relative to the Hellinger metric, given the model $\mathcal{P}$. Under certain conditions, such as likelihood ratios bounded away from zero and infinity, this is proved as a theorem by Birgé [22] and LeCam [122] and [120]. See also Wong and Shen [172]. From Birgé's work it is clear that condition (8.25) is a measure of the complexity of the model.

Condition (8.27) is the other main condition. It requires that the prior measures put a sufficient amount of mass near the true measure P_0. Here "near" is measured through a combination of the Kullback-Leibler divergence of p and p_0 and the $L_2(P_0)$-norm of $\log(p/p_0)$. Again, this condition is satisfied for $\epsilon_n' \geq \epsilon_n$ if it is satisfied for ϵ_n and thus is another restriction on a minimal value of ϵ_n.

The assertion of the theorem is an in-probability statement that the posterior mass outside a large ball of radius proportional to ϵ_n is approximately zero. The in-probability statement can be improved to an almost-sure assertion, but under stronger conditions, as indicated below.

Let h be the Hellinger distance and write $\log_+ x$ for $(\log x) \vee 0$.

Theorem 8.6.2. *Suppose that conditions (8.25) and (8.26) hold as in the preceding theorem and* $\sum_n e^{-Bn\epsilon_n^2} < \infty$ *for every* $B > 0$ *and*

$$\Pi_n\Big(P : h^2(P, P_0)\Big\|p_0/p\Big\|_\infty \leq \epsilon_n^2\Big) \geq e^{-n\epsilon_n^2 C}$$

Then for sufficiently large M, *we have that* $\Pi_n(P : d(P, P_0) \geq M\epsilon_n | X_1, \dots, X_n) \to 0$ *in* P_0^n*-almost surely.*

See also theorem 2.3 in [80].

These theorems are not tailored for finite-dimensional models. For such cases and for finite-dimensional sieves, they yield an extra logarithmic factor in addition to the correct rate of $1/\sqrt{n}$. Suitable refinements of (8.25) and (8.27) to address this issue are in [80].

Convergence of the posterior distribution at the rate ϵ_n implies the existence of point estimators, which are Bayes in that they are based on the posterior distribution, which converge at least as fast as ϵ_n in the frequentist sense. One possible construction is to define $\hat{P}_n$ as the (near) maximizer of

$$Q \to \Pi_n\big(P : d(P, Q) < \epsilon_n | X_1, \dots, X_n\big)$$

Theorem 8.6.3. *Suppose that* $\Pi_n(P : d(P, P_0) \geq \epsilon_n | X_1, \dots, X_n)$ *converges to* 0, *almost surely (respectively, in-probability) under* P_0^n *and let* $\hat{P}_n$ *maximize, up to* $o(1)$, *the function* $Q \mapsto \Pi_n\big(P : d(P,Q) < \epsilon_n | X_1, \ldots, X_n\big)$. *Then* $d(\hat{P}_n, P_0) \leq 2\epsilon_n$ *eventually almost surely (respectively, in-probability) under* P_0^n.

Proof. By definition, the ϵ_n-ball around $\hat{P}_n$ contains at least as much posterior probability as the ϵ_n-ball around P_0, both of which by posterior convergence at rate ϵ_n, has posterior probability close to unity. Therefore, these two balls cannot be disjoint. Now apply the triangle inequality. □

The theorem is well - known (See e.g. Le Cam ([120] or Le Cam and Yang [121]). If we use the Hellinger or total variation metric (or some other bounded metric whose square is convex), then an alternative is to use the posterior expectation, which typically has a similar property.

In order to state the next theorem we need a strengthening of the notion of entropy.

Given two functions $l, u : \mathcal{X} \to \mathbb{R}$ the bracket $[l, u]$ is defined as the set of all functions $f : \mathcal{X} \to \mathbb{R}$ such that $l \leq f \leq u$ everywhere. The bracket is said to be of size ϵ relative to the distance d if $d(l, u) < \epsilon$. In the following we use the Hellinger distance h for the distance d and the brackets to consist of nonnegative functions, integrable with respect to a fixed measure μ. Let $N_{[]}(\epsilon, \mathcal{P}, h)$ be the minimal number of brackets of size ϵ needed to cover $\mathcal{P}$. The corresponding bracketing entropy is defined as the logarithm of the bracketing number $N_{[]}(\epsilon, \mathcal{P}, h)$. It is easy to see that $N_{[]}(\epsilon, \mathcal{P}, h)$ is bigger than $N_{[]}(\epsilon/2, \mathcal{P}, h)$ and hence bigger than $D(\epsilon, \mathcal{P}, h)$. However, in many examples, bracketing and packing numbers lead to the same values of the entropy up to an additive constant.

In the spirit of Section 8.2.2 we now construct a discrete prior supported on densities constructed from minimal sets of brackets for the Hellinger distance. For a given number $\epsilon_n > 0$ let $\mathcal{P}i_n$ be the uniform discrete measure on the $N_{[]}(\epsilon_n, \mathcal{P}, h)$ densities obtained by covering $\mathcal{P}$ with a minimal set of ϵ_n-brackets and then renormalizing the upper bounds of the brackets to integrate to one. Thus if $[l_1, u_1], \ldots, [l_N, u_N]$ are the $N = N_{[]}(\epsilon_n, \mathcal{P}, h)$ brackets, then Π_n is the uniform measure on the N functions $u_j / \int u_j \, d\mu$. Finally, construct the hierarchical prior

$$\Pi = \sum_{n \in \mathcal{N}} \lambda_n \Pi_n$$

for a given sequence λ_n with $\lambda_n \geq 0$ and $\sum_n \lambda_n = 1$. This is essentially the third approach of Section 8.2.2. As before the rate at which $\lambda_n \to 0$ is important.

Theorem 8.6.4. *Suppose that* ϵ_n *are numbers decreasing in* n *such that*

$$\log N_{[]}(\epsilon_n, \mathcal{P}, h) \leq n\epsilon_n^2$$

for every n, and

$$n\epsilon_n^2/\log n \to \infty$$

. *Construct the prior* Π *as given previously for a sequence* λ_n *such that* $\lambda_n > 0$ *for all* n *and* $\log \lambda_n^{-1} = O(\log n)$. *Then the conditions of Theorem 8.6.2 are satisfied for* ϵ_n *a sufficiently large multiple of the present* ϵ_n *and hence the corresponding posterior converges at the rate* ϵ_n *almost surely, for every* $P_0 \in \mathcal{P}$, *relative to the Hellinger distance.*

There are many specific applications. The situation here is similar to that in several recent papers on rates of convergence of (sieved) maximum likelihood estimators, as in Birgé and Massart, (1996, 1997), Wong and Shen [172], or chapter 3.4 of van der Vaart and Wellner [161]. We consider again Example 8.5.1 of smooth densities of the previous section.

Example 8.6.1 (Smooth densities). Because upper and lower brackets can be constructed from uniform approximations, this shows that the bracketing Hellinger entropies grow like $\epsilon^{-1/r}$, so that we can take ϵ_n of the order $n^{-r/(2r+1)}$ to satisfy the relation $\log N_{[]}(\epsilon_n, \mathcal{P}, h) \leq n\epsilon_n^2$. This rate is known to be the frequentist optimal rate for estimators. From Theorem 8.6.3, we therefore conclude that for the prior constructed earlier, the posterior attains the optimal rate of convergence.

Since the lower bounds of the brackets are not really needed, the theorem can be generalized by defining $N_{]}(\epsilon, \mathcal{P}, h)$ as the minimal number of functions $u_1, \ldots, u_m$ such that for every $p \in \mathcal{P}$ there exist a function u_i such that both $p \leq u_i$ and $h(u_i, p) < \epsilon$. Next we construct a prior Π as before. These upper bracketing numbers are clearly smaller than the bracketing numbers $N_{[]}(\epsilon, \mathcal{P}, h)$, but we do not know any example where this generalization could be useful.

So far, we have implicitly required that the model $\mathcal{P}$ is totally bounded for the Hellinger metric. A simple modification works for countable unions of totally bounded models, provided that we use a sequence of priors. Suppose that the bracketing numbers of $\mathcal{P}$ are infinite, but there exist subsets $\mathcal{P}_n \uparrow \mathcal{P}$ with finite bracketing numbers. Let ϵ_n be numbers such that $\log N_{[]}(\epsilon_n, \mathcal{P}_n, h) \leq n\epsilon_n^2$ and be such that $n\epsilon_n^2$ is increasing with $n\epsilon_n^2/\log n \to \infty$. Then we construct Π_n as before with $\mathcal{P}$ replaced by $\mathcal{P}_n$, but we do not mix these uniform distributions. Instead, we consider Π_n itself as the sequence of prior distributions. Then the corresponding posteriors achieve the convergence rate ϵ_n.

It is worth observing that we use a condition on the entropies with bracketing, even though we apply Theorem 8.6.2, which demands control over metric entropies only.

This is necessary because the theorem also requires control over the likelihood ratios. If, for instance, the densities are uniformly bounded away from zero and infinity, so that the quotients p_0/p are uniformly bounded, then we can replace the bracketing entropy also by ordinary entropy. Alternatively, if the set of densities $\mathcal{P}$ possesses an integrable envelope function, then we can construct priors achieving the rate ϵ_n determined by the covering numbers up to logarithmic factors. Here we define ϵ_n as the minimal solution of the equation $\log N(\epsilon, \mathcal{P}, h) \leq n\epsilon^2$ and $N(\epsilon, \mathcal{P}, h)$ denotes the Hellinger covering number (without bracketing).

We assume that the set of densities $\mathcal{P}$ has a μ-integrable envelope function: a measurable function m with $\int m\, d\mu < \infty$ such that $p \leq m$ for every $p \in \mathcal{P}$. Given $\epsilon_n > 0$ let $\{s_{1,n}, \ldots, s_{N_n,n}\}$ be a minimal ϵ_n-net over $\mathcal{P}$ (hence $N_n = N(\epsilon_n, \mathcal{P}, h)$) and put

$$g_{j,n} = (s_{j,n}^{1/2} + \epsilon_n m^{1/2})^2 / c_{j,n}$$

where $c_{j,n}$ is a constant ensuring that $g_{j,n}$ is a probability density. Finally, let Π_n be the uniform discrete measure on $g_{1,n}, \ldots, g_{N_n,n}$ and let $\Pi = \sum_{n=1}^{\infty} \lambda_n \Pi_n$ be a convex combination of the Π_n as before. This is similar to the construction of sieved MLE in theorem 6 of Wong and Shen [172]. The following result guarantees an optimal rate of convergence.

Theorem 8.6.5. *Suppose that ϵ_n are numbers decreasing in n such that*

$$\log N(\epsilon_n, \mathcal{P}, h) \leq n\epsilon_n^2$$

for every n and $n\epsilon_n^2/\log n \to \infty$. Construct the prior $\Pi = \sum_{n=1}^{\infty} \lambda_n \Pi_n$ as given previously for a sequence λ_n such that $\lambda_n > 0$ for all n and $\log \lambda_n^{-1} = O(\log n)$. Assume m is a μ-integrable envelope. Then the corresponding posterior converges at the rate $\epsilon_n \log(1/\epsilon_n)$ in probability, relative to the Hellinger distance.

We omit the proof.

9 Survival Analysis—Dirichlet Priors

9.1 Introduction

In this chapter, our interest is in the distribution of a positive random variable X, which arises as the time to occurrence of an event. What makes the problem different from those considered so far is the presence of censoring. Typically, one does not always get to observe the value of X but only obtains some partial information about X, like $X \geq a$ or $a \leq X \leq b$. This loss of information is often modeled through various kinds of censoring mechanisms: left, right, interval, etc. See Andersen et al. [3] for a deep development of various censoring models. The earliest frequentist methods for censored data were in the context of right censored data, and it is this kind of censoring that we will study in this and in Chapter 10. Bayesian analysis of other kinds of censored data is still tentative, and much remains to be done.

Let X be a positive random variable with distribution F and let Y be independent of X with distribution G. The model studied in this section is: $F \sim \Pi$, given F; $X_1, X_2, \ldots, X_n$ are i.i.d F; given G; $Y_1, Y_2, \ldots, Y_n$ are i.i.d G and are independent of the X_is; the observations are $(Z_1, \delta_1), (Z_2, \delta_2), \ldots, (Z_n, \delta_n)$ where $Z_i = (X_i \wedge Y_i)$ and $\delta_i = I(X_i \leq Y_i)$.

Our interest is in the posterior distribution of F given $(Z_i, \delta_i) : 1 \leq i \leq n$.

Under the assumption that X and Y are independent, the posterior distribution of F given (Z, δ) is independent of G. If $Z_i = z_i$ and $\delta_i = 0$, the observation is referred to as (right) censored at z_i, and in this case it is intuitively clear that the information we have about X is just that $X_i > z_i$ and hence the posterior distribution of F given $(Z_i = z_i, \delta_i = 0)$ is $\Pi(\cdot|X_i > z_i)$. Similarly, the posterior distribution of F given $(Z_i = z_i, \delta_i = 1)$ is $\Pi(\cdot|X_i = z_i)$.

In Section 9.1, we study the case when the underlying prior for F is a Dirichlet process. This model was first studied by Susarla and Van Ryzin [154]. They obtained the Bayes estimate of F, and later Blum and Susarla [26] gave a mixture representation for the posterior. Here we develop a different representation for the posterior and show that the posterior is consistent.

In Section 9.2, we briefly discuss the notion of cumulative hazard function, describe some its properties, and use it to describe a result of Peterson who shows that, under mild assumptions, both F and G can be recovered from the distribution of (Z, δ). This result is used in Section 9.3.

In Section 9.3, we start with a Dirichlet prior for the distribution of (Z, δ) and through the map discussed in Section 9.2, transfer this to a prior for F. The properties discussed in Section 9.2 are used to study these priors.

In the last section, we look at Dirichlet process priors for interval censored data and note that some of the properties analogous to the right censored case do not hold here. Some of the material in this chapter is taken from [81] and [87].

9.2 Dirichlet Prior

Let α be a finite measure on $(0, \infty)$. The model that we consider here is $F \sim D_\alpha$; Given F; $X_1, X_2, \ldots, X_n$ are i.i.d F; Given G; $Y_1, Y_2, \ldots, Y_n$ are i.i.d G and are independent of the X_is; the observations are $(Z_1, \delta_1), (Z_2, \delta_2), \ldots, (Z_n, \delta_n)$ where $Z_i = (X_i \wedge Y_i)$ and $\delta_i = I(X_i \leq Y_i)$.

Our interest is in the posterior distribution of F given $(Z_i, \delta_i) : 1 \leq i \leq n$. Under the independence assumption the distribution of G plays no role in the posterior distribution of F.

The posterior can be represented in many ways. Susarla and Van Ryzin [154], who first investigated, obtained a Bayes estimate for F and showed that this Bayes estimate converges to the Kaplan-Meier estimate as $\alpha(\mathbb{R}^+) \to 0$. Blum and Susarla [26] complemented this result by showing that the posterior distribution is a mixture of Dirichlet processes. This mixture representation, while natural, is somewhat cumbersome.

Lavine [118] observed that the posterior can be realized as a Polya tree process. Under this representation computations are more transparent, and this is the representation that we use in this chapter. A more elegant approach comes from viewing a Dirichlet process as a neutral to right prior. This method is discussed in Chapter 10.

Since a Dirichlet process is also a Polya tree, we begin with a proposition that indicates that a Polya tree prior can be easily updated in the presence of partial information. The proof is straightforward and omitted.

Proposition 9.2.1. *Let μ be a $PT(\underline{\underline{T}}, \underline{\alpha})$. Given $P; X_1, X_2, \ldots, X_n$ are i.i.d P. The posterior given $I_{B_{\underline{\epsilon}_1}}(X_1), I_{B_{\underline{\epsilon}_2}}(X_2), \ldots, I_{B_{\underline{\epsilon}_n}}(X_n)$ is again a Polya tree with respect to $\underline{\underline{T}}$ and with parameters $\alpha'_{\underline{\epsilon}} = \alpha_{\underline{\epsilon}} + \#\{i : B_{\underline{\epsilon}_i} \subset B_{\underline{\epsilon}}\}$.*

Let $\boldsymbol{Z} = (Z_1, Z_2, \ldots, Z_n)$, where $Z_1 < \cdots < Z_n$. Consider the sequence of nested partitions $\{\pi_m(\boldsymbol{Z})\}_{m \geq 1}$ given by:

$$\begin{aligned} \pi_1(\boldsymbol{Z}) &: \; B_0 = (0, Z_1], B_1 = (Z_1, \infty) \\ \pi_2(\boldsymbol{Z}) &: \; B_{00}, B_{01}, B_{10} = (Z_1, Z_2], B_{11} = (Z_2, \infty) \end{aligned}$$

and for $l \leq (n-1), let$

$$\pi_{l+1}(\boldsymbol{Z}) \; : \; B_{0_l 0}, B_{0_l 1}, \ldots, B_{1_l, 0} = (Z_l, Z_{l+1}], B_{1_l 1} = (Z_{l+1}, \infty)$$

where 1_l is a string of 1s of length l, and 0_l is a string of 0s of length l. The remaining $B_{\boldsymbol{\epsilon}}$s are arbitrarily partitioned into two intervals such that $\{\pi_m(\boldsymbol{Z})\}_{m \geq 1}$ forms a sequence of nested partitions that generates $\mathcal{B}(\mathbb{R}^+)$.

Let $\alpha_{\epsilon_1, \ldots, \epsilon_l} = \alpha(B_{\epsilon_1, \ldots, \epsilon_l})$, and $C^n_{\epsilon_1, \ldots, \epsilon_l} = \sum_{\delta_i = 0} I[(Z_i, \infty) \subset B_{\epsilon_1, \ldots, \epsilon_l}]$. Also, let

$$U_i = \# \left\{ (Z_i, \delta_i) : Z_i > Z_{(i)}, \delta_i = 1 \right\}$$

be the number of uncensored observations strictly larger than $Z_{(i)}$.

Similarly denote by C_i the number of censored observations that are greater than or equal to $Z_{(i)}$, i.e.

$$C_i = \# \left\{ (Z_i, \delta_i) : Z_i \geq Z_{(i)}, \delta_i = 0 \right\}$$

where $n_i = C_i + U_{i-1}$ is the number of subjects alive at time $Z_{(i)}$ and $n_i^+ = C_i + U_i$ is the number of subjects who survived beyond $Z_{(i)}$. To evaluate the posterior given $(z_1, \delta_1), \ldots, (z_n, \delta_n)$, first look at the posterior given all the uncensored observations among $(z_1, \delta_1), \ldots, (z_n, \delta_n)$. Since the prior on $M(\mathcal{X})$—the space of all distributions for X–is a D_α, the posterior on $\mathbf{M}(\mathbf{X})$ is Dirichlet with parameter $\alpha + \sum_{(i:\Delta_i = 1)} \delta_{Z_i}$.

Because a Dirichlet process is a Polya tree with respect to every partition, it is so with respect to $\underline{\underline{T}}^*(\underline{Z}^*)$. Proposition 9.2.1 easily leads to the updated parameters $\alpha'_{\epsilon_1,\epsilon_2,\ldots,\epsilon_k}$. We summarize these observations in the following theorem.

Theorem 9.2.1. *Let $\mu = D_\alpha \times \delta_{G_0}$ be the prior on $M(\mathbb{R}^+)\times M(\mathbb{R}^+)$. Then the posterior distribution $\mu_1(\cdot \mid (z_1,\delta_1),\ldots,(z_n,\delta_n))$ is a Polya tree process with parameters $\boldsymbol{\pi}_n^{(\boldsymbol{Z},\boldsymbol{\delta})}$ and $\alpha_n^{(\boldsymbol{Z},\boldsymbol{\delta})} = \{\acute{\alpha}_{\epsilon_1,\ldots,\epsilon_l}\}$, where $\acute{\alpha}_{\epsilon_1,\ldots,\epsilon_l} = \alpha_{\epsilon_1,\ldots,\epsilon_l} + U_i] + C_i$.*

Remark 9.2.1. Note that if $B_{\epsilon_1,\ldots,\epsilon_l} = (Z_k,\infty)$ then

$$\alpha'_{\epsilon_1,\ldots,\epsilon_l} = \alpha(B_{\epsilon_1,\ldots,\epsilon_l}) + \text{ number of individuals surviving at time } Z_k$$

and for every other $B_{\epsilon_1,\ldots,\epsilon_l}$,

$$\alpha'_{\epsilon_1,\ldots,\epsilon_l} = \alpha(B_{\epsilon_1,\ldots,\epsilon_l}) + \text{ number of uncensored observations in } B_{\epsilon_1,\ldots,\epsilon_l}$$

The representation immediately allows us to find the Bayes estimate of the survival function $\bar{F} = 1 - F$. Fix $t > 0$ and let $Z_{(k)} \le t < Z_{(k+1)}$. Then, with $Z_{(0)} = 0$

$$\bar{F}(t) = \left[\prod_1^k \frac{\bar{F}(Z_{(i)})}{\bar{F}(Z_{(i-1)})}\right] \frac{\bar{F}(t)}{\bar{F}(Z_{(k)})} \tag{9.1}$$

A bit of reflection shows that Theorem 9.2.1 continues to hold if we change the partition to include t, i.e., partition B_{1_k} into $(Z_{(k)}, t]$ and (t,∞) and then continue as before. Thus the factors in (9.1) are independent beta variables and $\hat{\bar{F}}(t) = E(\bar{F}(t)|(Z_i,\delta_i) : 1 \le i \le n)$ is seen to be

$$\hat{\bar{F}}(t) = \left[\prod_1^k \frac{\alpha(Z_{(i)},\infty) + U_i + C_i}{\alpha(Z_{(i-1)},\infty) + U_{i-1} + C_i}\right] \frac{\alpha(t,\infty) + U_t + C_t}{\alpha(Z_{(k)},\infty) + U_k + C_t} \tag{9.2}$$

Rewrite expression (9.2) as

$$\left[\prod_1^k \frac{\alpha(Z_{(i)},\infty) + U_i + C_i}{\alpha(Z_{(i)},\infty) + U_i + C_{i+1}}\right] \frac{\alpha(t,\infty) + U_t + C_t}{\alpha(0,\infty) + n} \tag{9.3}$$

If the censored observations and the uncensored observations are distinct (as would be the case if F and G have no common discontinuity), then at any $Z_{(i)}$ that is an uncensored value, $C_i = C_{i+1}$ and the corresponding factor in (9.3) is 1. Thus (9.3) can be rewritten as

$$\left[\prod_{Z_{(i)} \leq t, \delta_i = 0} \frac{\alpha(Z_{(i)}, \infty) + U_i + C_i}{\alpha(Z_{(i)}, \infty) + U_i + C_{i+1}}\right] \frac{\alpha(t, \infty) + U_t + C_t}{\alpha(0, \infty) + n} \tag{9.4}$$

This is the expression obtained by Susarla and Van Ryzin [154]. The expression is a bit misleading because it appears that the estimate, unlike the Kaplan-Meier, is a product over censored values. Keeping in mind that $C_t = C_{k+1}$, it is easy to see that if t is a censored value, then the expression is left-continuous at t, and being a survival function it is hence continuous at t. Similarly, it can be seen that the expression has jumps at uncensored observations. Thus the expression can be rewritten as a product over censored observations times a continuous function. This form appears in the Chapter 10.

As $\alpha(0, \infty) \to 0$, (9.1) goes to

$$\left[\prod_1^k \frac{U_i + C_i}{U_{i-1} + C_i}\right] \frac{U_t + C_t}{U_k + C_k} \tag{9.5}$$

If $Z_{(i)}$ is uncensored then $U_i + C_i = N_i^+$ and $U_{i-1} + C_i = N_i$. If $Z_{(i)}$ is censored then $U_i + C_i = U_{i-1} + C_i$ and we get the usual Kaplan-Meier estimate.

We next turn to consistency.

Theorem 9.2.2. *Let F_0 and G have the same support and no common point of discontinuity. Then for any $t > 0$,*

(i) $E(\bar{F}(t)|(Z_i, \delta_i) : 1 \leq i \leq n) \to \bar{F}_0(t)$ a.e. $P^\infty_{F_0 \times G}$; and

(ii) $V(\bar{F}(t)|(Z_i, \delta_i) : 1 \leq i \leq n) \to 0$ a.e. $P^\infty_{F_0 \times G}$.

Hence the posterior of F is consistent $(F_0$.

Proof. Because F_0 and G have the same support and no common point of discontinuity, the censored and uncensored observations are distinct. Note that if $a, b, c \geq 0, a + b/a + c \geq b/c$. Using this fact, it is easy to see that (9.1) is larger than (9.5), and hence

$$\lim_{n \to \infty} E(\bar{F}(t)|(Z_i, \delta_i) : 1 \leq i \leq n) \geq \bar{F}_0(t) \text{ a.e. } P^\infty_{F_0 \times G}$$

On the other hand, writing (9.4) as $A_n(t)B_n(t)$ where

$$A_n(t) = \frac{\alpha(t,\infty) + U_t + C_t}{\alpha(0,\infty) + n}, B_n(t) = \left[\prod_{Z_{(i)} \le t, \delta_i = 0} \frac{\alpha(Z_{(i)},\infty) + U_i + C_i}{\alpha(Z_{(i)},\infty) + U_i + C_{i+1}} \right]$$

it is easy to see that $A_n(t) \to \bar{F}_0(t)\bar{G}_0(t)$ and

$$(B_n(t))^{-1} \ge \prod_{Z_{(i)} \le t, \delta_i = 0} \frac{U_i + C_i}{U_i + C_{i+1}}$$

The right side of the last expression is the Kaplan-Meier estimate of $\bar{G}$, and so

$$\lim_{n\to\infty} (B_n(t))^{-1} \ge \bar{G}(t)$$

and

$$\lim_{n\to\infty} B_n(t) \le \left(\bar{G}(t)\right)^{-1}$$

so that

$$\lim_{n\to\infty} A_n(t)B_n(t) \le \bar{F}_0(t)$$

Since the factors in (9.1) are beta variables, it is easy to write $E(\bar{F}^2(t)|(Z_i,\delta_i) : 1 \le i \le n)$. A bit of tedious calculation will show that

$$E(\bar{F}^2(t)|(Z_i,\delta_i) : 1 \le i \le n) \to \bar{F}_0^2(t)$$

We leave the details to the reader. □

9.3 Cumulative Hazard Function, Identifiability

Let F be a distribution function on $(0,\infty)$. So the survival function $\bar{F} = 1 - F$ is decreasing, right-continuous and $\lim_{t\to 0} \bar{F}(t) = 1, \lim_{t\to\infty} \bar{F}(t) = 0$. We will often write $F(A), \bar{F}(A)$ for the probability of a set A under the probability measure corresponding to the distribution function F. Thus $F\{t\} = \bar{F}\{t\} = F(t) - F(t-) = \bar{F}(t-) - \bar{F}(t)$ is the probability of $\{t\}$.

A concept of importance in survival analysis is failure rate and the related cumulative hazard function. For the distribution function F of a discrete probability, a natural expression for the hazard rate at s is $F\{s\}/\bar{F}(s-)$. Summing this over $s \le t$ gives a notion of cumulative hazard function for a discrete F at t as

$$\mathbf{H}(F)(t) = \sum_{s \le (t)} \frac{F\{s\}}{\bar{F}(s-)} = \int_0^{(\cdot)} \frac{dF(s)}{\bar{F}(s-)}$$

Extending this notion, cumulative hazard function for a general F is defined by

$$\mathbf{H}(F)(\cdot) = \int_0^{(\cdot)} \frac{dF(s)}{\bar{F}(s-)}$$

More precisely, let $F \in \mathcal{F}$ and let $T_F = \inf\{t : F(t) = 1\}$. Note that T_F may be ∞. Set

$$\mathbf{H}(F) = H_F(t) = \begin{cases} \int_{(0,t]} \frac{dF(s)}{F[s,\infty)}, & \text{for } t \le T_F \\ H_F(T_F) & \text{for } t > T_F \end{cases}$$

1. Let $\{s_1, s_2, \ldots\}$ be a dense subset of $(0, \infty)$. For each n, let $s_1^{(n)} < \cdots < s_n^{(n)}$ be an ordering of $\{s_1, \ldots, s_n\}$. Let $s_0^{(n)} = 0$ and define

$$H_F^n(t) = \begin{cases} \sum_{s_i^{(n)} < t} \frac{F(s_i^{(n)}, s_{i+1}^{(n)}]}{F(s_i^{(n)}, \infty)} & \text{for } t \le T_F \\ H_F^n(T_F) & \text{for } t > T_F \end{cases}$$

   Then, for all t, $H_F^n(t) \to H_F(t)$ as $n \to \infty$.

2. H_F is nondecreasing and right-continuous. The fact that H_F is nondecreasing follows trivially because F is nondecreasing. To see that H_F is right-continuous, fix a point t and note that if $j = \max\{i \le n : s_i^{(n)} < t\}$, then

$$H_F(t+) - H_F(t) = \lim_{n \to \infty} \frac{F(s_{j+1}^{(n)}, s_{j+2}^{(n)}]}{F(s_{j+1}^{(n)}, \infty)}$$

   where both $\{s_{j+1}^{(n)}\}$ and $\{s_{j+2}^{(n)}\}$ are nondecreasing sequences converging to t from above. Thus $F(s_{j+1}^{(n)}, s_{j+2}^{(n)}] \to 0$ as $n \to \infty$.

   If $t < T_F$, then the denominator of the right hand side of the equation is positive for some n, hence right-continuity follows. For $t \ge T_F$ it follows from the definition.

It is easy to see that $H_F(t) < \infty$ for every $t < T_F$. As with F, we will think of H_F simultaneously as a function and a measure. Thus the measure of any interval $(s,t]$ under H_F will be defined as $H_F(s,t] = H_F(t) - H_F(s)$. For $T_F < s < t$, define $H_F(s,t] = 0$.

3. For any t, H_F has a jump at t iff F has a jump at t, i.e. $\{t : H_F\{t\} > 0\} = \{t : F\{t\} > 0\}$.

4. It follows from preceding that

   (a) $T_F = \inf\{t : H_F(t) = \infty \text{ or } H_F\{t\} = 1\}$,

   (b) $H_F\{t\} \leq 1$ for all t,

   (c) $H_F(T_F) = \infty$ if T_F is a continuity point of F,and

   (d) $H_F\{T_F\} = 1$ if $F\{T_F\} > 0$.

These and other properties of **H** and details can be found in Gill and Johansen [90].

Let $\mathcal{A}'$ be the space of all functions on $[0,\infty)$ that are nondecreasing, right-continuous, and may, at any finite point, be infinity, but has jumps no greater than one, i.e.,

$$\mathcal{A}' = \{B \in \mathcal{H} \mid B\{t\} \leq 1 \text{ for all } t\}$$

Equip $\mathcal{A}'$ with the smallest σ-algebra under which, the maps $\{A \mapsto A(t), t \geq 0\}$ are measurable. **H** maps $\mathcal{F}$ into $\mathcal{A}'$ and **H** is measurable. The actual range of **H**, which we will now describe, is smaller.

For $A \in \mathcal{A}'$, let $T_A = \inf\{t : A(t) = \infty \text{ or } A\{t\} = 1\}$. Let $\mathcal{A}$ be the space of all cumulative hazard functions on $[0,\infty)$. Formally define $\mathcal{A}$ as

$$\mathcal{A} = \{A \in \mathcal{A}' \mid A(t) = A(T_A) \text{ for all } t \geq T_A\}$$

Endow $\mathcal{A}$ with the σ-algebra which is the restriction of the σ-algebra on $\mathcal{A}'$ to $\mathcal{A}$. The map **H** is a 1-1 measurable map from $\mathcal{F}$ onto $\mathcal{A}$ and, in fact, has an inverse [see Gill and Johansen [90]]. We consider this inverse map next and briefly summarize its properties.

Let $A \in \mathcal{A}'$. Let $\{s_1, s_2, \dots\}$ be dense in $(0,\infty)$. For each n, let $s_1^{(n)} < \cdots < s_n^{(n)}$ be as before. Fix $s < t$. If $A(t) < \infty$, define the product integral of A by

$$\prod_{(s,t]}(1 - dA) = \lim_{n\to\infty} \prod_{s < s_i^{(n)} \leq t} (1 - A(s_{i-1}^{(n)}, s_i^{(n)}])$$

where $A(a,b] = A(b) - A(a)$ for $a < b$. If $A(t) = \infty$ and $A(s) < \infty$, set $\prod_{(s,t]}(1-dA) = 0$. If $A(s) = \infty$, set $\prod_{(s,t]}(1-dA) = 1$.

Theorem 9.3.1. *Let $A \in \mathcal{A}$. Then F given by*

$$F(t) = 1 - \prod_{(0,t]}(1 - dA)$$

is an element of $\mathcal{F}$. Further,

$$A(t) = \int_{(0,t]} \frac{dF(s)}{F[s,\infty)}$$

The following properties of the product integral are included to lend the reader a better understanding of the nature of the map **H** and will be useful later. For details, we again refer to Gill and Johansen [90].

5. Like **H**, the product integral also has an explicit expression:

$$\prod_{(0,t]}(1 - dA) = \exp\left(-A^c(t)\right) \prod_{s \le t}(1 - A\{s\})$$

where A^c is the continuous part of A.

6. Let ρ_S denote the Skorokhod metric on $D[0,\infty)$ and let $\{H_n\}$ be a sequence in $\mathcal{A}$. Say that $\rho_S(H_n, A) \to 0$ for some $A \in \mathcal{A}$ as $n \to \infty$, if $\rho_S(H_n^T, A^T) \to 0$ for all $T > 0$ where H_n^T and A^T are restrictions of H_n and A to $[0,T]$. It may be shown, following Hjort [([100], Lemma A.2, pp. 1290–91), that if $\{H_n\}, A \in \mathcal{A}$ and $\rho_S(H_n, A) \to 0$, then $\mathbf{H}^{-1}(H_n) \stackrel{w}{\to} \mathbf{H}^{-1}(A)$. Thus, if $\mathcal{A}$ is endowed with the Skorokhod metric, then $\mathbf{H}^{-1}$ is a continuous map.

Let F be a distribution function. In the literature

$$\mathbf{A}(F) = -\log \bar{F}$$

is also used to formalize the notion of "cumulative hazard function of F." **A** arises by defining the *hazard rate* at s for a continuous random variable as

$$r(s) = \lim_{\Delta s \to 0} \frac{1}{\Delta s} \frac{P(s \le X < s + \Delta s)}{P(X \ge s}$$

If X has a distribution F with density f then $r(s) = f(s)/\bar{F}(s)$ and if the cumulative hazard function is defined as $\int_0^{(.)} r(s)ds$ then this gives $\mathbf{A}(F) = -\log \bar{F}(\cdot)$. One extends the definition for a discrete F formally to give $\mathbf{A}$.

It is easy to see that the two definitions coincide when F is continuous. However, in estimating a survival function or a cumulative hazard function one typically employs a discontinuous estimate. Further, priors like the Dirichlet sit on discrete distributions. The nature of the map, therefore, plays an important role in inference about lifetime distributions and hazard rates. For us, the cumulative hazard function of a distribution will be $\mathbf{H}(F)$.

We next turn to identifiability of (F, G) by (Z, δ). As before, let X and Y be independent with $X \sim F, Y \sim G$. Let $T(x, y) = (z, \delta) = (x \wedge y), I(x \leq y))$ and denote by $T^*(F, G)$ the distribution of T when $X \sim F, Y \sim G$.

$T^*(F, G)$ is thus a probability measure on $\mathcal{T} = (0, \infty) \times \{0, 1\}$. Any probability measure P on $\mathcal{T}$ gives rise to two subsurvival functions,

$$S^0(t) = P\left((t, \infty) \times \{0\}\right)$$

and

$$S^1(t) = P\left((t, \infty) \times \{1\}\right)$$

These satisfy

$$S^0(0+) + S^1(0+) = 1, \qquad S^i(t) \text{ decreasing in } t \qquad \lim_{t\to\infty} S^i(t) = 0 \tag{9.6}$$

Conversely, any pair of subsurvival functions satisfying (9.6) correspond to a probability on $\mathcal{T}$. The following proposition, due to Peterson [138], shows that under mild assumptions F and G can be recovered from $T^*(F, G)$.

Proposition 9.3.1. *Assume that F and G have no common points of discontinuity. Let $T^*(F, G) = (S^0, S^1)$. Then for any t such that $S^i(t) > 0, i = 0, 1$;*

1.

$$H_F(t) = \int_{(0,t]} \frac{dS^1(s)}{S^0(s-) + S^1(s-)} \tag{9.7}$$

2.

$$\bar{F}(t) = e^{-\int_0^t \frac{dS_c^1(s)}{S^0(s-)+S^1(s-)}} \prod_{s\leq t, S^1\{s\}>0} \left(1 - \frac{S^1\{s\}}{S^0(s-) + S^1(s-)}\right) \tag{9.8}$$

3.

$$\sup_t |F_n(t) - F(t)| + |G_n(t) - G(t)| \to 0 \textit{ iff}$$
$$\sup_t \left[|S_n^0(t) - S^0(t)| + |S_n^0(t) - S^0(t)|\right] \to 0 \quad (9.9)$$

A similar expression holds for $\bar{G}$. Thus, if we assume that F and G have no common points of discontinuity and have the same support, then both F and G can be recovered from $T^*(F, G)$.

9.4 Priors via Distributions of (Z, δ)

It might be argued that in the censoring context, subjective judgments such as exchangeability are to be made on the observables (Z, Δ) and would hence lead to priors for the distribution of (Z, Δ). The model of independent censoring can be used to transfer this prior to the distribution of the lifetime X.

Formally, let $\mathbf{M_0} \subset M(\mathcal{X}) \times \mathbf{M}(\mathbf{Y})$ be the class of all pairs of distribution functions (F, G) such that

1. F and G have no points of discontinuity in common, and
2. for all $t \geq 0, F(t) < 1$ and $G(t) < 1$.

Denote by T the function $T(x, y) = (x \wedge y, I_{x \leq y})$ and by T^* the function on $M(\S \times \mathcal{Y})$ defined by $T^*(P, Q) = (P, Q) \circ T^{-1}$, i.e., $T^*(P, Q)$ is the distribution of T under (P, Q). Let $\mathbf{M_0}^* = T^*(\mathbf{M_0})$. From the last section we know that on $\mathbf{M_0}$, T^* is 1-1. Note that every prior on $\mathbf{M_0}$ gives rise to a prior on $\mathbf{M_0}^*$ via T^* and every prior on $\mathbf{M_0}^*$ induces a prior on $\mathbf{M_0}$ through $(T^*)^{-1}$.

Theorem 9.4.1. *Let Π be a prior on $\mathbf{M_0}$ and $\Pi^* = \mu \circ \phi^{-1}$ be the induced prior on $\mathbf{M_0}^*$.*

(i) If $\Pi^*(\cdot|(Z_i, \delta_i) : 1 \leq i \leq n)$ on $\mathbf{M_0}^*$ is weakly consistent at $T^*(P_0, Q_0)$, and (P_0, Q_0) is continuous then the posterior $\Pi(\cdot|(Z_i, \delta_i) : 1 \leq i \leq n)$ on $\mathbf{M_0}$ is weakly consistent at (P_0, Q_0).

(ii) If $\Pi^*(U|(Z_i, \delta_i) : 1 \leq i \leq n) \to 1$ for U of the form

$$U = \{(S^0, S^1) : \sup_t [|S^0(t) - S_0^0(t)| + |S^1(t) - S_0^1(t)| < \epsilon]\}$$

(here (S_0^0, S_0^1) are the subsurvival functions corresponding to $T^*(P_0, Q_0)$), then the posterior $\Pi(\cdot|(Z_i, \delta_i) : 1 \le i \le n)$ on $\mathbf{M_0}$ is weakly consistent at P_0.

Proof. (i) immediately follows from the fact that for continuous distributions the neighborhoods arising from supremum metric and weak neighborhoods coincide (see Proposition 2.5.3). The second assertion follows from the continuity property described in Proposition 9.3.1 and by noting that $\Pi(.|(Z_i, \delta_i) : 1 \le i \le n)$ on $\mathbf{M_0}$ is just the distribution of $(T^*)^{-1}$ under $\Pi^*(.|(Z_i, \delta_i) : 1 \le i \le n)$. □

We have so far not demonstrated any prior on $\mathbf{M_0}^*$. We next argue that it is in fact possible to obtain a Dirichlet prior on $M(\mathcal{T})$ that gives mass 1 to $\mathbf{M_0}^*$.

Theorem 9.4.2. *Let α be probability measure on $\mathcal{T} = (0, \infty) \times \{0, 1\}$ and let (S_α^0, S_α^1) be the corresponding subsurvival functions. Assume*

(a) S_α^0 and S_α^1 have the same support and have no common points of discontinuity; and

(b) if for $i = 0, 1$, $H_i(t) = \int_{(0,t]} dS_\alpha^i(s)/((S_\alpha^0(s-) + S_\alpha^1(s-)))$ satisfies

$$\lim_{t\to\infty} H_i(t) = \infty \text{ for } i = 0, 1$$

then for any $c > 0$, $D_{c\alpha}(\mathbf{M_0}^*) = 1$.

Proof. We will work with pairs of random subsurvival functions than with random probabilities on $\mathcal{T}$. We will show that with $D_{c\alpha}$ probability 1,

(a) S^0 and S^1 have the same support and have no common points of discontinuity; and

(b) for $i = 0, 1$, $\int_{(0,\infty)} dS^i(s)/(S^0(s-) + S^1(s-)) = \infty$

That (a) holds with probability 1 is immediate from assumption (a). For (b), let $t_1, t_2, \ldots,$ continuity points of S_α^0, be such that

$$\sum_i \frac{S_\alpha^1(t_{i-1}, t_i]}{(S_\alpha^0(t_{i-1}) + S_\alpha^1(t_{i-1}))} = \infty$$

Such t_is can be chosen by first choosing s_i with $H_1(s_i) \uparrow \infty$ and then choosing t_is in $(s_i, s_{i+1}]$ with

$$\sum_{t_j \in (s_i, s_{i+1}]} \frac{S_\alpha^1(t_{i-1}, t_i]}{(S_\alpha^0(t_{i-1}) + S_\alpha^1(t_{i-1}))} \ge H_1(s_i) - H_1(s_{i-1}) + 2^{-i}$$

Let $Y_i = S^1(t_{i-1}, t_i]/((S^0(t_{i-1}) + S^1(t_{i-1})))$, clearly $\sum_i Y_i \geq \int dS^i(s)/(S^0(s-) + S^1(s-))$. Further, the Y_i's are bounded by 1 and under Dirichlet, are independent.

Note that $(S^0_\alpha(t_{i-1}) + S^1_\alpha(t_{i-1}))$ and Y_i are independent and hence

$$E(Y_i) = \frac{S^1_\alpha(t_{i-1}, t_i]}{(S^0_\alpha(t_{i-1}) + S^1_\alpha(t_{i-1}))}$$

Assumption (b) guarantees $\sum E(Y_i) = \infty$. This in turn gives $\sum E(Y_i) = \infty$ [See Loeve, [132] p 248)]. □

In addition to consistency, if the empirical distribution of (Z, Δ) is a limit of Bayes estimate on $\mathbf{M_0}^*$, then so is the Kaplan-Meier estimate. This method of constructing priors on $\mathbf{M_0}$ is appealing and merits further investigation—for instance the Dirichlet process on $\mathbf{M_0}^*$ arises through a Polya urn scheme, and it would be of interest to see the corresponding process for the induced prior.

9.5 Interval Censored Data

Susarla and Van Ryzin showed that the Kaplan-Meier estimate, which is also the nonparametric MLE, is the limit of Bayes estimates with a D_α prior for the distribution of X. The observations in this section show that this result does not carry over to other kinds of censored data.

Here our observation consists of n pairs $(L_i, R_i]; 1 \leq i \leq n$ where $L_i \leq R_i$ and corresponds to the information $X \in (L_i, R_i]$. We assume that $(L_i, R_i]; 1 \leq i \leq n$ are independent and that the underlying censoring mechanism is independent of the lifetime X so that the posterior distribution depends only on $(L_i, R_i]; 1 \leq i \leq n$. Let $t_1 < t_2 < \ldots, t_{k+1}$ denote the endpoints of $(L_i, R_i]; 1 \leq i \leq n$ arranged in increasing order and let $I_j = (t_j, t_{j+1}]$. For simplicity we assume that $t_1 = \min_i L_i$ and $t_{k+1} = \max_i R_i$.

Our starting point is a Dirichlet prior $D(c\alpha_1, c\alpha_2, \ldots, c\alpha_k)$ for $(p_1, p_2, \ldots, p_k)$ where $p_j = P\{X \in I_j\}$. Turnbull [159] suggested the use of the nonparametric maximum likelihood estimate obtained from the likelihood function

$$\prod_{i=1}^{n} \left(\sum_{I_j \subset (L_i, R_i]} p_j \right)$$

If $(p_1, p_2, \ldots, p_k)$ has a $D(c\alpha_1, c\alpha_2, \ldots, c\alpha_k)$ prior then the posterior distribution of $(p_1, p_2, \ldots, p_k)$ given $(L_i, R_i]; 1 \leq i \leq n$ is a mixture of Dirichlet distributions.

Call a vector $\underline{a} = (a_1, a_2, \ldots, a_n)$, where each a_i, is an integer, an *imputation of* $(L_i, R_i]; 1 \le i \le n$ if $I_{a_i} \subset (L_i, R_i]$. For an imputation $\underline{a}$, let $n_j(\underline{a})$ be the number of observations assigned to the interval I_j. Formally $n_j(\underline{a}) = \#\{i : a_i = j\}$.

Let the order $O(\underline{a})$ of an imputation be $\#\{j : n_j(\underline{a}) > 0\}$. Let $\mathbf{A}$ be the set of all imputations of $(L_i, R_i]; 1 \le i \le n$ and let $m = \min_{\underline{a} \in \mathbf{A}} O(\underline{a})$. Call an imputation $\underline{a}$ minimal if $O(\underline{a}) = m$.

It is not hard to see that the posterior distribution of $(p_1, p_2, \ldots, p_k)$ given $(L_i, R_i]; 1 \le i \le n$ is

$$\sum_{\underline{a} \in \mathbf{A}} C_{\underline{a}} D(c\alpha_1 + n_1(\underline{a}), c\alpha_2 + n_2(\underline{a}), \ldots, c\alpha_k + n_k(\underline{a}))$$

where

$$C_{\underline{a}} = \frac{\prod_1^k \Gamma(c\alpha_j + n_j(\underline{a}))}{\sum_{\underline{a}' \in \mathbf{A}} \prod_1^k \Gamma(c\alpha_j + n_j(\underline{a}'))}$$

The Bayes estimate of any p_j is

$$\hat{p_j} = \sum_{\underline{a} \in \mathbf{A}} C_{\underline{a}} \frac{c\alpha_j + n_j(\underline{a})}{c + n}$$

As $c \downarrow 0, (c\alpha_j + n_j(\underline{a}))/(c + n) \to n_j(\underline{a})/n$. The behavior of $C_{\underline{a}}$ is given by the next proposition.

Proposition 9.5.1. $\lim_{c \to 0} C_{\underline{a}} > 0$ *iff* $\underline{a}$ *is a minimal imputation.*

Proof. Suppose $\underline{a}$ is not minimal. Let $\underline{a}_0$ be an imputation with $O(\underline{a}) > O(\underline{a}_0)$:

$$C_{\underline{a}} \le \frac{\prod_1^k \Gamma\left(c\alpha_j + n_j(\underline{a})\right)}{\prod_1^k \Gamma\left(c\alpha_j + n_j(\underline{a}_0)\right)} = \frac{\prod_{j=1}^k \Gamma(c\alpha_j) \prod_{j:n_j(\underline{a}) \ne 0)} \left(\prod_0^{n_j(\underline{a})} (c\alpha_j + i)\right)}{\prod_{j=1}^k \Gamma(c\alpha_j) \prod_{j:n_j(\underline{a}_0) \ne 0)} \left(\prod_0^{n_j(\underline{a}_0)} (c\alpha_j + i)\right)}$$

Since $O(\underline{a}) > O(\underline{a}_0)$ the ratio goes to 0. Conversely, if $\underline{a}$ is minimal it is easy to see that

$$\frac{1}{C_{\underline{a}}} = \sum_{\underline{a}' \in \mathbf{A}} \frac{\prod_1^k \Gamma\left(c\alpha_j + n_j(\underline{a}')\right)}{\prod_1^k \Gamma\left(c\alpha_j + n_j(\underline{a})\right)}$$

converges to a positive limit. □

Thus the limiting behavior is determined by minimal imputations. A few examples clarify these notions.

Example 9.4.1. Consider the right censoring case, i.e., for each i either $L_i = R_i$ or $R_i = t_k$. Any minimal imputation is given by assigning compatible observations to the singletons corresponding to uncensored observations and I_k if the last(largest) observation is censored.

Example 9.4.2. Consider the case when we have current status or case I interval censored data. Here for each i, either $L_i = t_1$ or $R_i = t_{k+1}$ so that all we know is if X_i is to the right of L_i or to the left of R_i.

(i) If $\max_i L_i < \min_i R_i$ the minimal imputation is allocation of all the observations to the interval $(\max_i L_i, \min_i R_i]$.

(ii) In general, the minimal imputations have order 2. For example, a consistent assignment of the data to $(t_1, \min_i R_i], (\max_i L_i, t_{k+1}]$ would yield a minimal imputation.

A couple of simple numerical examples help clarify the different cases. In the following examples the prior of the distribution is $D_{c\alpha}$, where α is a probability measure. The limit is taken as $c \to 0$. Corresponding to any imputation $\underline{a}$, we will call the intervals I_js for which $n_j(\underline{a}) > 0$, an allocation, and an allocation corresponding to a minimal imputation will be called a minimal allocation.

Example (a): This example illustrates that the limit of Bayes estimates could be supported on a much bigger set than the NPMLE. The observed data consist of the four intervals $(1, \infty)$, $(2, \infty)$, $(0, 3]$, $(4, \infty)$.
The limit of Bayes estimates in this case turns out to be;
$\tilde{F}(0, 1] = 1/22$,
$\tilde{F}(1, 2] = 2/22$,
$\tilde{F}(2, 3] = 6/22$, and
$\tilde{F}(4, \infty] = 13/22$,
while the NPMLE is given by,
$\hat{F}(2, 3] = 1/2$ and
$\hat{F}(4, \infty] = 1/2$.

In the example, each minimal allocation consists of only two subntervals.
(i) (0,1], and $(4, \infty)$, with the corresponding numbers of X_is in the subintervals being 1 and 3, respectively, represents a minimal allocation.
(ii) (2, 3] and $(4, \infty)$ with the corresponding numbers of X_is in the subintervals being

1 and 3, respectively, represents another minimal allocation.
(iii) (2, 3] and (4, ∞) with the corresponding numbers of X_is in the subintervals being 2 and 2, respectively, represents yet another minimal allocation.

Example (b): This example shows that the limit of Bayes estimates could be supported on a smaller set than the NPMLE. The observed data consist of the intervals (0, 1], (2, ∞), (0, 3], (0, 4], and (5, ∞).
The limit of Bayes estimates in this case turns out to be:
$\tilde{F}(0,1] = 3/5$, and
$\tilde{F}(5,\infty) = 2/5$.
while the NPMLE is given by:
$\hat{F}(0,1] = 1/2$,
$\hat{F}(2,3] = 1/6$, and
$\hat{F}(5,\infty) = 1/3$.

As $c \to 0$, while Dirichlet priors lead to strange estimates for the current status data, the case $c = 1$ seems to present no problems. Even when $c \to 0$ we expect that the limiting behavior will be more reasonable when the data are case II interval censored, in the sense described in [91]. In this case, the tendency to push the observation to the extremes would be less pronounced.

In the current status data case the limit (as $c \downarrow 0$) of the posterior itself exhibits degeneracy. The following proposition is easy to establish.

Proposition 9.5.2. *Let* $R^* = \inf_{i:L_i=0} R_i$ *and* $L^* = \sup_{i:R_i=t_{k+1}} L_i$.

(i) If $R^* < L^*$ then as $c \downarrow 0$ the posterior distribution of $P(R^*, L^*)$ converges to the measure degenerate at 0

(ii) If $L^* < R^*$ then as $c \downarrow 0$ the posterior distribution of $P(L^*, R^*)$ converges to the measure degenerate at 1

10
Neutral to the Right Priors

10.1 Introduction

In Chapter 3, among other aspects, we looked at two properties of Dirichlet processes-the tail free property and the neutral to the right property. In this chapter we discuss priors that generalize Dirichlet processes via the neutral to the right property.

Neutral to the right priors are a class of nonparametric priors that were introduced by Doksum [48]. Historically, the concept of neutrality is due to Connor and Mosimann [34] who considered it in the multinomial context. Doksum extended it to distributions on the real line in the form of neutral to the right priors and showed that if Π is neutral to the right, then the posterior given n observations is also neutral to the right. This result was extended to the case of right-censored data by Ferguson and Phadia [64]. These topics are discussed in Section 10.2.

Doksum and Hjort showed that a prior is neutral to the right iff the cumulative hazard function has independent increments. Since independent increment processes are well understood, this connection provides a powerful tool for studying neutral to the right priors. In particular, independent increment processes have a canonical structure, the so-called Lévy representation. The associated Lévy measure can be used to elucidate properties of neutral to the right priors. For instance Hjort provides an explicit expression for the posterior given n independent observations in terms of the Lévy representation when the Lévy measure is of a specific form. In Section 10.3 we summarize these results.

In Section 10.4 we discuss beta processes. Hjort [100] and Walker and Muliere [166], respectively, developed beta processes and beta-Stacy processes, which provide concrete and useful classes of neutral to the right priors. These priors are analogous to the beta prior for the Bernoulli (θ), are analytically tractable, and are flexible enough to incorporate a wide variety of prior beliefs.

The rest of the chapter is devoted to consistency results for neutral to the right priors. These results center around an example of Kim and Lee [114] of a neutral to the right prior that is inconsistent at all continuous distributions.

10.2 Neutral to the Right Priors

For any $F \in \mathcal{F}$, as in the Chapter 9 $\bar{F}(\cdot) = 1 - F(\cdot)$ is the survival function corresponding to F. Let $\bar{F}(0) = 1$. We also continue to denote by $F(A)$ the measure of the set A under the probability measure corresponding to F.

Definition 10.2.1. A prior Π on $\mathcal{F}$ is said to be neutral to the right if, under Π, for all $k \geq 1$ and all $0 < t_1 < \ldots < t_k$,

$$\bar{F}(t_1), \frac{\bar{F}(t_2)}{\bar{F}(t_1)}, \ldots, \frac{\bar{F}(t_k)}{\bar{F}(t_{k-1})}$$

are independent.

If Π is neutral to the right, we will also refer to a random distribution function F with distribution Π as being neutral to right. Note that $(0/0)$ is defined here and throughout to be 1.

For a fixed F, if X is a random variable distributed as F, then for every $0 \leq s < t$, $\bar{F}(t)/\bar{F}(s)$ is simply the conditional probability $F(X > t|X > s)$. For $t > 0$, $\bar{F}(t)$ is viewed as the conditional probability $F(X > t|X > 0)$.

Example 10.2.1. Consider a finite ordered set $\{t_1, \ldots, t_n\}$ of points in $(0, \infty)$. To construct a neutral to right prior on the set $\mathcal{F}_{t_1,\ldots,t_n}$ of distribution functions supported by the points $t_1, \ldots, t_n$, we only need to specify $(n-1)$ independently distributed $[0,1]$-valued random variables $V_1, \ldots, V_{n-1}$, and then set $\bar{F}(t_i)/\bar{F}(t_{i-1}) = 1 - V_i$ for $1 \leq i \leq n-1$. Finally, set $\bar{F}(t_n)/\bar{F}(t_{n-1}) = 0$. Observe that $\bar{F}(t_n) = 0$ and, for

$1 \le i \le n-1$,

$$\bar{F}(t_i) = \prod_{j=1}^{i}(1 - V_j)$$

Example 10.2.2. In a similar fashion we can construct a neutral to right prior on the space $\mathcal{F}_{\underline{\underline{T}}}$ of all distribution functions supported by a countable subset $\underline{\underline{T}} = \{t_1 < t_2 < \ldots\}$ of $(0, \infty)$.

Let $\{V_i\}_{i \ge 1}$ be a sequence of independent $[0,1]$-valued random variables such that, for some $\eta > 0$,

$$\sum_{i \ge 1} \mathbf{P}(V_i > \eta) = \infty$$

This happens, for instance, when V_is are identically distributed with $\mathbf{P}(V_i > \eta) > 0$. As before, for $i \ge 1$, set $\bar{F}(t_i)/\bar{F}(t_{i-1}) = 1 - V_i$. In other words, $\bar{F}(t_k) = \prod_{i=1}^{k}(1 - V_i)$, for all $k \ge 1$. By the Borel-Cantelli lemma, we have

$$\mathbf{P}\left(\prod_{i \ge 1}(1 - V_i) = 0\right) = 1$$

This defines a neutral to right prior Π on $\mathcal{F}$ because

$$\lim_{t \to \infty} \bar{F}(t) = \lim_{k \to \infty} \prod_{i=1}^{k}(1 - V_i) = 0, \qquad \text{a.s. } \Pi$$

Dirichlet process priors of course provide a ready example of a family of neutral to the right priors. Other examples are the beta process and beta-Stacy process , to be discussed later.

As before, we consider the standard Bayesian set-up where Π is a prior and given F, $X_1, X_2, \ldots$ be i.i.d. F. For each $n \ge 1$, denote by $\Pi_{X_1,\ldots,X_n}$ a version of the posterior distribution, i.e. the conditional distribution of F given $X_1, \ldots, X_n$.

Following are some notations:

For $n \ge 1$, define the *observation process* $N_n(.)$ as follows:

$$N_n(t) = \sum_{i \le n} I_{(0,t]}(X_i) \qquad \text{for all } t > 0$$

For every $n \ge 1$, let $N_n(0) \equiv 0$. Observe that $N_n(.)$ is right-continuous on $[0, \infty)$. Let

$$\mathcal{G}_{t_1 \ldots t_k} = \sigma\left\{\bar{F}(t_1), \frac{\bar{F}(t_2)}{\bar{F}(t_1)}, \ldots, \frac{\bar{F}(t_k)}{\bar{F}(t_{k-1})}\right\}.$$

Thus $\mathcal{G}_{t_1 \ldots t_k}$ denotes the collection of all sets of the form

$$D = \left\{ (\bar{F}(t_1), \frac{\bar{F}(t_2)}{\bar{F}(t_1)}, \ldots, \frac{\bar{F}(t_k)}{\bar{F}(t_{k-1})}) \in C \right\}$$

where $C \in \mathcal{B}^k_{[0,1]}$.

Theorem 10.2.1 (Doksum). *Let Π be neutral to the right. Then $\Pi_{X_1,\ldots,X_n}$ is also neutral to the right.*

Proof. Fix $k \geq 1$ and let $t_1 < t_2 < \cdots < t_k$ be arbitrary points in $(0, \infty)$. Denote by $\mathbb{Q}$ the set of all rationals in $(0, \infty)$ and let $\mathbb{Q}' = \mathbb{Q} \cup \{t_1, \ldots, t_k\}$. Let $\{s_1, s_2, \ldots\}$ be an enumeration of $\mathbb{Q}'$. Observe that, for large enough m, $\{t_1, \ldots, t_k\} \subset \{s_1, \ldots, s_m\}$.

For such an m, let $s_1^{(m)} < \cdots < s_m^{(m)}$ be an ordering of $\{s_1, \ldots, s_m\}$. Let $Y_i^{(m)} = \bar{F}(s_i^{(m)})/\bar{F}(s_{i-1}^{(m)})$ and, under Π, let $\Pi_i^{(m)}$ denote the distribution of $Y_i^{(m)}$.

Let $n_1 \leq \cdots \leq n_m$. Then, given $\{N_n(s_1^{(m)}) = n_1, \ldots, N_n(s_m^{(m)}) = n_m\}$, the posterior density of $(Y_1^{(m)}, \ldots, Y_m^{(m)})$ is written as

$$\begin{aligned} f_{Y_1^{(m)},\ldots,Y_m^{(m)}}(y_1, \ldots, y_m) &= \frac{\prod_{i=1}^m (1 - y_i)^{n_i - n_{i-1}} y_i^{n-n_i}}{\int \prod_{i=1}^m (1 - y_i)^{n_i - n_{i-1}} y_i^{n-n_i} d\Pi_i^{(m)}(y_i)} \\ &= \prod_{i=1}^m \frac{(1 - y_i)^{n_i - n_{i-1}} y_i^{n-n_i}}{\int (1 - y_i)^{n_i - n_{i-1}} y_i^{n-n_i} d\Pi_i^{(m)}(y_i)} \end{aligned}$$

This shows that $(Y_1^{(m)}, \ldots, Y_m^{(m)})$ are independent under the posterior given $\{N_n(s_1^{(m)}), \ldots, N_n(s_m^{(m)})\}$. Hence,

$$\frac{\bar{F}(t_i)}{\bar{F}(t_{i-1})} = \prod_{t_{i-1} < s_j^{(m)} \leq t_i} \frac{\bar{F}(s_j^{(m)})}{\bar{F}(s_{j-1}^{(m)})}, \qquad i = 1, \ldots, k$$

are also independent under the posterior given the same information.

Now, by the right-continuity of $N_n(\cdot)$ we have, as $n \to \infty$,

$$\sigma\{N_n(s_j), j \leq m\} \uparrow \sigma\{N_n(t), t \geq 0\} \equiv \sigma(X_1, \ldots, X_n)$$

Hence, for any $A \in \mathcal{G}_{t_1 \ldots t_k}$, by the martingale Convergence theorem, we have

$$\Pi(A \mid N_n(s_1^{(m)}), \ldots, N_n(s_m^{(m)})) \to \Pi(A \mid X_1, \ldots, X_n) \quad \text{almost surely}$$

Since for each m, the random quantities $\bar{F}(t_1), \bar{F}(t_2)/\bar{F}(t_1) \ldots, \bar{F}(t_k)/\bar{F}(k_1)$ are independent given $\sigma(N_n(s_1^{(m)}), \ldots, N_n(s_m^{(m)}))$, independence also holds as $m \to \infty$. □

A perusal of the proof given above suggests that for any $t_1 < t_2$ the posterior distribution of $\bar{F}(t_2)/\bar{F}(t_1)$ depends on $\{N^n(s) : t_1 \leq s \leq t_2\}$. In words, the posterior depends on the number of observations less than t_1, the exact observations between t_1 and t_2 and the number of observations greater than t_2. This was observed by Doksum. The following theorem proved in [42] shows that this property essentially characterizes neutral to the right priors. Walker and Muliere [167] have also obtained characterizations of neutral to the right priors. Their results are presented in a different flavor.

Theorem 10.2.2. *Let Π be a prior on $\mathcal{F}$ such that $\Pi\{0 < F(t) < 1,$ for all $t\} = 1$. Then the following are equivalent:*

(i) Π is neutral to the right

(ii) for every t

$$\mathcal{L}\left(\bar{\mathbf{F}}(t)|\Pi(|\, X_1, X_2, \ldots, X_n)\right) = \mathcal{L}\left(\bar{\mathbf{F}}(t)|N^n(s) : 0 < s < t\right)$$

where $\mathcal{L}(.)$ stands for the Law of (.).

Thus, if one wants to estimate the probability that a subject survives beyond t years based on n samples of which n_1 fell below t, then a neutral to the right prior would lead to the same estimate if the remaining $n - n_1$ observations fell just above t or far beyond it. This is a property that is also shared by the empirical distribution function. This suggests that neutral to the right priors are appropriate when the interest is in all of F and inappropriate if the interest is in a local neighborhood of a fixed time point.

Ferguson and Phadia [64] extend Doksum's result in the case of inclusively and exclusively right censored observations. Let x be a real number in $(0, \infty)$. Given a distribution function $F \in \mathcal{F}$, an observation X from F is said to be *exclusively right censored* if we only know $X \geq x$ and *inclusively right-censored* if we know $X > x$. We state their result next. The proof is straightforward.

Theorem 10.2.3 (Ferguson and Phadia). *Let F be a random distribution function neutral to the right. Let X be a sample of size one from F, and let x be a number in $(0, \infty)$. Then*

(a) the posterior distribution of F given $X > x$ is neutral to the right, and

(b) the posterior distribution of F given $X \geq x$ is neutral to the right.

10.3 Independent Increment Processes

As mentioned in the introduction, neutral to the right priors relate to independent increment process via the cumulative hazard function. To recall from Chapter 9, the cumulative hazard function is given by

$$\mathbf{H}(F)(t) = H_F(t) = \begin{cases} \int_{(0,t]} \frac{dF(s)}{F[s,\infty)} & \text{for } t \le T_F \\ H_F(T_F) & \text{for } t > T_F \end{cases}$$

and discussed its properties.

The next result establishes the connection between neutral to the right priors and independent increment processes with nondecreasing paths via the map $\mathbf{H}$.

Theorem 10.3.1. *Let Π be a neutral to the right prior on $\mathcal{F}$. Then, under the measure Π^* on $\mathcal{A}$ induced by the map $\mathbf{H}$, $\{A(t) : t > 0\}$ has independent increments. Conversely, if Π^* is a probability measure on $\mathcal{A}$ such that the process $\{A(t) : t > 0\}$ has independent increments, then the measure induced on $\mathcal{F}$ by the map*

$$\mathbf{H}^{-1} : A \mapsto 1 - \prod_{(0,t]}(1 - dA)$$

is neutral to the right.

Proof. First suppose that Π is neutral to the right on $\mathcal{F}$ and let $t_1 < \cdots < t_k$ be arbitrary points in $(0, \infty)$. Consider, as before, a dense set $\{s_1, s_2, \ldots\}$ in $(0, \infty)$. Let, for each n, $s_1^{(n)} < \cdots < s_n^{(n)}$ be as before.

Suppose n is large enough that $s_n^{(n)} \ge t_k$. Then, for each $1 \le i \le k$, we have with A_F^n as

$$\begin{aligned} A_F^n(t_i) - A_F^n(t_{i-1}) &= \sum_{t_{i-1} < s_j^{(n)} \le t_i} \frac{F(s_{j-1}^{(n)}, s_j^{(n)}]}{F(s_{j-1}^{(n)}, \infty)} \\ &= \sum_{t_{i-1} < s_j^{(n)} \le t_i} \left(1 - \frac{\bar{F}(s_j^{(n)})}{\bar{F}(s_{j-1}^{(n)})}\right) \end{aligned}$$

Because for each n, $\bar{F}(s_1^{(n)})$, $\bar{F}(s_2^{(n)})/\bar{F}(s_1^{(n)})$, ..., $\bar{F}(s_n^{(n)})/\bar{F}(s_{n-1}^{(n)})$ are independent, $A_F^n(t_1)$, $A_F^n(t_2) - A_F^n(t_1)$, ..., $A_F^n(t_k) - A_F^n(t_{k-1})$ are also independent. Letting $n \to \infty$, we get that $A_F(t_1)$, $A_F(t_2) - A_F(t_1)$, ..., $A_F(t_k) - A_F(t_{k-1})$ are independent.

For the converse, suppose Π^* on $\mathcal{A}$ such that, under Π^*, $\{A(t) : t > 0\}$ is an independent increment process. Again, let $t_1 < \cdots < t_k$ be arbitrary points in $(0, \infty)$. Then with $s_1^{(n)} < \cdots < s_n^{(n)}$ as before, let, for $1 \le i \le k$,

$$\bar{F}_A^n(t_i) = \prod_{s_j^{(n)} \le t_i} (1 - A(s_{j-1}^{(n)}, s_j^{(n)}])$$

If $F_A = \mathbf{H}^{-1}(A)$, then it follows from the definition of the product integral that $\bar{F}_A^{(n)}(t) \to \bar{F}_A(t)$ for all t, as $n \to \infty$. Now, observe that, for $1 \le i \le k$,

$$\frac{\bar{F}_A^n(t_i)}{\bar{F}_A^n(t_{i-1})} = \prod_{t_{i-1} < s_j^{(n)} \le t_i} (1 - A(s_{j-1}^{(n)}, s_j^{(n)}])$$

Since $A(s_{j-1}^{(n)}, s_j^{(n)}], 1 \le j \le n$ are independent for each n so are $\bar{F}_A^n(t_i)/\bar{F}_A^n(t_{i-1}), 1 \le i \le k$. Consequently, we have independence in the limit, i.e., $\bar{F}_A(t_1)$, $\bar{F}_A(t_2)/\bar{F}_A(t_1)$, ..., $\bar{F}_A(t_k)/\bar{F}_A(t_{k-1})$ are independent. □

It is not hard to verify that for a neutral to the right prior Π,

$$E_\Pi \mathbf{H}(F) = \mathbf{H}(E_\Pi F)$$

Since the posterior given $X_1, X_2, \ldots, X_n$ is again neutral to the right, the above property continues to hold for $\Pi_{X_1, X_2, \ldots, X_n}$. It is shown in Dey etal. [43] that in the time discrete case the above property characterizes neutral to the right property. We expect a similar result to hold in general.

Doksum was the first to observe a connection between neutral to the right priors and independent increment processes. He, however, considered cumulative hazard function defined by $\mathbf{D}(F)(t) = D_F(t) = -\log \bar{F}(t)$. The proof of Theorem 10.3.2 is straightforward.

Theorem 10.3.2 (Doksum). *A prior Π on $\mathcal{F}$ is neutral to the right if and only if $\tilde{\Pi} = \Pi \circ \mathbf{D}^{-1}$ is an independent increment process measure such that $\tilde{\Pi}\{H \in \mathcal{H} : \lim_{t \to \infty} H(t) = \infty\} = 1$.*

The theory of neutral to the right priors owes much of its development and analytic elegance to its connection with independent increment processes. The principal examples of general families of neutral to the right priors have been constructed via this connection. Next, we briefly discuss the relevant theory of these processes in terms of a representation due to P. Lévy. Following is a brief description of the representation.

The following facts are wellknown and can be found in , for example, Ito [104] and Kallenberg [110].

Definition 10.3.1. A stochastic process $\{A(t)\}_{t\geq 0}$ is said to be an *independent increment process* if $A(0) = 0$ almost surely and if, for every k and every $\{t_0 < t_1 < \cdots < t_k\} \subset [0,\infty)$, the family $\{A(t_i) - A(t_{i-1})\}_{i=1}^{k}$ is independent.

Let $\mathcal{H}$ be a space of functions defined by

$$\mathcal{H} = \{H \mid H : [0,\infty) \mapsto [0,\infty], H(0) = 0, H \text{ non-decreasing, right-continuous}\} \tag{10.1}$$

Let $\mathbb{B}_{(0,\infty)\times[0,\infty]}$ be the Borel σ-algebra on $(0,\infty) \times [0,\infty]$.

Theorem 10.3.3. *Let Π^* be a probability on $\mathcal{H}$. Under Π^*, $\{A(t) : t > 0\}$ is an independent increment process if and only if the following three conditions hold. There exists*

1 *a finite or countable set $\mathbf{M} = \{t_1, t_2, \ldots\}$ of points in $(0,\infty)$ and, for each $t_i \in \mathbf{M}$, a positive random variable Y_i defined on $\mathcal{H}$ with density f_i;*

2 *a nonrandom continuous nondecreasing function b; and*

3 *a measure λ on $\left((0,\infty) \times [0,\infty], \mathbb{B}_{(0,\infty)\times[0,\infty]}\right)$ that for all $t > 0$, satisfies*

   (a) $\lambda(\{t\} \times [0,\infty]) = 0$ *and*

   (b) $\displaystyle\iint_{\substack{0<s\leq t\\ 0\leq u\leq\infty}} \frac{u}{1+u}\lambda(ds\,du) < \infty,$

such that

$$A(t) = b(t) + \sum_{t_i \leq t} Y_i(A) + \iint_{\substack{0<s\leq t\\ 0\leq u\leq\infty}} u\,\mu(ds\,du, A) \tag{10.2}$$

where, for each $A \in \mathcal{H}$, $\mu(\cdot, A)$ is a measure on $\left((0,\infty) \times [0,\infty], \mathbb{B}_{(0,\infty)\times[0,\infty]}\right)$ such that, under Π^*, $\mu(\cdot,\cdot)$ is a Poisson process with parameter $\lambda(\cdot)$, i.e., for arbitrary disjoint Borel subsets $E_1, \ldots, E_k$ of $(0,\infty)\times[0,\infty]$, $\mu(E_1,\cdot), \ldots, \mu(E_k,\cdot)$ are independent, and

$$\mu(E_i,\cdot) \sim Poisson(\lambda(E_i)) \qquad for\ 1 \leq i \leq k$$

Note the following facts about independent increment processes, which will be useful later and facilitate understanding of the remaining subject matter.

(1) The measure λ on $(0,\infty) \times [0,\infty]$ is often expressed as a family of measures $\{\lambda_t : t > 0\}$ where $\lambda_t(A) = \lambda((0,t] \times A)$ for Borel sets A.

(2) The representation may be expressed equivalently in terms of the moment-generating function of $A(t)$ as

$$\mathcal{E}(e^{-\theta A(t)}) = e^{-b(t)} \left[\prod_{t_i \le t} \mathcal{E}(e^{-\theta Y_i})\right] \exp\left[-\iint_{\substack{0<s\le t \\ 0\le u\le\infty}} (1 - e^{-\theta u})\, \lambda(ds\, du)\right]$$

(3) The random variables Y_i occurring in the decomposition arise from the jumps of the process at fixed points. Say that t is a fixed jump-point of the process if $\Pi^*(A\{t\} > 0) > 0$. It is known that there are at most countably many such fixed jump-points, and the set $\mathbf{M}$ is precisely the set of such points and that $Y_i = A\{t_i\}$.

(4) The random measure $A \mapsto \mu(\cdot, A)$ also has an explicit description. For any Borel subset E of $(0,\infty) \times [0,\infty]$,

$$\mu(E, A) = \#\left\{(t, A\{t\}) \in E : A\{t\} > 0\right\}$$

(5) Let $A^c(t) = A(t) - b(t) - \sum_{t_i \le t} A\{t_i\}$. Then

$$A^c(t) = \iint_{\substack{0<s\le t \\ 0\le u\le\infty}} u\, \mu(du\, ds, A)$$

(6) The countable set $\mathbf{M}$, the set of densities $\{f_i : i \ge 1\}$, the measure λ, and the nonrandom function b are known as the four components of the process $\{A(t) : t > 0\}$, or, equivalently, of the measure Π^*. The measure λ is known as the Lévy measure of Π^*.

(7) A Lévy process Π^* without any non-random component, i.e., for which $b(t) = 0$, for all $t > 0$, has sample paths that increase only in jumps almost surely Π^*. Most of the Lévy processes that we encounter here will be of this type.

10.4 Basic Properties

Let Π be a neutral to the right prior on $\mathcal{F}$. From what we have seen so far, the maps $\mathbf{D}$ and $\mathbf{H}$ yield independent increment process measures $\tilde{\Pi}$ and Π^*, respectively. Let the Lévy measures of $\tilde{\Pi}$ and Π^* be denoted $\tilde{\lambda}$ and λ^*, respectively. The next proposition establishes a simple relationship between $\tilde{\lambda}$ and λ^*.

Proposition 10.4.1. *Suppose $\tilde{\lambda}$ and λ^* are as earlier. Then*

1 for each t, $\tilde{\lambda}_t$ is the distribution of $x \mapsto -\log(1-x)$ under the measure λ_t^*, and

2 for each t, λ_t^* is the distribution of $x \mapsto 1 - e^{-x}$ under $\tilde{\lambda}_t$

Proof. The proposition is an easy consequence of the following easy fact.

If $\omega \mapsto \mu(\cdot, \omega)$ is an $\mathbf{M}(\mathfrak{X})$-valued random measure which is a Poisson process with parameter measure λ, then for any measurable function $g : \mathfrak{X} \to \mathfrak{X}$, the random measure $\omega \mapsto \mu(g^{-1}(\cdot), \omega)$ is a Poisson process with parameter measure $\lambda \circ g^{-1}$.

Note that

$$\begin{aligned}
\mathbf{D}(F)(t) - \mathbf{D}(F)(t-) &= -\log \frac{F(t,\infty)}{F[t,\infty)} \\
&= -\log\left\{1 - \frac{F\{t\}}{F[t,\infty)}\right\} \\
&= -\log[1 - (\mathbf{H}(F)(t) - \mathbf{H}(F)(t-))]
\end{aligned}$$

□

It is of interest to know if we can choose neutral to the right priors with large support. The next proposition gives a sufficient condition that will ensure that the support is all of $\mathcal{F}$. Recall that the (topological) support E of a measure μ on a metric space $\mathcal{X}$ is the smallest closed set E with $\mu(E^c) = 0$. We view $\mathcal{F}$ as a metric space under convergence in distribution.

Proposition 10.4.2. *If the support of the Lévy measure λ_H is all of $[0,\infty) \times [0,1]$ then the support of Π is all of $\mathcal{F}$.*

Proof. We need to show that every open set (in the topology given by convergence in distribution) has positive Π measure. Since the set of continuous distributions is dense in $\mathcal{F}$, it is enough to show that neighborhoods of continuous distributions have positive Π measure. We will establish a stronger fact, namely, that every uniform neighborhood has positive prior probability.

Let F_0 be a continuous distribution , $A_0 = \mathbf{H}(F_0)$ be the hazard function of F_0 and let $U = \{F : \sup_{0<s\leq t} |F(s) - F_0(s)| < \epsilon\}$. In view of the last section, U contains a set $\mathbf{H}^{-1}(V)$, where V is of the form $V = \{A : \sup_{0<s\leq t} |A(s) - A_0(s)| < \delta\}$. We will show that $\Pi(U) > 0$ by showing that $\Pi \circ \mathbf{H}^{-1}(V) > 0$.

To see this, set $\delta_0 = \delta/3$ and choose $0 = t_0 < 0 < t_1 < t_2 \ldots < t_k < t_{k+1} = t$ such that for $i = 1, 2, \ldots, (k+1)$; $A_0(t_i) - A_0(t_{i-1}) < \delta_0$.

Recall the definition of $\mu(.; A)$. Let

$$W = \{A : \mu(E_i; A) = 1, i = 1, 2, \ldots, k\}$$

where $E_i = (t_{i-1}, t_i] \times (A_0(t_i) - A_0(t_{i-1}) - \delta_0/k, A_0(t_i) - A_0(t_{i-1}) + \delta_0/k)$.

If $t_i < s \leq t_{i+1}$,

$$\begin{aligned} &|A(s) - A_0(s)| \\ &\leq \sum_1^i |(A_0(t_j) - A_0(t_{j-1})) - (A(t_j) - A(t_{j-1}))| + |(A_0(s) - A_0(t_i)) - (A(s) - A(t_i))| \end{aligned}$$

The first term on the right-hand side is less than $i\delta_0/k$ and the second term is less than $2\delta_0$ so that for every $s \in (0, t], |A(s) - A_0(s)| < \delta$. Hence $W \subset V$.

Under the measure induced by $\mathbf{H}^{-1}$, the random variables $\mu(E_i; A) = 1, i = 0, 1, 2, \ldots, k-1$ are independent Poisson random variables with parameters $\lambda(E_i), i = 1, 2, \ldots, k$. These are positive by assumption and hence V has positive $\Pi \circ \mathbf{H}^{-1}$ measure. □

Let A^* be a right continuous function increasing to ∞. A convenient class of neutral to the right priors are those with Lévy measure λ_H of the form

$$d\lambda_H(x, s) = a(x, s) dA^*(x) ds \qquad 0 < x < \infty, 0 < s < 1 \tag{10.3}$$

with $\int_0^1 s a(x, s) ds < \infty$ for all x. Without loss of generality we assume that for all x, $\int_0^1 s a(x, s) ds = 1$. This ensures that the prior expectation of $A(t)$ is $A_0(t)$.

Every neutral to the right prior gives rise to a Lévy measure via λ_H. Is every Lévy measure on $\mathbb{R}^+ \times [0, 1]$ obtainable as λ_H of a neutral to the right prior? The next proposition answers the question for the class of measures just discussed.

Proposition 10.4.3. *Let A^* be $H(F^*)$ for some distribution function F^* and*

$$d\lambda_H(x, s) = a(x, s) dA^*(x) ds \qquad 0 < x < \infty, 0 < s < 1$$

such that for all x, $\int_0^1 sa(x,s)ds = 1$ so that $E(A(t) = A^*(t)$.

The function $A \mapsto \prod_{(0,t]}(1 - dA(s))$ (where $\prod_{(0,t]}$ stands for the product integral) defines a neutral to the right prior on $\mathcal{F}$.

Proof. It can be easily deduced from the basic properties of the product integral that the function $A \mapsto \prod_{(0,t]}(1 - dA(s))$ induces a probability measure on the set of all functions which are right continuous and decreasing. In order to show that this is a prior on $\mathcal{F}$ we need to verify that if $\bar{F}(t) = \prod_{(0,t]}(1 - dA(s))$, then with probability 1 $\lim_{t\to\infty} \bar{F}(t) = 0$. This follows because the property of independent increments gives

$$E\prod_{(0,t]}(1 - dA(s)) = \prod_{(0,t]}(1 - dE(A)(s)) = \bar{F}^*$$

Each $\bar{F}(t)$ is decreasing in t and $\lim_{t\to\infty} E(\bar{F}(t) = \lim_{t\to\infty} \bar{F}^*(t) = 0$. □

Lévy representation plays a central role in the study of posteriors of neutral to the right priors. When the prior is neutral to the right, since the posterior given $X_1, X_2, \ldots, X_n$ is again neutral to the right, this posterior has a Lévy representation. An expression for the posterior in terms of λ_D can be found in Ferguson [62] and in terms of λ_H can be found in Hjort [100]. There is another proof due to Kim [113]. James [105] has a some what different approach, an approach we believe is promising and deserves further study. We will give a result from [100] without proof.

Our setup consists of random variables $X_1, X_2, \ldots, X_n$ that are independent identically distributed F and $Y_1, Y_2, \ldots, Y_n$, which are independent of the X_is and are independent identically distributed as G_0. The observations are $Z_i = X_i \wedge Y_i$ and $\delta_i = I(X_i \le Y_i)$. Let

$$N^n(t) = \sum_1^n I(Z_i > t) \text{ be the number of observations greater than } t$$

and

$$M^n(t) \text{ be the number of } Z_i s \text{ equal to } t$$

Theorem 10.4.1 (Hjort). *Let Π be a neutral to the right prior with Lévy measure of the form (10.3). When all the uncensored values—the Z_is with $\delta_i = 1$—are distinct among themselves, and from the values of the censored observations, the posterior has the Lévy representation given by*

1 M_u^n*: the set of uncensored values are points of fixed jumps. The distribution of the jump at* Z_i *has the density*

$$\frac{(1-s)^{N^n(Z_i)} sa(Z_i,s)}{\int_0^1 (1-s)^{N^n(Z_i)} sa(Z_i,s)ds}$$

2 the Lévy measure of the continuous part has

$$\hat{a}(x,s) = (1-s)^{N^n(x)+M^n(x)}$$

Remark 10.4.1. Consequently

$$E\left(\frac{\bar{\mathbf{F}}(t_2)}{\bar{\mathbf{F}}(t_1)}|\Pi(\middle| (Z_i,\delta_i) :\le i \le n\right)$$

$$= \left[\prod_{Z_i \in M_u^n : t_1 < Z_i \le t_2} \frac{\int_0^1 (1-s)^{N^n(Z_i)+1} sa(Z_i,s)ds}{\int_0^1 (1-s)^{N^n(Z_i)} sa(Z_i,s)ds}\right] e^{-\int_{t_1}^{t_2}\int_0^1 (1-s)^{N^n(z)+M^n(z)} sa(z,s)dsd\hat{A}(z)} \tag{10.4}$$

10.5 Beta Processes

Beta processes, introduced by Hjort [100] are continuous analogs of a time-discrete case where (see Example 10.2.2) the V_is are independent beta random variables. The continuous case is obtained as a limit of the time-discrete case. However, in order to ensure that the limit exists, the parameters of the beta random variables have to be chosen carefully. In addition to introducing beta processes and elucidating their properties for right censored data, Hjort [100] studied extensions to situations more general than right censored data. This chapter only deals with a part of [100].

10.5.1 Definition and Construction

Let A^* be a hazard function with finitely many jumps. Let $t_1, \ldots, t_k$ be the jump-points of A^*. Let $c(\cdot)$ be a piecewise continuous non-negative function on $[0,\infty)$ and let $A^{*,c}$ denote the continuous part of A^*. Let $A^*(t) < \infty$ for all t.

Definition 10.5.1. An independent increment process A is said to be a beta process with parameters $c(.)$ and A^*, written $A \sim \text{beta}(c, A^*)$, if the following holds: A has Lévy representation as in Theorem 10.3.3 with

1 $\mathbf{M} = \{t_1, \ldots, t_k\}$ and the jump-size at any t_j given by

$$Y_j \equiv A\{t_j\} \sim beta(c(t_j)A^*\{t_j\}, c(t_j)(1 - A^*\{t_j\}))$$

2 Lévy measure given by

$$\lambda(ds\, du) = c(s)u^{-1}(1-u)^{c(s)-1} du\, dA^{*,c}(s)$$

for $0 \le s < \infty, 0 < u < 1$; and

3 $b(t) \equiv 0$ for all $t > 0$.

The existence of such a process is guaranteed by Proposition 10.4 but this existence result does not give any insight into the prior. A better understanding of the prior comes from the construction of Hjort who obtained these priors as weak limits of time-discrete processes on $\mathcal{A}'$ and showed that the sample paths are almost surely in $\mathcal{A}$. In a very similar spirit, we construct the prior on $\mathcal{F}$ as a weak limit of priors sitting on a discrete set of points on $(0, \infty)$.

Let $F^* \in \mathcal{F}$ and, to begin, assume that it is continuous. Let $A^* = \mathbf{H}(F^*)$ be the cumulative hazard function corresponding to F^*.

Let $\mathbb{Q}$ be a countable dense set in $(0, \infty)$, enumerated as $\{s_1, s_2, \ldots\}$. For each $n \ge 1$, let $\{\, s_1^{(n)} < \cdots < s_n^{(n)} \,\}$ be an ordering of $s_1, \ldots, s_n$. Construct a prior Π_n on $\mathcal{F}_{s_1,\ldots,s_n}$ as in Example 10.2.1 by requiring that, under Π_n,

$$V_i^{(n)} \sim \text{beta}\left(c(s_{i-1}^{(n)}) \frac{\bar{F^*}(s_i^{(n)})}{\bar{F^*}(s_{i-1}^{(n)})}, c(s_{i-1}^{(n)}) \left(1 - \frac{\bar{F^*}(s_i^{(n)})}{\bar{F^*}(s_{i-1}^{(n)})} \right) \right) \quad \text{for } 1 \le i \le n-1. \tag{10.5}$$

Let $V_n^{(n)} \equiv 1$ and let F be a random distribution function , such that, under Π_n,

$$\mathcal{L}(\bar{F}(t)) = \mathcal{L}\left(\prod_{s_i^{(n)} \le t} (1 - V_i^{(n)}) \right) \quad \text{for all } t > 0$$

Theorem 10.5.1. *$\{\Pi_n\}_{n \ge 1}$ converges weakly to a neutral to the right prior Π on $\mathcal{F}$, which corresponds to a beta process.*

Proof. First observe that, as $n \to \infty$,

$$\begin{aligned}
\mathcal{E}_{\Pi_n}(\bar{F}(t)) &= \prod_{s_i^{(n)} \le t} \mathcal{E}_{\Pi_n}(1 - V_i^{(n)}) \\
&= \prod_{s_i^{(n)} \le t} \left(1 - \frac{F^*(s_{i-1}^{(n)}, s_i^{(n)}]}{F^*(s_{i-1}^{(n)}, \infty)}\right) \\
&\to \prod_{(0,t]} (1 - d\mathbf{H}(F^*)) \\
&= \prod_{(0,t]} (1 - dA^*) = \bar{F^*}(t)
\end{aligned}$$

for all $t \ge 0$. Thus $\mathcal{E}_{\Pi_n}(F) = F_n \overset{w}{\to} F^*$ as $n \to \infty$. Hence, by Theorem 2.5.1, $\{\Pi_n\}$ is tight.

We now follow Hjort's calculations to show that the finite-dimensional distributions of the process F, under the prior Π_n, converges weakly to those under the prior induced by a beta process with parameters c and A_0 on $\mathcal{H}$.

Consider, for each $n \ge 1$, an independent increment process A_n^c with process measure Π_n^* on $\mathcal{A}$ such that, for each fixed $t > 0$,

$$\mathcal{L}(A_n^c(t)) = \mathcal{L}(\sum_{s_i^{(n)} \le t} V_i^{(n)})$$

Thus, for each $n \ge 1$, A_n^c is a pure jump-process with fixed jumps at $s_1^{(n)}, \dots, s_{n-1}^{(n)}$ and with random jump sizes given by $V_i^{(n)}, \dots, V_{n-1}^{(n)}$ at these sites. Clearly, Π_n^* induces the prior Π_n on $\mathcal{F}$.

Now, for any fixed $t > 0$, repeating computations as in Hjort [[100], Theorem 3.1, pp. 1270-72] with

$$c_{n,i} = c(s_{i-1}^{(n)}), \qquad b_{n,i} = c_{n,i} \frac{\bar{F^{*,c}}(s_i^{(n)})}{\bar{F^{*,c}}(s_{i-1}^{(n)})} \quad \text{and} \quad a_{n,i} = c_{n,i} - b_{n,i}$$

one concludes that, for each θ, as $n \to \infty$,

$$\mathcal{E}[e^{-\theta A_n^c(t)}] \to \exp\left\{\int_0^1 \int_0^t (1 - e^{-\theta u}) \lambda(ds\, du)\right\}$$

and, similarly,

$$\mathcal{E} \exp - \sum_{j=1}^{m} \theta_j A_n^c(a_{j-1}, a_j] \to \exp\left\{ - \sum_{j=1}^{m} \int_0^1 \int_{a_{j-1}}^{a_j} (1 - e^{-\theta_j u}) \lambda(ds\, du) \right\}$$

Thus the finite-dimensional distributions of the independent increment processes A_n converge to the finite-dimensional distributions of an independent increment process with Lévy measure as in Definition 10.5.1. If the process measure is denoted by Π^* and the corresponding induced measure on $\mathcal{F}$ is denoted by Π, then considering the Skorokhod topology on $\mathcal{A}$ and by the continuity of $\mathbf{H}^{-1}$, we conclude that, for all $a_1, \ldots, a_m$,

$$\mathcal{L}(\bar{F}(a_1), \ldots, \bar{F}(a_m) \mid \Pi_n) \stackrel{w}{\to} \mathcal{L}(\bar{F}(a_1), \ldots, \bar{F}(a_m) \mid \Pi)$$

Therefore, $\{\Pi_n\}$ converges weakly to Π, a neutral to the right prior on $\mathcal{F}$. □

10.5.2 Properties

The following properties of beta processes are from Hjort [100].

1 Let $A^* \in \mathcal{A}$ be a hazard function with finitely many points of discontinuity and let c be a piecewise continuous function on $(0, \infty)$.

If $A \sim \text{beta}(c, A^*)$ then $\mathcal{E}(A(t)) = A^*(t)$. In other words $F = \mathbf{H}^{-1}(A)$ follows a beta(c, F^*) prior distribution and we have $\mathcal{E}(F(t)) = F^*(t)$ where $F^* = \mathbf{H}^{-1}(A^*)$.

The function c enters the expression for the variance. If $\mathbf{M} = \{t_1, \ldots, t_k\}$ is the set of discontinuity points of A_0 then

$$\mathbf{V}(A(t)) = \sum_{t_j \le t} \frac{A^*\{t_j\}(1 - A^*\{t_j\})}{c(t_j) + 1} + \int_0^t \frac{dA^{*,c}(s)}{c(s) + 1}$$

where $A^{*,c}(t) = A^*(t) - \sum_{t_i \le t} A^*\{t_i\}$.

2 Let $A \sim \text{beta}(c, A^*)$ where, as before, A^* has discontinuities at points in $\mathbf{M}$. Given F, let $X_1, \ldots, X_n$ be i.i.d. F. Then the posterior distribution of F given $X_1, \ldots, X_n$ is again a beta process, i.e., the corresponding independent increment process is again beta.

To describe the posterior parameters, let $\mathbf{X_n}$ be the set of distinct elements of $\{x_1, \ldots, x_n\}$. Define

$$Y_n(t) = \sum_{i=1}^{n} I_{(X_i \ge t)} \quad \text{and} \quad \bar{Y}_n(t) = \sum_{i=1}^{n} I_{(X_i > t)}$$

With $N_n(t)$ as before, note that $\bar{Y}_n(t) = n - N_n(t)$ and $Y_n(t) = n - N_n(t-)$.

Using this notation, the posterior beta process has parameters

$$c_{X_1 \ldots X_n}(t) = c(t) + Y_n(t)$$
$$A^*_{X_1 \ldots X_n}(t) = \int_0^t \frac{c(z)\, dA^*(z) + dN_n(z)}{c(z) + Y_n(z)}$$

More explicitly, $A^*_{X_1 \ldots X_n}$ has discontinuities at points in $\mathbf{M}^* = \mathbf{M} \cup \mathbf{X_n}$, and for $t \in \mathbf{M}^*$,

$$A^*_{X_1 \ldots X_n}\{t\} = \frac{c(t).A^*\{t\} + N_n\{t\}}{c(t) + Y_n(t)}$$
$$A^{*,c}_{X_1 \ldots X_n}(t) = \int_0^t \frac{c(z)\, dA^{*,c}(z)}{c(z) + Y_n(z)}$$

Note that if $t \in \mathbf{M}^*$,

$$A\{t\} \sim \text{beta}\left(c(t)\, A^*\{t\} + N_n\{t\}, c(t)(1 - A^*\{t\}) + Y_n(t) - N_n\{t\}\right).$$

3 Our interest is in the following special case of 2. Suppose $A \sim$ beta(c, A^*) and t A^* is continuous. Then the posterior given $X_1, \ldots, X_n$ is again a beta process with parameters

$$c_{X_1 \ldots X_n}(t) = c(t) + Y_n(t)$$

and

$$A^* X_1 \ldots X_n(t) = A^{*,d} X_1 \ldots X_n(t) + A^{*,c} X_1 \ldots X_n(t)$$

where

$$A^{*,d}_{X_1 \ldots X_n}(t) = \sum_{\substack{t_i \in \mathbf{X}_n \\ t_i \le t}} \frac{N_n\{t_i\}}{c(t_i) + Y_n(t_i)}$$

and

$$A^{*,c}_{X_1 \ldots X_n}(t) = \int_0^t \frac{c(z)\, dA^*(z)}{c(z) + Y_n(z)}$$

As a consequence, if $t \in \mathbf{X}_n$, then under the posterior $\Pi_{X_1,\dots,X_n}$ we have

$$A\{t\} \sim \text{beta}(N_n\{t\}, c(t) + \bar{Y}_n(t)).$$

Also note that the Bayes estimates are

$$\mathcal{E}_{\Pi_{X_1,\dots,X_n}}(A(t)) = A^* X_1 \dots X_n(t)$$

and

$$\mathcal{E}_{\Pi_{X_1,\dots,X_n}}(\bar{F}(t)) = \prod_{\substack{t_i \in \mathbf{X}_n \\ t_i \le t}} \left(1 - \frac{N_n\{t_i\}}{c(t_i) + Y_n(t_i)}\right) \exp\left\{- \int_0^t \frac{c(z)\, dA^*(z)}{c(z) + Y_n(z)}\right\} \tag{10.6}$$

4 A neat expression for the posterior and the Bayes estimate for right censored data can be easily obtained using Theorem 10.4.1. We leave the details to the reader.

Using these explicit expressions it is not very difficult to show that beta processes lead to consistent posteriors. However since we take up the consistency issue more generally in the next section we do not pursue it here.

Like the Dirichlet, any two beta processes tend to be mutually singular. This is proved in [43].

Walker and Muliere [167] started with a positive function D on $(0, \infty)$ and a distribution function $\hat{F}$ and constructed a class of priors on $\mathcal{F}$ called beta-Stacy processes. We again consider the simple case when $\hat{F}$ is continuous. The beta-Stacy process is the neutral to the right prior with

$$d\lambda_D(s,x) = D(x)\frac{e^{-sD(x)\bar{F}(x)}}{1 - e^{-s}} ds d\hat{A}x; \qquad 0 < x < \infty, 0 < s < \infty$$

The beta process prior thus relates to an independent increment process via $\mathbf{H}$ and the beta -Stacy via $\mathbf{D}$. Viewing the processes as measures on $\mathcal{F}$ provides a mean to calibrate the prior information in $\mathbf{H}$ in terms of that in $\mathbf{D}$ and vice versa. Though not explicitly formulated in the following form, the relationship between the two priors is already implicit in remark 2 and remark 4 of [167].

Theorem 10.5.2. *Π is a Beta Stacy $(D, \hat{F})$ process iff Π is a Beta $(C, \hat{A})$ process prior where $C = D\bar{\hat{F}}$ and $\hat{A}$ is the cumulative hazard function of $\hat{F}$.*

Proof. Because Beta Stacy process has λ_D given above, we can compute its λ_H using Proposition 10.4.1. This immediately yields the assertion. □

10.6 Posterior Consistency

Since neutral to the right priors, like tail free priors, possess nice independence and conjugacy properties it appeared that they would always yield consistent posteriors. However, Kim and Lee [114] gave an example of a neutral to the right prior which is inconsistent. Their elegant example is constructed with a homogeneous Lévy measure and is inconsistent at every continuous distribution.

Recall from Theorem 4.2.1 that to establish posterior consistency at F_0, it is enough to show that with F_0^∞ probability 1, for all t

(i) $\lim_{n\to\infty} E(\mathbf{F}(t)|X_1, X_2, \ldots, X_n) = F_0(t)$ and

(ii) $\lim_{n\to\infty} V(\mathbf{F}(t)|X_1, X_2, \ldots, X_n) = 0.$

The next theorem shows that for neutral to the right priors consistency of Bayes estimates ensures consistency of the posterior.

Theorem 10.6.1. *Let Π be a neutral to the right prior of the form (10.3). If*

$$\lim_{n\to\infty} E(\mathbf{F}(t)|X_1, X_2, \ldots, X_n) = F_0(t)$$

then

$$\lim_{n\to\infty} V(\mathbf{F}(t)|X_1, X_2, \ldots, X_n) = 0$$

Proof. Let $X_{[1]} < X_{[2]} \ldots X_{[k]}$ be the ordering of the observations $X_1, X_2, \ldots, X_n$ which are less than t. Then, apart from an exponential factor going to 1,

$$E(\bar{\mathbf{F}}(t)^2|X_1, X_2, \ldots, X_n) = \prod_2^k \frac{\int_0^1 (1-s)^{j+2} a(s, X_{[j]})ds}{\int_0^1 (1-s)^j a(s, X_{[j]})ds}$$

multiplying each term by $\int_0^1 (1-s)^{j+1} a(s, X_{[j]})ds / \int_0^1 (1-s)^{j+1} a(s, X_{[j]})ds$, we get

$$= \prod_2^k \frac{\int_0^1 (1-s)^{j+2} a(s, X_{[j]})ds}{\int_0^1 (1-s)^{j+1} a(s, X_{[j]})ds} \prod_1^k \frac{\int_0^1 (1-s)^{j+1} a(s, X_{[j]})ds}{\int_0^1 (1-s)^j a(s, X_{[j]})ds} \to (\bar{F}_0(t))^2$$

□

There is another structural aspect of neutral to the right priors. Consistency for the censored case follows from consistency for the uncensored case. Following is the result. For a proof, see Dey et al. [43]

Theorem 10.6.2. *Suppose X is a survival time with distribution F and Y is a censoring time distributed as G. $X_1, X_2, \ldots,$ are given $\mathbf{F} = F$, i.i.d. F and $Y_1, Y_2, \ldots,$ be i.i.d. G, where G is continuous and has support all of $\mathbb{R}^+$. We also assume that the X_is and Y_is are independent. Let $Z_i = X_i \wedge Y_i$ and $\Delta_i = I(X_i \leq Y_i)$. If Π is a neutral to the right prior for $\mathbf{F}$ whose posterior is consistent at all continuous distributions F_0, then the posterior given $(Z_i, \Delta_i) : i \geq 1$ is also consistent at all continuous F_0.*

Proof. Fix $t_1 < t_2$. since the exponential term in 10.4 goes to 0 as $n \to \infty$, our assumption on consistency translates into: for any continuous distribution F, if $X_1, X_2, \ldots, X_n$ are i.i.d. F, then

$$\lim_{n\to\infty} \prod_{X_i \in (t_1,t_2]} \frac{\int_{0,1} s(1-s)^{N_n(X_i)+1} a(X_i,s)ds}{\int_{0,1} s(1-s)^{N_n(X_i)} a(X_i,s)ds} = \frac{\bar{F}(t_2)}{\bar{F}(t_1)}$$

Fix F_0 continuous. Let $X_1, X_2, \ldots, X_n$be i.i.d. F_0 and $Y_1, \ldots, Y_n$ be i.i.d. G, and let (Z_i, Δ_i) be as above. We will first show that

$$\lim_{n\to\infty} \prod_{Z_i \in M_n^* \cap (0,t]} \frac{\int_{0,1} s(1-s)^{N_n(X_i)+1} a(X_i,s)ds}{\int_{0,1} s(1-s)^{N_n(X_i)} a(X_i,s)ds} = \bar{F}(t) \text{ a.s. } (F_0 \times G)^\infty$$

where $M_n^* = \{Z_j : \Delta_j = 0\}$.

With $t_1 < t_2$ fixed, let ϕ be an increasing continuous mapping of (t_1, ∞) into (t_2, ∞) and define

$$Z_i^* = Z_i I(\Delta_i = 1) + \phi(Z_i) I(|Delta_i = 0)$$

Then Z_i^* are again i.i.d. with a continuous distribution F_0^* such that

$$\frac{F_0^*(t_2)}{F_0^*(t_1)} = \frac{\bar{J}(t_1,1) - \bar{J}(t_2,1)}{\bar{J}^*(t_1)}$$

where $\bar{J}^*(t) = P(Z > t)$ and $\bar{J}(t_1 = P(Z > t, \Delta = 1)$.

Now using our assumption, if $N_*^n(t) = \sum_{i=1}^n I(Z_i^* > t)$ then

$$\lim_{n\to\infty} \prod_{Z_i^* \in (t_1,t_2]} \frac{\int_{0,1} s(1-s)^{N_*^n(Z_i^*)+1} a(Z_i^*,s)ds}{\int_{0,1} s(1-s)^{N_*^n(Z_i^*} a(Z_i^*,s)ds} = \frac{\bar{J}(t_1,1) - \bar{J}(t_2,1)}{\bar{J}^*(t_1)} \text{ a.s}$$

Note that the above product is only over the uncensored Z_is and that, for each $t_1 < t_2$ with $\Delta_i = 1$, $N^n(Z_i) \leq N_*^n(Z_i)$. Now using the Cauchy-Schwarz inequality we get

$$\left[\int_0^1 (1-s)^{n+2} sa(x,s)ds\right]\left[\int_0^1 (1-s)^n sa(x,s)ds\right]$$
$$= \left[\int_0^1 [(1-s)^{(n+2)/2}]^2 sa(x,s)ds\right]\left[\int_0^1 [(1-s)^{(n)/2}]^2 sa(x,s)ds\right]$$
$$\geq \left[\int_0^1 (1-s)^{n+1} sa(x,s)ds\right]^2$$

and consequently $\int_0^1 (1-s)^{n+1} sa(x,s)/\int_0^1 (1-s)^n sa(x,s)ds$ is decreasing in n. Hence, we have

$$\lim_{n\to\infty} \prod_{Z_i \in M_n^* \cap (t_1,t_2]} \frac{\int_{0,1} s(1-s)^{N_n(Z_i)+1} a(Z_i,s)ds}{\int_{0,1} s(1-s)^{N_n(Z_i)} a(Z_i,s)ds}$$
$$\leq \lim_{n\to\infty} \prod_{Z_i^* \in (t_1,t_2]} \frac{\int_{0,1} s(1-s)^{N_*^n(Z_i^*)+1} a(Z_i^*,s)ds}{\int_{0,1} s(1-s)^{N_*^n(Z_i^*)} a(Z_i^*,s)ds}$$
$$= \frac{\bar{J}(t_1,1) - \bar{J}(t_2,1)}{\bar{J}^*(t_1)}$$

Let $0 = t_0 < t_1 < t_2 < \ldots < t_k = t$ be a partition of $(0,t]$. Then

$$\lim_{n\to\infty} \prod_{Z_i \in M_n^* \cap (0,t]} \frac{\int_{0,1} s(1-s)^{N_n(Z_i)+1} a(Z_i,s)ds}{\int_{0,1} s(1-s)^{N_n(Z_i)} a(Z_i,s)ds} \leq \prod_1^k \frac{\bar{J}(t_{i-1},1) - \bar{J}(t_i,1)}{\bar{J}^*(t_{i-1})}$$

As the width of the partition $\max|t_i - t_{i-1}$ goes to 0, the right-hand side converges to the product integral $\prod_{(0,t]}(1 - J(ds,1)/\bar{J}(s))$, which from Peterson [138] is equal to $\bar{F}(t)$.

Let $\hat{\bar{F}}_n$ denote the Bayes estimate of $\bar{F}$ given $X_1, X_2, \ldots, X_n$ and let $\bar{F}_n^*$ denote the Bayes estimate of $\bar{F}$ given $(Z_i, \delta_i) : 1 \leq i \leq n$. we have shown that for all t,

$$\bar{F}_n^*(t) \leq \hat{\bar{F}}_n(t) \text{ and hence } \liminf_n F_n^* \geq \bar{F}_0$$

Similarly, by considering the "Bayes" estimate for G, with $M_0^n = \{(Z_j, \Delta_j : \Delta_j = 0)\}$,

$$\liminf_n \prod_{Z_i \leq t : Z_i \in M_0^n} \frac{\int_0^1 (1-s)^{N^n(Z_i)+1} a(Z_i,s)ds}{\int_0^1 (1-s)^{N^n(Z_i)} a(Z_i,s)ds} \geq \bar{G}$$

Consider,

$$\prod_{Z_i \le t: Z_i \in M_u^n} \frac{\int_0^1 (1-s)^{N^n(Z_i)+1} a(Z_i,s)ds}{\int_0^1 (1-s)^{N^n(Z_i)} a(Z_i,s)ds} \prod_{Z_i \le t: Z_i \in M_0^n} \frac{\int_0^1 (1-s)^{N^n(Z_i)+1} a(Z_i,s)ds}{\int_0^1 (1-s)^{N^n(Z_i)} a(Z_i,s)ds} \tag{10.7}$$

but this is equal to

$$\prod_{Z_i \le t} \frac{\int_0^1 (1-s)^{N^n(Z_i)+1} a(Z_i,s)ds}{\int_0^1 (1-s)^{N^n(Z_i)} a(Z_i,s)ds}$$

But this is just the Bayes estimate based on i.i.d. observations from the continuous survival distribution $\bar{F}_0(t)\bar{G}(t)$ and by assumption (10.7) converges to $\bar{F}_0(t)\bar{G}(t)$. The conclusion follows easily. □

Thus, as far as consistency issues are concerned, we only need to study the uncensored case. We begin looking at the simple case when the Lévy measure is homogeneous. In the sequel for any $a, b > 0$, we denote by $B(a,b)$,the usual beta function given by

$$B(a,b) = \frac{\Gamma(a)\Gamma(b)}{\Gamma(a+b)} = \left[\int_0^1 (1-s)^{a-1} s^{b-1} ds\right]$$

If f is an integrable function on $(0,1)$ we set

$$K(n,f) = \int_0^1 (1-s)^n f(s)ds$$

We will repeatedly use the fact that

$$\text{for any } p, q;\ \lim_{n\to\infty} n^{q-p} \frac{\Gamma(n+p)}{\Gamma(n+q)} = 1$$

Lemma 10.6.1. *Suppose f is a nonnegative function on $(0,1)$ such that*

(a) $0 < \int_0^1 f(s)ds < \infty$ and

(b) for some $\alpha < 1$, $0 < \lim_{s\to 0} s^\alpha f(s) = b < \infty$.

Then

$$\lim_{n\to\infty} \frac{K(n,f)}{B(n+1, 1-\alpha)} = b$$

Proof. Since

$$\int_{\epsilon}^{1}(1-s)^n f(s)ds \le (1-\epsilon)^n \int_0^1 f(s)ds = o(n^{-(1-\alpha)})$$

and as $n \to \infty, n^{1-\alpha}B(n, 1-\alpha) \to \Gamma(1-\alpha)$, we have

$$\lim_{n\to\infty} \frac{\int_{\epsilon}^{1}(1-s)^n f(s)ds}{B(n+1, 1-\alpha)} = 0 \tag{10.8}$$

Similarly, because $\alpha < 1$,

$$\int_{\epsilon}^{1}(1-s)^n s^{-\alpha}ds \le (1-\epsilon)^n \int_0^{\epsilon} s^{-\alpha}ds \le (1-\epsilon)^n \frac{1-\epsilon^{1-\alpha}}{1-\alpha} = o(n^{-(1-\alpha)})$$

which in turn yields

$$\lim_{n\to\infty} \frac{\int_{\epsilon}^{1}(1-s)^n s^{-\alpha}ds}{B(n+1, 1-\alpha)} = 0 \tag{10.9}$$

Given δ, use assumption (b) to choose $\epsilon > 0$ such that for $s < \epsilon$

$$(b-\delta)s^{-\alpha} < f(s) < (b+\delta)s^{-\alpha}$$

Then

$$K(n, f) \le (b+\delta)B(n+1, 1-\alpha) + \int_{\epsilon}^{1}(1-s)^n f(s)ds$$

and by (10.8) we have

$$\lim_{n\to\infty} \frac{K(n, f)}{B(n+1, 1-\alpha)} \le (b+\delta)$$

A similar argument using (10.9) shows that

$$\lim_{n\to\infty} \frac{K(n, f)}{B(n+1, 1-\alpha)} \ge (b-\delta)$$

Since δ is arbitrary, the lemma follows. □

Theorem 10.6.3. *Let A^* be a cumulative hazard function which is continuous and finite for all x. Suppose that a neutral to the right prior with no fixed jumps has the expected hazard function A^* and the Lévy measure*

$$d\lambda_H(x, s) = a(s)dA^*(x)ds \qquad 0 < x < \infty, 0 < s < 1$$

such that

$$\text{for some } \alpha < 1, \qquad 0 < \lim_{s \to 0} s^{1+\alpha} a(s) = b < \infty \tag{10.10}$$

If F_0 is a continuous distribution with $F_0(t) > 0$ for all t, then with F_0^∞-probability 1, the posterior converges weakly to the measure degenerate at $F_0^{1-\alpha}$. In particular, if (10.10) holds with $\alpha = 0$ then the posterior is consistent at F_0.

Proof. Set $f(s) = sa(s)$. We have $\int_0^1 f(s)ds = 1$. Using (10.4),

$$E(\bar{\mathbf{F}}(t)|X_1, X_2, \ldots, X_n) = \left[\prod_{X_i \le t} \frac{K(N^n(X_i) + 1, f)}{K(N^n(X_i), f)} \right] e^{-\psi_n(t)} \tag{10.11}$$

where $\psi_n(t) = \int_0^t \int_0^1 (1-s)^{N^n(x)+M^n(x)} sa(s) ds dA^*(x)$.

For any $x < t, (1-s)^{N^n(x)+M^n(x)} < (1-s)^{N^n(t)}$ and hence $\psi_n(t)$ is bounded above by $(\int_0^1 (1-s)^{N^n(t)} ds) A^*(t)$. Since $N^n(t) \to \infty$ as $n \to \infty$, it follows that $\psi_n(t) \to 0$ as $n \to \infty$. Hence the exponential factor goes to 1.

If $X_{(1)} < X_{(2)} \ldots < X_{(n-N^n(t))}$ is an ordering of the $n - N^n(t)$ samples that are less than t, then, since with F_0 probability 1 the $X_1, X_2, \ldots, X_n$ are all distinct, $N^n(X_{(1)}) = n-1, N^n(X_{(2)}) = n-2$, and so on. Thus the first term in (10.11) reduces to

$$\prod_{i=0}^{(i=n-N^n(t))} \frac{K(n-i, f)}{K((n-i-1), f)} = \frac{K(n, f)}{K(N^n(t) - 1, f)}$$

It follows from Lemma 10.6.1 that

$$\begin{aligned}
\lim_{n \to \infty} \frac{K(n, f)}{K(N^n(t) - 1, f)} &= \lim_{n \to \infty} \frac{B(N^n(t) - 1, 1 - \alpha)}{B(n, 1 - \alpha)} \\
&= \lim_{n \to \infty} \frac{\Gamma(N^n(t) - \alpha)}{\Gamma(N^n(t) - 1)} \frac{\Gamma(n)}{\Gamma(n + 1 - \alpha)} \\
&= \lim_{n \to \infty} \left(\frac{N^n(t)}{n+1} \right)^{1-\alpha} = \bar{\mathbf{F}}_0(t)^{1-\alpha} \text{ a.s. } F_0^\infty
\end{aligned}$$

□

Remark 10.6.1. The Kim-Lee example had the homogeneous Lévy measure given by $a(s) = 2s^{-3/2}$. In this case the conditions of the Theorem 10.6.3 are satisfied with $\alpha = 1/2$ so that the posterior converges to $F_0^{1/2}$.

We next turn to a sufficient condition for consistency in the general case. We begin with an extension of Lemma 10.6.1.

For each x in a set $\mathcal{X}$ let $f(x,.)$ be a non negative function on $(0,1)$. Let m_n be a sequence of integers such that $\lim_{n\to\infty} \frac{m_n}{n} = c, 0 < c < 1$.

Lemma 10.6.2. *Suppose*

(a) $0 < \sup_x \int_0^1 f(x,s)ds = I < \infty$ *and*

(b) As $s \to 0$, $f(x,s)$ *converges uniformly (in* x*), to the constant function 1, i.e., as* $\epsilon \to 0$,

$$\delta_\epsilon = \sup_x \sup_{s<\epsilon} |f(s,x) - 1| \to 0$$

Then

$$\lim_{n\to\infty} \prod_{m_n}^{n} \left(\frac{i+2}{i+1}\right) \frac{\int_0^1 (1-s)^{i+1} f(x_i,s)ds}{\int_0^1 (1-s)^i f(x_i,s)ds} = 1$$

and the convergence is uniform in the x_i'*s.*

Proof. To avoid unpleasant expressions involving fractions of integrals, set

$$K_{i,x} = \int_0^1 (1-s)^i f(x,s)ds \text{ and } L_{i,x} = \int_0^1 s(1-s)^i f(x,s)ds$$

We will show that for any x, given δ small, there is an m_0 such that, for $i > m_0$,

$$\frac{i+1-2\delta}{i+2} \le \frac{K_{i+1,x}}{K_{i,x}} \le \frac{i+1+2\delta}{i+2} \tag{10.12}$$

The bounds in inequality 10.12 do not depend on the x_is. Consequently, we have uniformly in the x_is,

$$\left(1 - \frac{2\delta}{m_n+1}\right)^{n-m_n} \le \prod_{m_n}^{n} \frac{i+2}{i+1} \frac{K_{i+1,x}}{K_{i,x}} \le \left(1 + \frac{2\delta}{m_n+1}\right)^{n-m_n}$$

For small positive y, $e^{-2y} < 1 - y < 1 + y < e^y$. Hence, as $n \to \infty$, the left-hand side converges to $e^{-4\delta(1-c)/c}$ and the right side to $e^{2\delta(1-c)/c}$. Letting δ go to 0 we have the result.

To prove (10.12) note that

$$\frac{K_{i+1,x}}{K_{i,x}} = 1 - \frac{L_{i,x}}{K_{i,x}}$$

For any $0 < \epsilon = 1 - \alpha < 1$,

$$(1 - \delta_\epsilon H_{i,\epsilon}) \leq K_{i,x} \leq (1 + \delta_\epsilon) H_{i,\epsilon} + \alpha^i I$$

and

$$(1 - \delta_\epsilon J_{i,\epsilon}) \leq L_{i,x} \leq (1 + \delta_\epsilon) J_{i,\epsilon} + \alpha^i I$$

where

$$H_{i,\epsilon} = \int_0^\epsilon (1 - s)^i ds = \frac{1 - \alpha^{i+1}}{i + 1}$$

and

$$J_{i,\epsilon} = \int_0^\epsilon s(1 - s)^i ds = \frac{1 - \alpha^{i+1}(1 + \epsilon + i\epsilon)}{(i + 1)(i + 2)}$$

Now

$$(i + 2)\frac{L_{i,x}}{K_{i,x}} \leq (i + 2) \left\{ \frac{(1 + \delta_\epsilon) J_{i,\epsilon}}{(1 - \delta_\epsilon) H_{i,\epsilon}} + \frac{\alpha^i I}{(1 - \delta_\epsilon) H_{i,\epsilon}} \right\}$$

which goes to $(1+\delta_\epsilon)/(1-\delta_\epsilon)$ as $i \to \infty$. Further, the right-hand side does not involve x, and hence this convergence is uniform in x.

On the other hand,

$$(i + 2)\frac{L_{i,x}}{K_{i,x}} \geq (i + 2) \left\{ \frac{(1 - \delta_\epsilon) J_{i,\epsilon}}{(1 + \delta_\epsilon)(H_{i,\epsilon} + \alpha^i I)} \right\}$$

which goes to $(1 - \delta_\epsilon)/(1 + \delta_\epsilon)$, again uniformly in x.

Because $\delta_\epsilon \to 0$ as ϵ goes to 0, given any $\delta > 0$, for sufficiently small ϵ, $(1-\delta_\epsilon)/(1+\delta_\epsilon)$ is larger than $(1 - \delta)$ and $(1 + \delta_\epsilon)/(1 - \delta_\epsilon)$ is smaller than $(1 + \delta)$.

Thus given any $\delta > 0$, there is an n_ϵ such that for $i > n_\epsilon$,

$$1 - \delta \leq \frac{(i + 2)L_{i,x}}{K_{i,x}} \leq 1 + \delta$$

Using $K_{i+1,x}/K_{i,x} = 1 - (L_{i,x}/K_{i,x})$, we get

$$1 - \frac{1 + \delta}{i + 2} < \frac{K_{i+1,x}}{K_{i,x}} < 1 + \frac{1 - \delta}{i + 2}$$

and this is (10.12)

□

Remark 10.6.2. In the Lemma 10.6.2, assumption (a) can be replaced by:
(a') $0 < \sup_x \int_0^1 (1-s)f(x,s)ds < \infty$.

This follows from setting $g(s,x) = (1-s)f(x,s)$ and noting that g satisfies assumptions (a) and (b) and that

$$\frac{\int_0^1 (1-s)^{n+1} f(x,s)ds}{\int_0^1 (1-s)^n f(x,s)ds} = \frac{\int_0^1 (1-s)^n g(x,s)ds}{\int_0^1 (1-s)^{n-1} g(x,s)ds}$$

and observing that $(n+2)/(m_n+2) = \prod_{m_n}^n [(i+2)/(i+1)$ and $(n+1)/(m_n+1) = \prod_{m_n}^n [(i+1)/i$ both converge to the same limit $1/c$.

Theorem 10.6.4. *Let Π be a neutral to the right prior with*

$$d\lambda_H(x,s) = c(x)a(x,s)dA^*(x)ds \qquad 0 < x < \infty, 0 < s < 1$$

If $f(x,s) = sa(x,s)$ satisfies the assumption of the Lemma 10.6.2 (or the remark following it) then the posterior is consistent at any continuous distribution F_0.

Proof. Since the exponential factor in equation (10.4) goes to 1, it follows immediately from Lemma 10.6.2 that for each t with $\bar{F_0}(t) > 0$,

$$E(\bar{\mathbf{F}}(t)|\Pi()|X_1, X_2, \ldots, X_n) \to \bar{F_0}(t)$$

□

Theorem 10.6.5. *The posterior of the beta(C, A^*) prior is consistent at all continuous distribution F_0.*

Proof. Since the Lévy measure satisfies the conditions of Remark 10.6.2, this is an immediate consequence of Theorem 10.6.4. □

Remark 10.6.3. Kim and Lee [114] have shown consistency when

1 $(1-s)f(x,s) \leq 1$ and

2 as $x \to 0, f(x,s)$ converges uniformly in x to a positive continuous function $b(x)$.

The result is marginally more general than that of Kim and Lee. The methods that we have used are more elementary.

To summarize, neutral to priors are an elegant class of priors that can, in terms of mathematical tractability, conveniently handle right censored data. We have also seen that some caution is required if one wants consistent posteriors. As with the Dirichlet, mixtures of neutral to priors would yield more flexibility in terms of prior opinions and posteriors that are amenable to simulation. These remain to be explored.

11

Exercises

11.0.1. If two probability measures on $\mathbb{R}^K$ agree on all sets of the form $(a_1, b_1] \times (a_2, b_2], \ldots \times (a_k, b_k]$ then they agree on all Borel sets in $\mathbb{R}^k$.

11.0.2. Let M_t be the median of $Beta(ct, c(1-t))$ where $0 < t < 1$. Show that $M_t \geq \frac{1}{2}$ iff $t \geq \frac{1}{2}$. [Hint: If $x \geq \frac{1}{2}$ show that $x^{ct-1}(1-x)^{c(1-t)-1}$ is increasing in t. Suppose $t \geq \frac{1}{2}$ and $M_t < \frac{1}{2}$. Then $\int_0^{1/2} x^{ct-1}(1-x)^{c(1-t)-1} dx \geq \frac{1}{2}$. Make the change of variable $x \mapsto (1-x)$ to obtain a contradiction]

11.0.3. Suppose $\alpha$is a finite measure. Define $X_1, X_2, \ldots$ by

$$X_1 \text{ is distributed as } \bar{\alpha}$$

, for any $n \geq 1$,

$$P(X_{n+1} \in B | X_1, X_2, \ldots, X_n) = \frac{\alpha(B) + \sum_1^n \delta_{X_i}(B)}{\alpha(\mathbb{R} + n)}$$

Show that $X_1, X_2, \ldots$ form an exchangeable sequence and the corresponding DeFinneti measure on $M(\mathbb{R})$ is D_α

11.0.4. Assume a Dirichlet prior and show that the predictive distribution of X_{n+1} given $X_1, X_2, \ldots, X_n$, converges to P_0 weakly almost surely P_0. Examine what happens when the prior is a mixture of Dirichlet processes.

11.0.5. Show that if $P \in M_\alpha$ and U is a neighborhood of P in set-wise convergence then $D_\alpha(U) > 0$. However M_α is not the smallest closed set with this property.

11.0.6. Show that a Polya tree prior is a Dirichlet process iff for any $\underline{\epsilon} \in E_i^*, \alpha_{\underline{\epsilon}0}+\alpha_{\underline{\epsilon}1} = \alpha_{\underline{\epsilon}}$.

11.0.7. Let L_μ be the set of all probability measures dominated by a $\sigma-$finite measure μ. Verify that, when restricted to L_μ all the three $\sigma-$algebras discussed in section 2.2 coincide.

11.0.8. Let E be a measurable subset of $\Theta \times \mathcal{X}$ such that $\theta \neq \theta', E_\theta \cap E_{\theta'} = \emptyset$ and for all θ, $P_\theta(E_\theta) = 1$. For any two priors Π_1, Π_2 on Θ show that $\|\Pi_1 - \Pi_2\| = \|\lambda_1 - \lambda_2\|$, where λ_i are the respective marginals on $\mathcal{X}$.

Derive the Blackwell- Dubins merging result from Doob's theorem

11.0.9. Consider $f_\theta = U(0, \theta); 0 < \theta < 1$. Show that the Schwartz condition fails at $\theta = 1$ but posterior consistency holds. Can you use the results in Section 4.3 to prove consistency?

11.0.10. Suppose $X_1, X_2, \ldots, X_n$ are i.i.d. Ber(p), i.e.,

$$\Pr(X_i = 1) = p = 1 - \Pr(X_i = 0)$$

A prior for p may be elicited by asking for a rule for predicting X_{n+1}. Suppose for all $n \geq 1$, one is given the rule

$$\Pr(X_{n+1} = 1 | X_1, X_2, \ldots, X_n) = \frac{a + \sum_1^n X_i}{a + b + n}$$

Assuming that the prediction loss is squared error, show that there is a unique prior corresponding to this rule and identify the prior

11.0.11. With X_is as in Exercise11.0.11, consider a conjugate prior and a realization of the X_is such that $\hat{p} = \sum_1^n X_i/n$ is bounded away from 0 and 1 as $n \to \infty$. Show directly (without using the results established in the text) that as $n \to \infty$, the posterior distribution of $\sqrt{n}(\hat{p} - p)/(\hat{p}(1 - \hat{p})$ converges weakly to $N(0, 1)$

11.0.12. Let $X_1, X_2 \ldots, X_n$ be i.i.d. $N(0, 1)$. Consider a Bayesian who does not know the true density and who uses the model, $\theta \sim N(\mu, \eta)$ and given θ, $X_1, X_2 \ldots, X_n$ be i.i.d. $N(\theta, 1)$. Calculate the posterior of θ given $X_1, X_2 \ldots, X_n$ and verify that with probability 1 under the joint distribution under $N(0, 1)$, the density of $\sqrt{n}(\theta - \bar{X})$ converges in L_1 distance to $N(0, 1)$.

11.0.13. Consider X_is as in Exercise11.0.11. Consider a beta prior, i.e., a prior with density

$$\Pi(p) = cp^{\alpha-1}(1-p)^{\beta-1}, \alpha \geq, \beta \geq 0$$

a Discuss why relatively small values of $\alpha+\beta$ indicate relative lack of prior information

b Consider a sequence of hyperparameters α_i, β_i such that $\alpha_i + \beta_i \to 0$ but $\alpha_i/\beta_i \to C, 0 < C < 1$. Show that the corresponding sequence of priors converge weakly, and determine the limiting prior. Would you call this prior noninformative? Reconcile your answer with the discussion in (a)

11.0.14. (1). For a multinomial with probabilities $p_1, p_2, \ldots, p_k$ for k classes,calculate the Jeffreys prior. [Hint: Use the following well known identity (see [144]): Let B be a positive definite matrix. Let $A = B + xx^T$. Then $\det A = \det B(1 + x^T B^{-1} x)$]

(2). In the above problem calculate the reference prior for (p_1, p_2) assuming $k = 3$.

For the next four problems $P \sim D_\alpha$ and given P,$X_1, X_2, \ldots, X_n$ are i.i.d. P.

11.0.15. Assume $\int_{-\infty}^{\infty} x^2 d\alpha < \infty$. Calculate the prior variance of the population mean$\int xdP$

11.0.16. Assuming α has the Cauchy density

$$\frac{1}{\pi}\frac{1}{1+x^2}$$

and $\int xdP = T(P)$ is well defined for almost all P, show that $T(P)$ has the same Cauchy distribution.

[Hint: Use Sethuraman's construction]

11.0.17. For $\bar{\alpha}$ Cauchy, show that $\int xdP = T(P)$ is well defined for almost all P.

[Hint: If Y_i is a sequence of independent random variables such that $\sum_1^n Y_i$ converges in distribution, then $\sum_1^\infty Y_i$ is finite a.s. Alternatively, use methods of Doss and Selke [55]]

11.0.18. Let $\alpha_\theta = N(\theta, 1)$ and $\theta \sim N(\mu, \eta)$. Given $X_1, X_2, \ldots, X_n$ are all distinct, calculate the posterior distribution of θ.

For the next three problems, let $P' \sim D_\alpha$, P a convolution of P' and $N(0, h^2)$ and h have the prior density $\Pi(h)$. Given P, let $X_1, X_2, \ldots, X_n$ be i.i.d. f, where f is the density of P

11.0.19. Let C_n be the information that all the X_is are distinct. For any fixed x calculate $E(f(x)|X_1, X_2, \ldots, X_n, C_n)$ assuming the X_is are all distinct.

11.0.20. Let the true density f_0 be uniform on $(0, 1)$. Verify if the Bayes estimate $E(f|X_1, X_2, \ldots, X_n, h)$ is consistent in the L_1 distance

11.0.21. Let f_0 be Normal or Cauchy with location and scale parameters chosen by you but not equal to 0 and 1. Set $n = 50$ from f_0, draw a sample of size n, namely, $X_1, X_2, \ldots, X_n$. Simulate the Bayes estimate of $f(x)$ when the prior is a Dirichlet mixture of normal and $\bar{\alpha} = N(0, 1)$ or $N(\mu, \sigma^2)$ with μ and σ^2 independent, μ normal and σ^2 is inverse gamma truncated above.

Plot f_0 and the Bayes estimate. Discuss whether putting a prior on μ, σ^2 leads to a Bayes estimate that is closer to f_0 than the Bayes estimate under a prior with fixed values of μ and σ^2. (Base your comments from 10 simulations on each case).

11.0.22. Let f_0 be normal or Cauchy. Using the Polya tree prior recommended in Chapter6 and a normal or Cauchy prior for the location parameter, calculate numerically the posterior for θ, for various values of n and various choices of $X_1, X_2, \ldots, X_n$.

11.0.23. (a) Assume the regression model discussed in Chapter7 with a prior for the random density f that is Dirichlet mixture of Normal or Cauchy . Calculate and plot the posterior for β for the different priors listed in Exercise11.0.21.

(b) Do the same but symmetrize f around 0. Discuss whether the behavior of the posterior for β is similar o that in (a)

11.0.24. Examine Doob's theorem in the regression set up considered in Chapter 7

11.0.25. Show that the Bayes estimate for survival function under a Dirichlet prior with censored data has a representation as a product of survival probabilities and that it converges to the Kaplan-Meier estimate as $\alpha(\mathbb{R}) \to 0$.

11.0.26. Show that the Bayes estimate for the bivariate survival function is inconsistent in the following example (due to R.Pruitt):
$(T_1, T_2) \sim F$ and $F \sim D_\alpha$ where α is the uniform distribution on $(0, 2) \times (0, 2)$. The censoring random variable (C_1, C_2) takes the values $(0, 2), (2, 0)$ and $(2, 2)$ with equal probability of 1/3. The Bayes estimator for F is inconsistent when F_0 is the uniform distribution on $(1, 2) \times (0, 1)$.

11.0.27. Show, in the context of Chapter9 that if one starts with a Dirichlet prior for the distribution of (Z, Δ) (i.e., a prior for probability measures on $\{0, 1\} \times \mathbb{R}^+$), then the induced prior for F-the distribution of the survival time X is a Beta process.

References

[1] James H. Albert and Siddhartha Chib. Bayesian analysis of binary and polychotomous response data. *J. Amer. Statist. Assoc.*, 88(422):669–679, 1993.

[2] S.-I. Amari, O. E. Barndorff-Nielsen, R. E. Kass, S. L. Lauritzen, and C. R. Rao. *Differential geometry in statistical inference*. Institute of Mathematical Statistics, Hayward, CA, 1987.

[3] Per Kragh Andersen, Ørnulf Borgan, Richard D. Gill, and Niels Keiding. *Statistical models based on counting processes*. Springer-Verlag, New York, 1993.

[4] Charles E. Antoniak. Mixtures of Dirichlet processes with applications to Bayesian nonparametric problems. *Ann. Statist.*, 2:1152–1174, 1974.

[5] Andrew. Barron, Mark J. Schervish, and Larry Wasserman. The consistency of posterior distributions in nonparametric problems. *Ann. Statist.*, 27(2):536–561, 1999.

[6] Andrew R. Barron. The strong ergodic theorem for densities: generalized Shannon-McMillan-Breiman theorem. *Ann. Probab.*, 13(4):1292–1303, 1985.

[7] Andrew R. Barron. Uniformly powerful goodness of fit tests. *Ann. Statist.*, 17(1):107–124, 1989.

[8] ANDREW R. BARRON. Information-theoretic characterization of Bayes performance and the choice of priors in parametric and nonparametric problems. In *Bayesian statistics, 6 (Alcoceber, 1998)*, pages 27–52. Oxford Univ. Press, New York, 1999.

[9] D. BASU. Statistical information and likelihood. *Sankhyā Ser. A*, 37(1):1–71, 1975. Discussion and correspondance between Barnard and Basu.

[10] D. BASU AND R. C. TIWARI. A note on the Dirichlet process. In *Statistics and probability: essays in honor of C. R. Rao*, pages 89–103. North-Holland, Amsterdam, 1982.

[11] S. BASU AND S. MUKHOPADHYAY. Binary response regression with normal scale mixture links. In *Generalized Linear Models- A Bayesian Perspective*, pages 231–241. Marcel-Dekker, New York, 1998.

[12] S. BASU AND S. MUKHOPADHYAY. Bayesian analysis of binary regression using symmetric and asymmetric links. *Sankhyā Ser. B*, 62:372–387, 2000.

[13] JAMES O. BERGER. *Statistical decision theory and Bayesian analysis*. Springer-Verlag, New York, 1993. Corrected reprint of the second (1985) edition.

[14] JAMES O. BERGER AND JOSE-M. BERNARDO. Estimating a product of means: Bayesian analysis with reference priors. *J. Amer. Statist. Assoc.*, 84(405):200–207, 1989.

[15] JAMES O. BERGER AND JOSÉ M. BERNARDO. On the development of reference priors. In *Bayesian statistics, 4 (Peñíscola, 1991)*, pages 35–60. Oxford Univ. Press, New York, 1992.

[16] JAMES O. BERGER AND LUIS R. PERICCHI. The intrinsic Bayes factor for model selection and prediction. *J. Amer. Statist. Assoc.*, 91(433):109–122, 1996.

[17] ROBERT H. BERK AND I. RICHARD SAVAGE. Dirichlet processes produce discrete measures: an elementary proof. In *Contributions to statistics*, pages 25–31. Reidel, Dordrecht, 1979.

[18] JOSE-M. BERNARDO. Reference posterior distributions for Bayesian inference. *J. Roy. Statist. Soc. Ser. B*, 41(2):113–147, 1979. With discussion.

[19] P. J. Bickel. On adaptive estimation. *Ann. Statist.*, 10(3):647–671, 1982.

[20] P. J. Bickel and J. A. Yahav. Some contributions to the asymptotic theory of Bayes solutions. *Z. Wahrscheinlichkeitstheorie und Verw. Gebiete*, 11:257–276, 1969.

[21] Patrick Billingsley. *Convergence of probability measures.* John Wiley & Sons Inc., New York, second edition, 1999. A Wiley-Interscience Publication.

[22] Lucien Birgé. Approximation dans les espaces métriques et théorie de l'estimation. *Z. Wahrsch. Verw. Gebiete*, 65(2):181–237, 1983.

[23] David Blackwell. Discreteness of Ferguson selections. *Ann. Statist.*, 1:356–358, 1973.

[24] David Blackwell and Lester Dubins. Merging of opinions with increasing information. *Ann. Math. Statist.*, 33:882–886, 1962.

[25] David Blackwell and James B. MacQueen. Ferguson distributions via Pólya urn schemes. *Ann. Statist.*, 1:353–355, 1973.

[26] J. Blum and V. Susarla. On the posterior distribution of a Dirichlet process given randomly right censored observations. *Stochastic Processes Appl.*, 5(3):207–211, 1977.

[27] J. Borwanker, G. Kallianpur, and B. L. S. Prakasa Rao. The Bernstein-von Mises theorem for Markov processes. *Ann. Math. Statist.*, 42:1241–1253, 1971.

[28] Olaf Bunke and Xavier Milhaud. Asymptotic behavior of Bayes estimates under possibly incorrect models. *Ann. Statist.*, 26(2):617–644, 1998.

[29] Burr, D, Cooke G.E., Doss H. and P.J. Goldschmidt-Clermont. A meta analysis of studies on the association of the platlet pla polymorphism of glycoprotein iiia and risk of coronary heart disease. Technical report, 2002.

[30] N. N. Čencov. *Statistical decision rules and optimal inference.* American Mathematical Society, Providence, R.I., 1982. Translation from the Russian edited by Lev J. Leifman.

[31] Ming-Hui Chen and Dipak K. Dey. Bayesian modeling of correlated binary responses via scale mixture of multivariate normal link functions. *Sankhyā Ser. A*, 60(3):322–343, 1998. Bayesian analysis.

[32] Ming-Hui Chen, Qi-Man Shao, and Joseph G. Ibrahim. *Monte Carlo methods in Bayesian computation*. Springer-Verlag, New York, 2000.

[33] Bertrand S. Clarke and Andrew R. Barron. Information-theoretic asymptotics of Bayes methods. *IEEE Trans. Inform. Theory*, 36(3):453–471, 1990.

[34] Robert J. Connor and James E. Mosimann. Concepts of independence for proportions with a generalization of the Dirichlet distribution. *J. Amer. Statist. Assoc.*, 64:194–206, 1969.

[35] Harald Cramér. *Mathematical Methods of Statistics*. Princeton University Press, Princeton, N. J., 1946.

[36] Harald Cramér and M. R. Leadbetter. *Stationary and related stochastic processes. Sample function properties and their applications*. John Wiley & Sons Inc., New York, 1967.

[37] Sarat Dass and Jayeong Lee. A note on the consistency of bayes factors for testing point null versus nonparametric alternatives.

[38] G. S. Datta and J. K. Ghosh. Noninformative priors for maximal invariant parameter in group models. *Test*, 4(1):95–114, 1995.

[39] A. P. Dawid, M. Stone, and J. V. Zidek. Marginalization paradoxes in Bayesian and structural inference. *J. Roy. Statist. Soc. Ser. B*, 35:189–233, 1973. With discussion by D. J. Bartholomew, A. D. McLaren, D. V. Lindley, Bradley Efron, J. Dickey, G. N. Wilkinson, A. P.Dempster, D. V. Hinkley, M. R. Novick, Seymour Geisser, D. A. S. Fraser and A. Zellner, and a reply by A. P. Dawid, M. Stone, and J. V. Zidek.

[40] William A. Dembski. Uniform probability. *J. Theoret. Probab.*, 3(4):611–626, 1990.

[41] Luc Devroye and László Györfi. No empirical probability measure can converge in the total variation sense for all distributions. *Ann. Statist.*, 18(3):1496–1499, 1990.

[42] J. Dey, L. Drăghici, and R. V. Ramamoorthi. Characterizations of tail free and neutral to the right priors. In *Advances on methodological and applied aspects of probability and statistics*, pages 305–325. Gordon and Breach science publishers.

[43] J. Dey, R.V. Erickson, and R.V. Ramamoorthi. Some aspects of neutral to right priors. *submitted*. Bayesian analysis.

[44] P. Diaconis and D. Freedman. Partial exchangeability and sufficiency. In *Statistics: applications and new directions (Calcutta, 1981)*, pages 205–236. Indian Statist. Inst., Calcutta, 1984.

[45] P. Diaconis and D. Freedman. On inconsistent Bayes estimates of location. *Ann. Statist.*, 14(1):68–87, 1986.

[46] Persi Diaconis and David Freedman. On the consistency of Bayes estimates. *Ann. Statist.*, 14(1):1–67, 1986. With a discussion and a rejoinder by the authors.

[47] Persi Diaconis and Donald Ylvisaker. Conjugate priors for exponential families. *Ann. Statist.*, 7(2):269–281, 1979.

[48] Kjell Doksum. Tailfree and neutral random probabilities and their posterior distributions. *Ann. Probability*, 2:183–201, 1974.

[49] J. L. Doob. Application of the theory of martingales. In *Le Calcul des Probabilités et ses Applications.*, pages 23–27. Centre National de la Recherche Scientifique, Paris, 1949. Colloques Internationaux du Centre National de la Recherche Scientifique, no. 13,.

[50] Hani Doss. Bayesian estimation in the symmetric location problem. *Z. Wahrsch. Verw. Gebiete*, 68(2):127–147, 1984.

[51] Hani Doss. Bayesian nonparametric estimation of the median. I. Computation of the estimates. *Ann. Statist.*, 13(4):1432–1444, 1985.

[52] Hani Doss. Bayesian nonparametric estimation of the median. II. Asymptotic properties of the estimates. *Ann. Statist.*, 13(4):1445–1464, 1985.

[53] Hani Doss. Bayesian nonparametric estimation for incomplete data via successive substitution sampling. *Ann. Statist.*, 22(4):1763–1786, 1994.

[54] HANI DOSS AND B. NARASIMHAN. Dynamic display of changing posterior in Bayesian survival analysis. In *Practical nonparametric and semiparametric Bayesian statistics*, pages 63–87. Springer, New York, 1998.

[55] HANI DOSS AND THOMAS SELLKE. The tails of probabilities chosen from a Dirichlet prior. *Ann. Statist.*, 10(4):1302–1305, 1982.

[56] L. DRĂGHICI AND R. V. RAMAMOORTHI. A note on the absolute continuity and singularity of Polya tree priors and posteriors. *Scand. J. Statist.*, 27(2):299–303, 2000.

[57] R. M. DUDLEY. Measures on non-separable metric spaces. *Illinois J. Math.*, 11:449–453, 1967.

[58] RICHARD M. DUDLEY. *Real analysis and probability*. Wadsworth & Brooks/Cole Advanced Books & Software, Pacific Grove, CA, 1989.

[59] MICHAEL D. ESCOBAR AND MIKE WEST. Bayesian density estimation and inference using mixtures. *J. Amer. Statist. Assoc.*, 90(430):577–588, 1995.

[60] MICHAEL D. ESCOBAR AND MIKE WEST. Computing nonparametric hierarchical models. In *Practical nonparametric and semiparametric Bayesian statistics*, pages 1–22. Springer, New York, 1998.

[61] THOMAS S. FERGUSON. A Bayesian analysis of some nonparametric problems. *Ann. Statist.*, 1:209–230, 1973.

[62] THOMAS S. FERGUSON. Prior distributions on spaces of probability measures. *Ann. Statist.*, 2:615–629, 1974.

[63] THOMAS S. FERGUSON. Bayesian density estimation by mixtures of normal distributions. In *Recent advances in statistics*, pages 287–302. Academic Press, New York, 1983.

[64] THOMAS S. FERGUSON AND ESWAR G. PHADIA. Bayesian nonparametric estimation based on censored data. *Ann. Statist.*, 7(1):163–186, 1979.

[65] THOMAS S. FERGUSON, ESWAR G. PHADIA, AND RAM C. TIWARI. Bayesian nonparametric inference. In *Current issues in statistical inference: essays in honor of D. Basu*, pages 127–150. Inst. Math. Statist., Hayward, CA, 1992.

[66] J.-P. FLORENS, M. MOUCHART, AND J.-M. ROLIN. Bayesian analysis of mixtures: some results on exact estimability and identification. In *Bayesian statistics, 4 (Peñíscola, 1991)*, pages 127–145. Oxford Univ. Press, New York, 1992.

[67] SANDRA FORTINI, LUCIA LADELLI, AND EUGENIO REGAZZINI. Exchangeability, predictive distributions and parametric models. *Sankhyā Ser. A*, 62(1):86–109, 2000.

[68] DAVID A. FREEDMAN. Invariants under mixing which generalize de Finetti's theorem: Continuous time parameter. *Ann. Math. Statist.*, 34:1194–1216, 1963.

[69] DAVID A. FREEDMAN. On the asymptotic behavior of Bayes' estimates in the discrete case. *Ann. Math. Statist.*, 34:1386–1403, 1963.

[70] DAVID A. FREEDMAN. On the asymptotic behavior of Bayes estimates in the discrete case. II. *Ann. Math. Statist.*, 36:454–456, 1965.

[71] MARIE GAUDARD AND DONALD HADWIN. Sigma-algebras on spaces of probability measures. *Scand. J. Statist.*, 16(2):169–175, 1989.

[72] J. K. GHORAI AND H. RUBIN. Bayes risk consistency of nonparametric Bayes density estimates. *Austral. J. Statist.*, 24(1):51–66, 1982.

[73] S. GHOSAL, J. K. GHOSH, AND R. V. RAMAMOORTHI. Non-informative priors via sieves and packing numbers. In *Advances in statistical decision theory and applications*, pages 119–132. Birkhäuser Boston, Boston, MA, 1997.

[74] S. GHOSAL, J. K. GHOSH, AND R. V. RAMAMOORTHI. Posterior consistency of Dirichlet mixtures in density estimation. *Ann. Statist.*, 27(1):143–158, 1999.

[75] SUBHASHIS GHOSAL. Normal approximation to the posterior distribution for generalized linear models with many covariates. *Math. Methods Statist.*, 6(3):332–348, 1997.

[76] SUBHASHIS GHOSAL. Asymptotic normality of posterior distributions in high-dimensional linear models. *Bernoulli*, 5(2):315–331, 1999.

[77] SUBHASHIS GHOSAL. Asymptotic normality of posterior distributions for exponential families when the number of parameters tends to infinity. *J. Multivariate Anal.*, 74(1):49–68, 2000.

[78] SUBHASHIS GHOSAL, JAYANTA K. GHOSH, AND R. V. RAMAMOORTHI. Consistent semiparametric Bayesian inference about a location parameter. *J. Statist. Plann. Inference*, 77(2):181–193, 1999.

[79] SUBHASHIS GHOSAL, JAYANTA K. GHOSH, AND TAPAS SAMANTA. On convergence of posterior distributions. *Ann. Statist.*, 23(6):2145–2152, 1995.

[80] SUBHASHIS GHOSAL, JAYANTA K. GHOSH, AND AAD W. VAN DER VAART. Convergence rates of posterior distributions. *Ann. Statist.*, 28(2):500–531, 2000.

[81] J. K. GHOSH, R. V. RAMAMOORTHI, AND K. R. SRIKANTH. Bayesian analysis of censored data. *Statist. Probab. Lett.*, 41(3):255–265, 1999. Special issue in memory of V. Susarla.

[82] J. K. GHOSH, B. K. SINHA, AND S. N. JOSHI. Expansions for posterior probability and integrated Bayes risk. In *Statistical decision theory and related topics, III, Vol. 1 (West Lafayette, Ind., 1981)*, pages 403–456. Academic Press, New York, 1982.

[83] JAYANTA K. GHOSH. *Higher Order Asymptotics*, volume 4. NSF-CBMS Regional Conference Series in probability and Statistics, 1994.

[84] JAYANTA K. GHOSH, SUBHASHIS GHOSAL, AND TAPAS SAMANTA. Stability and convergence of the posterior in non-regular problems. In *Statistical decision theory and related topics, V (West Lafayette, IN, 1992)*, pages 183–199. Springer, New York, 1994.

[85] JAYANTA K. GHOSH, SHRIKANT N. JOSHI, AND CHIRANJIT MUKHOPADHYAY. Asymptotics of a Bayesian approach to estimating change-point in a hazard rate. *Comm. Statist. Theory Methods*, 25(12):3147–3166, 1996.

[86] JAYANTA K. GHOSH AND RAHUL MUKERJEE. Non-informative priors. In *Bayesian statistics, 4 (Peñíscola, 1991)*, pages 195–210. Oxford Univ. Press, New York, 1992.

[87] JAYANTA K. GHOSH AND R. V. RAMAMOORTHI. Consistency of Bayesian inference for survival analysis with or without censoring. In *Analysis of censored data (Pune, 1994/1995)*, pages 95–103. Inst. Math. Statist., Hayward, CA, 1995.

[88] JAYANTA K. GHOSH AND TAPAS SAMANTA. Nonsubjective Bayes testing—an overview. *J. Statist. Plann. Inference*, 103(1-2):205–223, 2002. C. R. Rao 80th birthday felicitation volume, Part I.

[89] GHOSH.J.K. Review of approximation theorems in statistics by serfling. *Journal of Ameri. Stat.*, 78(383):732, September 1983.

[90] RICHARD D. GILL AND SØREN JOHANSEN. A survey of product-integration with a view toward application in survival analysis. *Ann. Statist.*, 18(4):1501–1555, 1990.

[91] PIET GROENEBOOM. Nonparametric estimators for interval censoring problems. In *Analysis of censored data (Pune, 1994/1995)*, volume 27 of *IMS Lecture Notes Monogr. Ser.*, pages 105–128. Inst. Math. Statist., Hayward, CA, 1995.

[92] J. HANNAN. Consistency of maximum likelihood estimation of discrete distributions. In *Contributions to probability and statistics*, pages 249–257. Stanford Univ. Press, Stanford, Calif., 1960.

[93] J. A. HARTIGAN. *Bayes theory*. Springer-Verlag, New York, 1983.

[94] J. A. HARTIGAN. Bayesian histograms. In *Bayesian statistics, 5 (Alicante, 1994)*, pages 211–222. Oxford Univ. Press, New York, 1996.

[95] DAVID HEATH AND WILLIAM SUDDERTH. De Finetti's theorem on exchangeable variables. *Amer. Statist.*, 30(4):188–189, 1976.

[96] DAVID HEATH AND WILLIAM SUDDERTH. On finitely additive priors, coherence, and extended admissibility. *Ann. Statist.*, 6(2):333–345, 1978.

[97] DAVID HEATH AND WILLIAM SUDDERTH. Coherent inference from improper priors and from finitely additive priors. *Ann. Statist.*, 17(2):907–919, 1989.

[98] N. L. HJORT. Bayesian approaches to non- and semiparametric density estimation. In *Bayesian statistics, 5 (Alicante, 1994)*, pages 223–253. Oxford Univ. Press, New York, 1996.

[99] NILS LID HJORT. *Application of the Dirichlet Process to some nonparametric estimation problems(in Norwegian)*. Ph.D. thesis, University of TromsØ.

[100] NILS LID HJORT. Nonparametric Bayes estimators based on beta processes in models for life history data. *Ann. Statist.*, 18(3):1259–1294, 1990.

[101] N.L HJORT AND D POLLARD. Asymptotics of minimisers of convex processes. *Statistical Research Report, Department of Mathematics. University of Oslo*, 1994.

[102] I. A. IBRAGIMOV AND R. Z. HASMINSKIĬ. *Statistical estimation*. Springer-Verlag, New York, 1981. Asymptotic theory, Translated from the Russian by Samuel Kotz.

[103] HEMANT ISHWARAN. Exponential posterior consistency via generalized Pólya urn schemes in finite semiparametric mixtures. *Ann. Statist.*, 26(6):2157–2178, 1998.

[104] K. ITO. *Stochastic processes*. Matematisk Institut, Aarhus Universitet, Aarhus, 1969.

[105] LANCELOT JAMES. Poisson process partition calculus with applications to exchangeable models and bayesian nonparametrics. .

[106] HAROLD JEFFREYS. An invariant form for the prior probability in estimation problems. *Proc. Roy. Soc. London. Ser. A.*, 186:453–461, 1946.

[107] R. A. JOHNSON. An asymptotic expansion for posterior distributions. *Ann. Math. Statist.*, 38:1899–1906, 1967.

[108] RICHARD A. JOHNSON. Asymptotic expansions associated with posterior distributions. *Ann. Math. Statist.*, 41:851–864, 1970.

[109] JOSEPH B. KADANE, JAMES M. DICKEY, ROBERT L. WINKLER, WAYNE S. SMITH, AND STEPHEN C. PETERS. Interactive elicitation of opinion for a normal linear model. *J. Amer. Statist. Assoc.*, 75(372):845–854, 1980.

[110] OLAV KALLENBERG. *Foundations of modern probability*. Springer-Verlag, New York, 1997.

[111] ROBERT E. KASS AND LARRY WASSERMAN. The selection of prior distributions by formal rules. *Journal of the American Statistical Association*, 91:1343–1370, 1996.

[112] J. H. B. KEMPERMAN. On the optimum rate of transmitting information. *Ann. Math. Statist.*, 40:2156–2177, 1969.

[113] YONGDAI KIM. Nonparametric Bayesian estimators for counting processes. *Ann. Statist.*, 27(2):562–588, 1999.

[114] YONGDAI KIM AND JAEYONG LEE. On posterior consistency of survival models. *Ann. Statist.*, 29(3):666–686, 2001.

[115] A. N. KOLMOGOROV AND V. M. TIHOMIROV. ε-entropy and ε-capacity of sets in functional space. *Amer. Math. Soc. Transl. (2)*, 17:277–364, 1961.

[116] RAMESH M. KORWAR AND MYLES HOLLANDER. Contributions to the theory of Dirichlet processes. *Ann. Probability*, 1:705–711, 1973.

[117] STEFFEN L. LAURITZEN. *Extremal families and systems of sufficient statistics*. Springer-Verlag, New York, 1988.

[118] MICHAEL LAVINE. Some aspects of Pólya tree distributions for statistical modelling. *Ann. Statist.*, 20(3):1222–1235, 1992.

[119] MICHAEL LAVINE. More aspects of Pólya tree distributions for statistical modelling. *Ann. Statist.*, 22(3):1161–1176, 1994.

[120] LUCIEN LE CAM. *Asymptotic methods in statistical decision theory*. Springer-Verlag, New York, 1986.

[121] LUCIEN LE CAM AND GRACE LO YANG. *Asymptotics in statistics*. Springer-Verlag, New York, 1990. Some basic concepts.

[122] L. LECAM. Convergence of estimates under dimensionality restrictions. *Ann. Statist.*, 1:38–53, 1973.

[123] E. L. LEHMANN. *Testing statistical hypotheses*. Springer-Verlag, New York, second edition, 1997.

[124] E. L. LEHMANN. *Theory of point estimation*. Springer-Verlag, New York, 1997. Reprint of the 1983 original.

[125] PETER J. LENK. The logistic normal distribution for Bayesian, nonparametric, predictive densities. *J. Amer. Statist. Assoc.*, 83(402):509–516, 1988.

[126] Tom Leonard. A Bayesian approach to some multinomial estimation and pretesting problems. *J. Amer. Statist. Assoc.*, 72(360, part 1):869–874, 1977.

[127] Tom Leonard. Density estimation, stochastic processes and prior information. *J. Roy. Statist. Soc. Ser. B*, 40(2):113–146, 1978. With discussion.

[128] D. V. Lindley. On a measure of the information provided by an experiment. *Ann. Math. Statist.*, 27:986–1005, 1956.

[129] D. V. Lindley. The use of prior probability distributions in statistical inference and decisions. In *Proc. 4th Berkeley Sympos. Math. Statist. and Prob., Vol. I*, pages 453–468. Univ. California Press, Berkeley, Calif., 1961.

[130] Albert Y. Lo. Consistency in the location model: the undominated case. *Ann. Statist.*, 12(4):1584–1587, 1984.

[131] Albert Y. Lo. On a class of Bayesian nonparametric estimates. I. Density estimates. *Ann. Statist.*, 12(1):351–357, 1984.

[132] Michel Loève. *Probability theory. II*. Springer-Verlag, New York, fourth edition, 1978. Graduate Texts in Mathematics, Vol. 46.

[133] R. Daniel Mauldin, William D. Sudderth, and S. C. Williams. Pólya trees and random distributions. *Ann. Statist.*, 20(3):1203–1221, 1992.

[134] Amewou-Atisso Messan, Subhashis Ghosal, Jayanta K. Ghosh, and R. V. Ramamoorthi. Posterior consistency for semiparametic regression problems. *Bernoulli*, To appear(2), 2002.

[135] Radford M. Neal. Markov chain sampling methods for Dirichlet process mixture models. *J. Comput. Graph. Statist.*, 9(2):249–265, 2000.

[136] Michael A. Newton, Claudia Czado, and Rick Chappell. Bayesian inference for semiparametric binary regression. *J. Amer. Statist. Assoc.*, 91(433):142–153, 1996.

[137] Michael A. Newton, Fernando A. Quintana, and Yunlei Zhang. Nonparametric Bayes methods using predictive updating. In *Practical nonparametric and semiparametric Bayesian statistics*, pages 45–61. Springer, New York, 1998.

[138] ARTHUR V. PETERSON, JR. Expressing the Kaplan-Meier estimator as a function of empirical subsurvival functions. *J. Amer. Statist. Assoc.*, 72(360, part 1):854–858, 1977.

[139] DAVID POLLARD. *Convergence of stochastic processes*. Springer-Verlag, New York, 1984.

[140] DAVID POLLARD. *A user's guide to measure theoretic probability*. Cambridge University Press, Cambridge, 2002.

[141] KATHRYN ROEDER AND LARRY WASSERMAN. Practical Bayesian density estimation using mixtures of normals. *J. Amer. Statist. Assoc.*, 92(439):894–902, 1997.

[142] DONALD B. RUBIN. The Bayesian bootstrap. *Ann. Statist.*, 9(1):130–134, 1981.

[143] GABRIELLA SALINETTI. Consistency of statistical estimators: the epigraphical view. In *Stochastic optimization: algorithms and applications (Gainesville, FL, 2000)*, volume 54 of *Appl. Optim.*, pages 365–383. Kluwer Acad. Publ., Dordrecht, 2001.

[144] MARK J. SCHERVISH. *Theory of statistics*. Springer-Verlag, New York, 1995.

[145] LORRAINE SCHWARTZ. On Bayes procedures. *Z. Wahrscheinlichkeitstheorie und Verw. Gebiete*, 4:10–26, 1965.

[146] GIDEON SCHWARZ. Estimating the dimension of a model. *Ann. Statist.*, 6(2):461–464, 1978.

[147] ROBERT J. SERFLING. *Approximation theorems of mathematical statistics*. John Wiley & Sons Inc., New York, 1980. Wiley Series in Probability and Mathematical Statistics.

[148] JAYARAM SETHURAMAN. A constructive definition of Dirichlet priors. *Statist. Sinica*, 4(2):639–650, 1994.

[149] JAYARAM SETHURAMAN AND RAM C. TIWARI. Convergence of Dirichlet measures and the interpretation of their parameter. In *Statistical decision theory and related topics, III, Vol. 2 (West Lafayette, Ind., 1981)*, pages 305–315. Academic Press, New York, 1982.

[150] Xiaotong Shen and Larry Wasserman. Rates of convergence of posterior distributions. *Ann. Statist.*, 29(3):687–714, 2001.

[151] B. W. Silverman. *Density estimation for statistics and data analysis*. Chapman & Hall, London, 1986.

[152] Richard L. Smith. Nonregular regression. *Biometrika*, 81(1):173–183, 1994.

[153] S. M. Srivastava. *A course on Borel sets*. Springer-Verlag, New York, 1998.

[154] V. Susarla and J. Van Ryzin. Nonparametric Bayesian estimation of survival curves from incomplete observations. *J. Amer. Statist. Assoc.*, 71(356):897–902, 1976.

[155] V. Susarla and J. Van Ryzin. Large sample theory for a Bayesian nonparametric survival curve estimator based on censored samples. *Ann. Statist.*, 6(4):755–768, 1978.

[156] Henry Teicher. Identifiability of finite mixtures. *Ann. Math. Statist.*, 34:1265–1269, 1963.

[157] Daniel Thorburn. A Bayesian approach to density estimation. *Biometrika*, 73(1):65–75, 1986.

[158] Luke Tierney and Joseph B. Kadane. Accurate approximations for posterior moments and marginal densities. *J. Amer. Statist. Assoc.*, 81(393):82–86, 1986.

[159] Bruce W. Turnbull. The empirical distribution function with arbitrarily grouped, censored and truncated data. *J. Roy. Statist. Soc. Ser. B*, 38(3):290–295, 1976.

[160] A. W. van der Vaart. *Asymptotic statistics*. Cambridge University Press, Cambridge, 1998.

[161] Aad W. van der Vaart and Jon A. Wellner. *Weak convergence and empirical processes*. Springer-Verlag, New York, 1996. With applications to statistics.

[162] Richard von Mises. *Probability, statistics and truth*. Dover Publications Inc., New York, english edition, 1981.

[163] ABRAHAM WALD. Note on the consistency of the maximum likelihood estimate. *Ann. Math. Statistics*, 20:595–601, 1949.

[164] A. M. WALKER. On the asymptotic behaviour of posterior distributions. *J. Roy. Statist. Soc. Ser. B*, 31:80–88, 1969.

[165] STEPHEN WALKER AND NILS LID HJORT. On Bayesian consistency. *J. R. Stat. Soc. Ser. B Stat. Methodol.*, 63(4):811–821, 2001.

[166] STEPHEN WALKER AND PIETRO MULIERE. Beta-Stacy processes and a generalization of the Pólya-urn scheme. *Ann. Statist.*, 25(4):1762–1780, 1997.

[167] STEPHEN WALKER AND PIETRO MULIERE. A characterization of a neutral to the right prior via an extension of Johnson's sufficientness postulate. *Ann. Statist.*, 27(2):589–599, 1999.

[168] MIKE WEST. Modelling with mixtures. In *Bayesian statistics, 4 (Peíscola, 1991)*, pages 503–524. Oxford Univ. Press, New York, 1992.

[169] MIKE WEST. Approximating posterior distributions by mixtures. *J. Roy. Statist. Soc. Ser. B*, 55(2):409–422, 1993.

[170] MIKE WEST, PETER MÜLLER, AND MICHAEL D. ESCOBAR. Hierarchical priors and mixture models, with application in regression and density estimation. In *Aspects of uncertainty*, pages 363–386. Wiley, Chichester, 1994.

[171] E. T. WHITTAKER AND G. N. WATSON. *A course of modern analysis*. Cambridge University Press, Cambridge, 1996. An introduction to the general theory of infinite processes and of analytic functions; with an account of the principal transcendental functions, Reprint of the fourth (1927) edition.

[172] WING HUNG WONG AND XIAOTONG SHEN. Probability inequalities for likelihood ratios and convergence rates of sieve MLEs. *Ann. Statist.*, 23(2):339–362, 1995.

[173] MICHAEL WOODROOFE. Very weak expansions for sequentially designed experiments: linear models. *Ann. Statist.*, 17(3):1087–1102, 1989.

Index

Springer Series in Statistics

(continued from p. ii)

Knottnerus: Sample Survey Theory: Some Pythagorean Perspectives.
Kotz/Johnson (Eds.): Breakthroughs in Statistics Volume I.
Kotz/Johnson (Eds.): Breakthroughs in Statistics Volume II.
Kotz/Johnson (Eds.): Breakthroughs in Statistics Volume III.
Küchler/Sørensen: Exponential Families of Stochastic Processes.
Le Cam: Asymptotic Methods in Statistical Decision Theory.
Le Cam/Yang: Asymptotics in Statistics: Some Basic Concepts, 2nd edition.
Liu: Monte Carlo Strategies in Scientific Computing.
Longford: Models for Uncertainty in Educational Testing.
Mielke/Berry: Permutation Methods: A Distance Function Approach.
Pan/Fang: Growth Curve Models and Statistical Diagnostics.
Parzen/Tanabe/Kitagawa: Selected Papers of Hirotugu Akaike.
Politis/Romano/Wolf: Subsampling.
Ramsay/Silverman: Applied Functional Data Analysis: Methods and Case Studies.
Ramsay/Silverman: Functional Data Analysis.
Rao/Toutenburg: Linear Models: Least Squares and Alternatives.
Reinsel: Elements of Multivariate Time Series Analysis, 2nd edition.
Rosenbaum: Observational Studies, 2nd edition.
Rosenblatt: Gaussian and Non-Gaussian Linear Time Series and Random Fields.
Särndal/Swensson/Wretman: Model Assisted Survey Sampling.
Schervish: Theory of Statistics.
Shao/Tu: The Jackknife and Bootstrap.
Simonoff: Smoothing Methods in Statistics.
Singpurwalla and Wilson: Statistical Methods in Software Engineering: Reliability and Risk.
Small: The Statistical Theory of Shape.
Sprott: Statistical Inference in Science.
Stein: Interpolation of Spatial Data: Some Theory for Kriging.
Taniguchi/Kakizawa: Asymptotic Theory of Statistical Inference for Time Series.
Tanner: Tools for Statistical Inference: Methods for the Exploration of Posterior Distributions and Likelihood Functions, 3rd edition.
van der Laan: Unified Methods for Censored Longitudinal Data and Causality.
van der Vaart/Wellner: Weak Convergence and Empirical Processes: With Applications to Statistics.
Verbeke/Molenberghs: Linear Mixed Models for Longitudinal Data.
Weerahandi: Exact Statistical Methods for Data Analysis.
West/Harrison: Bayesian Forecasting and Dynamic Models, 2nd edition.